Synthetic Aperture Radar for Disaster Management

Scrivener Publishing
100 Cummings Center, Suite 541J
Beverly, MA 01915-6106

Publishers at Scrivener
Martin Scrivener (martin@scrivenerpublishing.com)
Phillip Carmical (pcarmical@scrivenerpublishing.com)

Synthetic Aperture Radar for Disaster Management

Bridging Science and Solutions

Edited by

Pooja Joshi
Swami Rama Himalayan University, Dehradun, Uttarakhand, India

Ashutosh Bhatt
Swami Rama Himalayan University, Dehradun, Uttarakhand, India

Neelam Sharma
Banasthali Vidyapith University, Jaipur, Rajasthan, India

and

Kapil Joshi
Dept. of Computer Science & Engineering, Uttaranchal Institute of Technology, Uttaranchal University, Dehradun, Uttarakhand, India

Scrivener Publishing

WILEY

This edition first published 2026 by John Wiley & Sons, Inc., 111 River Street, Hoboken, NJ 07030, USA and Scrivener Publishing LLC, 100 Cummings Center, Suite 541J, Beverly, MA 01915, USA

For more information about Scrivener publications please visit www.scrivenerpublishing.com.

Wiley Global Headquarters
111 River Street, Hoboken, NJ 07030, USA

For details of our global editorial offices, customer services, and more information about Wiley products visit us at www.wiley.com.

The manufacturer's authorized representative according to the EU General Product Safety Regulation is Wiley-VCH GmbH, Boschstr. 12, 69469 Weinheim, Germany, e-mail: Product_Safety@wiley.com.

Library of Congress Cataloging-in-Publication Data

ISBN 978-1-394-39893-5

Front cover images courtesy of Wikimedia Commons
Cover design by Russell Richardson

Set in size of 11pt and Minion Pro by Manila Typesetting Company, Makati, Philippines

Printed in the USA

10 9 8 7 6 5 4 3 2 1

Dedicated to

To the resilient communities worldwide facing the brunt of disasters, and to the scientists, engineers, and policymakers working tirelessly to turn data into solutions.

Contents

Foreword

The present century has witnessed an unprecedented escalation in the frequency and severity of disasters. This trend is primarily driven by climate change, rapid urbanization, and growing ecological imbalances, underscoring the urgent need for proactive and integrated disaster risk management strategies. In this context, the deployment of cutting-edge technologies for risk mitigation, vulnerability assessment, and resilience building has become increasingly critical. Among these, Synthetic Aperture Radar (SAR) has emerged as a transformative tool, offering high-resolution, all-weather, day-and-night observations that are indispensable for effective disaster preparedness, timely response, and efficient recovery.

This edited volume is a timely and significant contribution to the global scientific and applied knowledge landscape. By integrating foundational principles of SAR with advanced computational approaches, including machine learning, deep learning, and multi-sensor data fusion, the book presents a comprehensive framework that is both scientifically rigorous and practically relevant. What distinguishes this work is its holistic perspective, which seamlessly transitions from radar physics to real-world disaster case studies, from Earth observation missions to planetary applications, and from technical innovations to policy and governance considerations.

The editors—Dr. Pooja Joshi, Dr. Ashutosh Bhatt, Dr. Neelam Sharma, and Dr. Kapil Joshi—bring together diverse expertise spanning computer science, radar remote sensing, geoinformatics, and disaster management. Their collective effort, complemented by contributions from leading researchers and practitioners, ensures that this book serves as more than a technical manual; it acts as a bridge between science and actionable solutions. By weaving together theoretical depth with practical insights, the volume empowers researchers, disaster managers, policymakers, and students to leverage SAR technologies for building safer, more resilient societies.

This book serves as a valuable reference for the academic and professional community and will inspire future innovations in the application of space-borne and airborne radar technologies for sustainable disaster resilience.

As humanity confronts increasingly complex challenges, works such as this reaffirm our capacity to translate science into solutions that genuinely make a difference.

I commend the authors and editors for their courage in addressing both the technical and humanitarian dimensions of SAR. Their work stands as a testament to the power of science in the service of society. To the reader, I extend this invitation: Let these pages not only inform you but also inspire and galvanize you. Whether you are a student, policymaker, or technologist, you hold a piece of the puzzle. Together, let us harness SAR's potential to transform vulnerability into resilience, data into action, and despair into hope.

The storms ahead are formidable, but with tools like SAR and the collective will to use them wisely, humanity need not face them alone.

Prof. (Dr.) Anil Gupta
Professor & PI – ICARS (CoE of DST, Govt. of India) IIT Roorkee
Gr. Noida Campus, NCR, India
& Senior ICIMOD Fellow - DRR

Preface

As humanity navigates the escalating challenges of a rapidly changing world—marked by intensifying climate extremes, environmental degradation, and urbanization—the imperative to mitigate disasters and protect vulnerable communities has never been more urgent. Natural and human-induced catastrophes, from devastating floods and earthquakes to glacial collapses and urban crises, demand not only swift response but also proactive, data-driven strategies to minimize loss of life, economic disruption, and ecological harm. This book is a critical resource at this pivotal juncture, uniting the frontiers of SAR technology with the pressing demands of disaster resilience. It illustrates how science, engineering, and policy can converge to build a safer, more sustainable future, offering a transformative lens through which to address the complex interplay of global vulnerabilities and technological innovation.

Synthetic Aperture Radar (SAR) is a cornerstone of modern remote sensing, transcending the limitations of traditional optical sensors by operating independently of sunlight and weather conditions. Its unparalleled ability to penetrate clouds, monitor surface deformations at millimetre scales, and map dynamic environments—from melting glaciers to floodplains—positions SAR as an indispensable tool in disaster management. Yet, despite its transformative potential, SAR's full capabilities remain underexplored in practical applications. This book bridges that gap. It offers a holistic exploration of SAR's technical foundations, advanced methodologies, and ethical imperatives to empower stakeholders across disciplines. From the physics and engineering underpinning SAR systems to the integration of artificial intelligence (AI) and machine learning (ML) for rapid disaster prediction and response, the chapters within cover a spectrum of knowledge that will inspire innovation while emphasizing actionable solutions.

The volume begins by demystifying the core principles of SAR technology, including signal processing, polarimetry, and interferometry (InSAR), revealing how these techniques decode complex terrain changes—from reservoir dynamics to shifts in snow water equivalent. It then transitions to the AI revolution, showcasing how machine learning accelerates disaster preparedness by automating damage assessment, enhancing early-warning systems, and addressing challenges like algorithmic bias. Beyond Earth, the book explores SAR's role in planetary exploration through missions such as NASA's Magellan and VERITAS, as well as ESA's Envision, which map celestial bodies to advance geological and climatic understanding. Ethical and policy considerations are central to this discourse, with dedicated chapters confronting dilemmas around data privacy, equitable access, and geopolitical tensions, ensuring that technological advancements align with societal needs.

Interwoven with real-world case studies—from Arctic glacier monitoring to urban disaster recovery—the book underscores SAR's versatility in addressing environmental, climatic, and humanitarian crises. It highlights the engineering marvels behind SAR payload design, emphasizing sustainability and cost-efficiency in next-generation instrumentation, while championing global collaboration through initiatives like the Sentinel-1 mission. Disaster management, inherently interdisciplinary, demands synergy among engineers, policymakers, scientists, and communities. By integrating geospatial insights with AI-driven analytics, SAR enables not only reactive responses but proactive strategies—monitoring infrastructure stability, tracking deforestation-linked landslides, or assessing post-disaster recovery in conflict zones.

Yet, as we harness SAR's immense potential, ethical deployment remains paramount. Questions of data transparency, algorithmic accountability, and equitable resource distribution remind us that technology must serve humanity with wisdom and equity. This book is a technical guide and a rallying cry—for researchers refining InSAR algorithms, policymakers drafting disaster protocols, and students passionate about geospatial innovation. It is a testament to the power of collaboration, offering hope in the face of escalating climate predictions. Together, let us transform data into solutions, and solutions into lasting resilience.

As editors, we are deeply grateful to the global experts who contributed their insights, making this collaborative effort a reality. It is our hope that this book will serve as a catalyst for advancing SAR technology, fostering resilience, and safeguarding communities in an increasingly vulnerable world. Finally, we extend our thanks to Martin Scrivener and the team at Scrivener Publishing for their effort and guidance in bringing this work to publication.

December 2025

Part 1

FOUNDATIONS OF SYNTHETIC APERTURE RADAR (SAR) AND DISASTER MANAGEMENT

1

Fundamentals of Synthetic Aperture Radar Technology: The Intersection of Physics, Engineering, and Geospatial Analysis for Advanced Imaging Systems

R. Venkatesh[1]* and Pooja Joshi[2]

[1]Department of Physics, PSNA College of Engineering and Technology (Autonomous), Dindigul, Tamil Nadu, India
[2]Swami Rama Himalayan University, Dehradun, Uttarakhand, India

Abstract

The synthetic aperture radar (SAR) is the most powerful Earth observation remote sensing tool, used to capture imagery in extremely high resolutions digitally, regardless of weather conditions or day or night. The chapter elucidates the fundamentals of SAR technology in an interdisciplinary manner that bridges physics, engineering, and geospatial analysis. It begins by describing the core physical principles of SAR, such as wave propagation, signal processing, and image formation. The perturbing effects of satellite-aided radar imaging on radar hardware design, signal transmission, and data acquisition are also shown. The role of geospatial analysis as a reference for interpreting satellite and aerial camera data will be discussed, particularly since it applies to mapping, surveillance, and disaster management. The study will thus be an extensive reader-and-learner in the technical position of SAR within all emerging imaging systems with extensive applications, most of which are defense applications in environmental monitoring and infrastructure development.

***Keywords*:** Synthetic aperture radar, remote sensing, signal processing, radar technology, geospatial analysis, earth observation, radar imaging systems

**Corresponding author*: sukatesh@gmail.com

Pooja Joshi, Ashutosh Bhatt, Neelam Sharma and Kapil Joshi (eds.) Synthetic Aperture Radar for Disaster Management: Bridging Science and Solutions, (3–46) © 2026 Scrivener Publishing LLC

1.1 Introduction to Synthetic Aperture Radar Technology

Synthetic aperture radar (SAR) is a robust remote sensing technology employed to obtain high-resolution images of the Earth's surface. In contrast to optical or infrared imaging systems, which depend on visible light or thermal radiation, SAR technology utilizes radio waves (microwaves) to collect information, enabling data acquisition regardless of weather conditions, day or night. SAR finds extensive applications in a range of fields, including Earth observation, military surveillance, disaster monitoring, and environmental studies.

Principles of synthetic aperture radar technology

Synthetic aperture radars work by emitting a radar wave from a platform (often an aircraft or satellite) toward the Earth's surface. The radar waves strike objects on the surface and are reflected to the radar receiver. The signal return time provides information on the distance to the object, while the Doppler shift (frequency change) can be used to infer the object's velocity. The "synthetic aperture" aspect of the technology results from the fact that physically long antennas are not needed for the radar. Instead, the motion of the platform carrying the radar is utilized to "simulate" a long antenna [1]. During its flight along its path, the platform accumulates radar information at various points, which is then processed to create a high-resolution image of the target region.

Key elements of a synthetic aperture radar technology system

Transmitter (Radar): The transmitter emits a burst of radio waves toward the target. The waves are generally in the microwave frequency band (1–100 GHz).

Receiver: The receiver records the reflected signals that bounce back from the Earth's surface.

Antenna: The radar system generally employs a side-looking antenna that emits and receives signals at a low angle, usually 30–60 degrees to the ground, for optimal mapping of terrain features.

Platform: SAR systems are usually installed on satellites, aircraft, or unmanned aerial vehicles. The platform travels along a fixed route, continuously transmitting and receiving radar waves.

Signal Processor: SAR data is complex and involves considerable signal processing. The processor combines radar signals with data related to the motion and position of the platform (GPS) and produces detailed images of the Earth's surface.

Data Storage and Analysis System: SAR data is frequently archived for subsequent analysis and may be processed into alternative forms of imagery (e.g., amplitude, phase, or interferometric SAR data) according to the requirement.

1.1.1 Overview of a Synthetic Aperture Radar

The SAR is a type of remote sensing technology that uses radar waves for high-resolution Earth surface imaging (Figure 1.1). It does not depend on visible light to function, but instead, SAR works in the microwave region of the electromagnetic spectrum. This provides the SAR with a possibility to be functional in all weather conditions and under all-time regimes [2].

The concept of SAR was developed in the 1950s, and since then, this technology has become the foundation of military and civilian applications (Figure 1.2). A SAR system sends radar signals to the area to be imaged and then reflects them to the sensor [3]. The innovation of SAR technology lies in utilizing the motion of the radar platform, such as an aircraft or satellite, to simulate a much larger "aperture" than is physically present. This enables the system to obtain high-resolution images.

Generally, SAR systems are mounted on satellites or aircraft, providing comprehensive high-resolution coverage over large areas. These radar

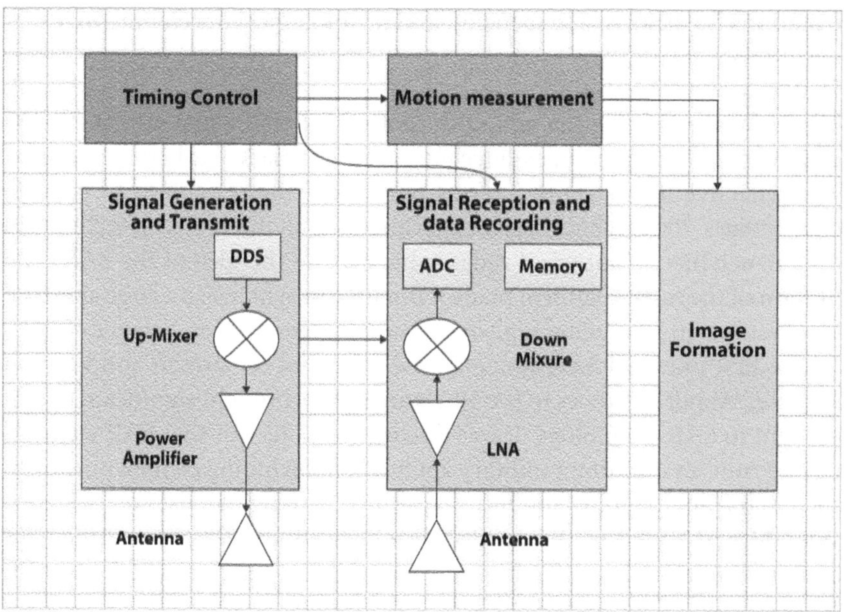

Figure 1.1 Synthetic aperture radar functional block diagram.

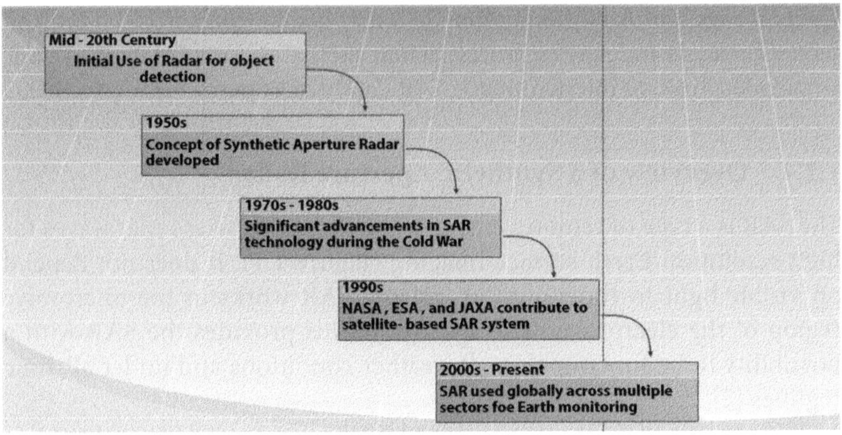

Figure 1.2 A journey through synthetic aperture radar: Key milestones and uses.

systems are designed to detect the time delay and intensity of the returned signal, generating images that depict physical features of the Earth's surface, topography, vegetation, and infrastructure, among others. The SAR is unique in its capacity to "look" through clouds, smoke, and darkness to perform environmental monitoring and evaluation of natural catastrophes or surveillance [4]. This table (Table 1.1) indicates some major milestones in the development of SAR, major applications realized over time, key contributions by major space agencies, and technological developments.

Historical background

The SAR originated in the mid-20th century, when the first military application of radar for reconnaissance purposes was made. Early radar systems were effective for detecting objects but lacked the resolution to provide clear images. Researchers began experimenting in the 1950s on techniques to enhance the resolution of radar, which led to the idea of the SAR. The motion of the radar platform made it possible to synthesize a large aperture that resulted in producing high-resolution images even at distances [5].

The first practical SAR systems were mounted on aircraft and later on satellites. Major advances in the 1970s and 1980s brought significant development to SAR technology. It was during the Cold War that SAR technology became an important military application, including civilian purposes, such as terrain mapping and reconnaissance. The civilian uses of SAR considerably expanded in the 1990s. More aspects of disaster management and environmental monitoring were focused on, with organizations such

Table 1.1 Timeline of SAR development and applications.

Time period	Key developments	Applications	Significant contributions
Mid-20th Century	The radar was used for military reconnaissance. Early radar systems lacked resolution for detailed imaging.	Military reconnaissance	Initial use of radar for object detection.
1950s	Researchers experimented with techniques to improve radar resolution; this led to the introduction of the SAR concept.	Early radar applications, research	The concept of SAR developed.
1970s-1980s	The first practical SAR systems were deployed on aircraft and satellites. Major advances in resolution.	Military reconnaissance, terrain mapping	Significant advancements in SAR technology during the Cold War.
1990s	Civilian applications expanded; focus on disaster management and environmental monitoring.	Disaster management and environmental monitoring	Agencies NASA, ESA, and JAXA contribute to satellite-based SAR systems.
Today (2000s–Present)	Ongoing advancements in signal processing, computational power, and sensor technology.	Defense, agriculture, forestry, land-use monitoring, and climate studies	Synthetic aperture radars are used globally across multiple sectors for Earth monitoring.

as NASA, ESA, and JAXA, which have made contributions to developing and operationally deploying satellite-based SAR systems.

Today, SAR technology is extensively used in defense, agriculture, forestry, land-use monitoring, and climate studies. As developments are made in signal processing, computational power, and sensor technology, the capacity of SAR systems continues to improve to unprecedented levels to monitor and map the Earth's surface [6].

1.1.2 Key Principles of Synthetic Aperture Radar Imaging

Figure 1.3 depicts the geometric configuration of a SAR system. Here, R denotes the distance between the sensor and the target, θ refers to the angle of incidence, α is the look angle, γ represents the depression angle, and β signifies the grazing angle. The radar sensor is positioned at a specific altitude above the nadir point (the vertical projection directly beneath the sensor). Spatial resolution in the SAR is governed by two orthogonal directions: the azimuth direction (aligned with the sensor's flight path) and the range direction (perpendicular to the flight path). These directions define the resolution cell's dimensions. The angle of incidence is formed between the incoming electromagnetic wave emitted by the SAR sensor and the normal vector (a line perpendicular) to the Earth's surface at the point where the wave strikes the target.

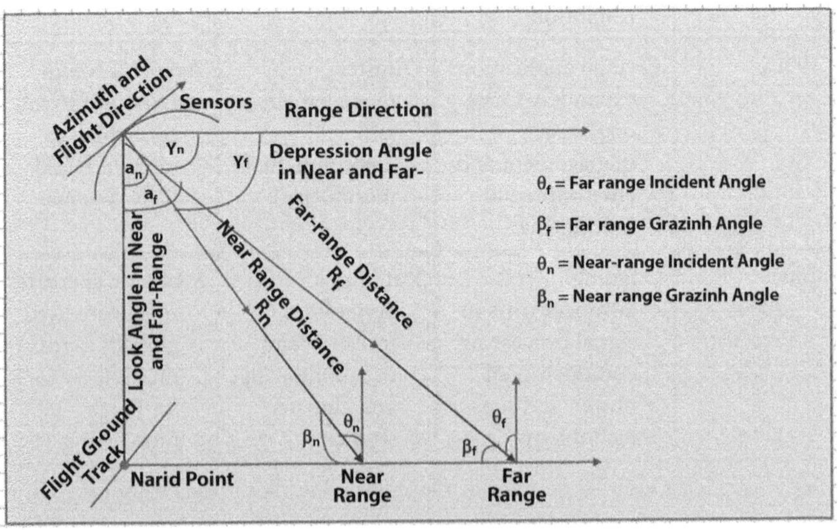

Figure 1.3 (a) The geometry of the radar system for data collection [7].

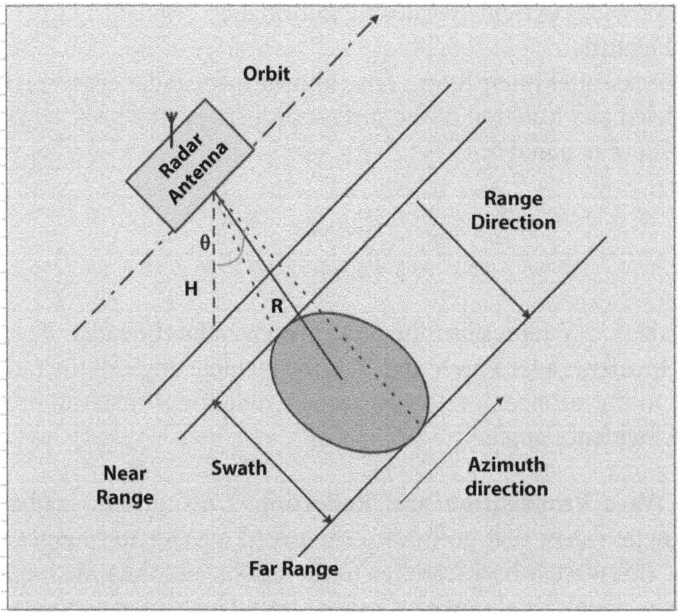

Figure 1.3 (b) Fundamentals of SAR imaging geometry [7].

The radar sensor emits microwave energy toward the Earth's surface, illuminating an elliptical area known as the footprint. This footprint determines the swath width of the SAR image. The segment of the illuminated region closest to the nadir is termed the near range, while the farthest segment is the far range. The dimensions of the footprint, and thus the swath, depend on the antenna's physical width (W) and the wavelength (λ) of the transmitted microwave signal [7].

In Figure 1.3(b), SAR imaging relies on several fundamental physical principles that enable the system to create high-resolution images despite the radar's physical aperture being quite small [8, 9].

Backscatter: It is the portion of the radar signal reflected or scattered in a direction back toward the radar sensor after it strikes a surface or target. When radar waves are incident on a target, they are reflected from the surface, and some of the energy is scattered to the radar receiver (Figure 1.3). The backscatter level is influenced by several variables, including surface roughness, material properties, and radar beam incidence angle [10]. Backscatter in SAR images is commonly utilized to decipher surface features such as vegetation density, soil type, or human-made structures, since it provides extremely valuable information related to target properties and texture.

The backscatter intensity σ_0 (also known as the radar cross section) can be modeled as a function of the surface properties. The basic equation for the backscatter signal is:

$$\sigma_0 = \frac{P_{received}}{P_{incident}} \cdot \frac{1}{\cos\theta} \tag{1.1}$$

Where $P_{received}$ represents the power received by the radar, $P_{incident}$ represents incident radar power and θ is the incident angle of the radar wave relative to the surface. It relies on surface roughness, material properties, and the incidence angle.

Radar Wave Propagation and Reflection: SAR systems radiate electromagnetic waves that go toward the earth's surface to interact with all objects. This wave is backscattered to the sensor, and the time required for the return signal, also known as "range," would give information regarding the distance to the reflecting surface.

The range R to the target is calculated using the time of flight and the speed of light c:

$$R = \frac{c \cdot t_{flight}}{2} \tag{1.2}$$

Where, c is the speed of light in a vacuum (3 x 10^8 m/s), t_{flight} is the round-trip time for the radar wave to reach the target and return.

Doppler Effect and Motion Compensation: The Doppler shift is a critical component of the SAR, as it occurs due to the relative motion between the radar platform and the target. The movement of the platform changes the frequency of the reflected waves. By analyzing such frequency shifts, SAR systems can determine velocities at the Earth's surface, enabling precise image formation. The Doppler shift also contributes to achieving high resolution along the flight path of the radar and simulates a much larger aperture.

The Doppler shift Δf due to the relative motion between the target and the radar platform is expressed as:

$$\Delta f = \frac{2v \cdot f_{radar}}{c} \tag{1.3}$$

Where, v is the relative of the target with respect to the radar platform (in the direction of the radar beam), f_{radar} is the radar signal frequency, c is the speed of light.

Synthetic Aperture: In SAR, unlike conventional radar systems where a fixed antenna is utilized, the movement of the radar platform creates an "effective aperture" or a "synthetic" one. The radar system collects multiple signals from the same area over time as the platform moves. It then combines or processes those signals to form a high-resolution image that represents the target area. It increases the size of the effective aperture and has fine resolution both in the range and azimuth direction along the flight path.

The synthetic aperture length $L_{aperture}$ depends on the platform's velocity $v_{platform}$ and the time duration T over which the radar collects data:

$$L_{aperture} = v_{platform} \cdot T \tag{1.4}$$

The azimuth direction resolution $\Delta_{x_{azimuth}}$ is inversely proportional to the synthetic aperture length:

$$\Delta_{x_{azimuth}} = \frac{\lambda}{L_{aperture}} \tag{1.5}$$

Where λ is the wavelength of the radar signal.

Image Formation Algorithms: After collecting radar returns, advanced signal processing algorithms are used to transform raw data into images. It corrects the distortions introduced because of platform motion, enhancing the signal-to-noise ratio and improving the final SAR image's resolution. Commonly employed image formation algorithms include range-Doppler and chirp scaling algorithms, which also consider the geometry and velocity of the radar platform.

Resolution and Pixel Size: Factors such as radar frequency, the size of the synthetic aperture, and velocity of the platform determine the resolution of a SAR image. Higher-frequency radars allow finer resolution, but they tend to be much more sensitive to atmospheric interference. Combining these factors with the synthetic aperture technique results in SAR achieving resolution on the order of a meter or less.

1.1.3 Applications and Importance of Synthetic Aperture Radar

Due to the unique capabilities of SAR technology, it is now widely applied in various areas of military and civilian life.

Earth Observation and Remote Sensing: The SAR is widely used for Earth observation, which provides an opportunity to monitor land surfaces, vegetation, and infrastructure. Its capacity to pierce through clouds and work during nighttime can be highly invaluable in monitoring environmental changes, such as deforestation, desertification, land-use changes, or even crops.

Disaster Management and Emergency Response: The SAR now plays a crucial role in disaster management, especially after natural phenomena such as earthquakes, floods, and landslides. Their systems allow urgent and updated images of the disaster-stricken area, which is important in assessing damage, identifying areas that require urgent relief, and monitoring the recovery process. Slight ground movements, such as surface displacements following an earthquake, are measurable using SAR and are therefore vital in measuring the event size.

Military and Defense: The SAR is utilized in defense to aid reconnaissance, surveillance, and target detection. Since it can create high-resolution images, the military can utilize SAR to monitor enemy movements, monitor infrastructure development, and identify strategic sites. Its ability to operate day and night, as well as its moving target indication, only make SAR a valuable tool in military operations.

Ocean and Ice Monitoring: The SAR is used to track oceans, glaciers, and ice sheets. The SAR tracks sea ice movements, locates oil spills, and gives insight into the study of ocean currents. SAR can be used to map the surface of ice-covered regions. Such information is important for climate change research and monitoring polar environments.

Urban Planning and Infrastructure Monitoring: The SAR is of great importance for monitoring urban infrastructure because of its precision in mapping and measuring surface displacements. The stability of buildings, roads, bridges, and dams can be evaluated using SAR data. This helps identify potential risks, especially in areas prone to subsidence or seismic activity. Moreover, the SAR's capability for monitoring changes in the urban environment effectively supports many urban planning efforts. This can be seen from monitoring a construction project to tracking population growth.

Topographic and Geophysical Mapping: The SAR is also utilized for highly detailed topographic mapping, especially in inaccessible areas like dense forests or deserts. In the field of SAR interferometry, one can

Table 1.2 The applications and significance of SAR technology.

Application area	Description	Importance
Earth Observation and Remote Sensing	Monitoring land surfaces, vegetation, infrastructure, and environmental changes like deforestation and land-use changes.	Crucial for tracking environmental changes, crop monitoring, and providing data for agriculture.
Disaster Management and Emergency Response	Providing rapid imagery of disaster-affected areas (e.g., earthquakes, floods). Detecting ground movements and damage.	Essential for assessing damage, aiding disaster response, and tracking recovery efforts.
Military and Defense	Used for reconnaissance, surveillance, and target identification.	Valuable for monitoring enemy movements, infrastructure, and detecting moving targets under all weather conditions.
Ocean and Ice Monitoring	Tracking sea ice movements, monitoring oil spills, analyzing ocean currents, and mapping ice-covered regions.	Key for climate change research, polar monitoring, and tracking environmental changes in oceans and ice sheets.
Urban Planning and Infrastructure Monitoring	Mapping and measuring surface displacements to assess the stability of infrastructure, such as buildings, roads, and bridges.	Vital for ensuring urban infrastructure stability and supporting urban planning and risk management.
Topographic and Geophysical Mapping	Creating detailed topographic maps and studying tectonic plate movements, volcanic activity, and land subsidence.	Useful for studying geographical changes, tectonic activity, and other geological phenomena in remote areas.

measure with high precision surface motion over a span of time, and in that study, tectonic plate motion, volcanic activity, or even land subsidence.

SAR technology can produce clear images in any type of weather condition and in diverse applications. Hence, the need for a high-quality modern geospatial analysis and remote sensing tool continues to grow through diverse applications like resource monitoring and tracking of natural resources and environmental change [11, 12]. Military surveillance, disaster management, and several others also continue to need SAR in this line of research. This table (Table 1.2) demonstrates how SAR has an application in various fields and plays a vital role in monitoring environmental changes, defense, and even urban planning.

1.2 Physics behind Synthetic Aperture Radar Systems

The physics of the SAR helps one understand how this technology generates such a clear image of the Earth's surface. Obviously, it is nothing more than the principles behind the interaction between electromagnetic waves and objects located on ground level, and the motion of the radar platform, combined with sophisticated signal processing algorithms to reconstruct such a detailed image [13].

1.2.1 Electromagnetic Wave Propagation

Synthetic aperture radar systems use electromagnetic waves to penetrate the Earth's surface and acquire data. It follows that electromagnetic waves propagate through space and interact with matter, and that SAR systems map and analyze the physical properties of the objects being imaged. To obtain SAR data, one needs to know how electromagnetic waves propagate. This table (Table 1.3) compares the SAR technology critical physics principles involved. These include the basic wave propagation mechanisms, frequency utilization, and ways the radar waves can interact with a surface to develop high-resolution images.

Electromagnetic Spectrum: The SAR works within the microwave portion of the electromagnetic spectrum (Figure 1.4) at the frequency range of 1–40 GHz (wavelength between 30 cm and 0.75 cm). Microwaves suit the SAR perfectly because they can pass through clouds, smoke, and rain, while visible light cannot. This allows the SAR to take images regardless of the time of day or weather.

Table 1.3 Key physics concepts behind SAR systems.

Concept	Description	Key details
Electromagnetic Spectrum	The SAR operates in the microwave region of the spectrum, typically between 1–40 GHz.	Microwaves are ideal for penetrating clouds, smoke, and rain, enabling the SAR to work in all weather conditions and at any time of day.
Wave Propagation	The SAR uses electromagnetic waves to probe the Earth's surface, transmitting and receiving radar waves.	Maxwell's equations describe how electric and magnetic fields interact and propagate, guiding the transmission and reception of radar waves.
Attenuation and Scattering	As radar waves travel, they lose energy (attenuation) and scatter depending on the surface.	Higher frequencies (shorter wavelengths) are sensitive to small features, such as vegetation. Lower frequencies (longer wavelengths) penetrate deeper.
Reflection and Backscatter	When radar waves hit an object, they reflect, carrying information about the object's surface.	The backscatter intensity depends on surface properties, such as dielectric material, roughness, and the angle of incidence.

Wave Propagation: Electromagnetic waves, such as radar waves, travel through free space based on Maxwell's equations. The electric and magnetic fields propagate and interact as waves, as depicted in these equations. The SAR systems send these radar waves, and energy gets transmitted to the Earth's surface [14]. When such waves interact with objects, they reflect toward the sensor. The delay in time between reception and transmission enables the radar system to calculate the distance to the reflective surface, which is crucial to image formation.

Attenuation and Scattering: While propagating through the atmosphere or interacting with various surface conditions, radar waves experience attenuation and scattering. The degree to which these take place varies based on the radar wave frequency as well as the nature of the surface encountered. For instance, higher frequencies (shorter wavelengths) are more sensitive to small surface roughness, such as vegetation, whereas lower frequencies (longer wavelengths) penetrate deeper in the ground or through vegetation layers.

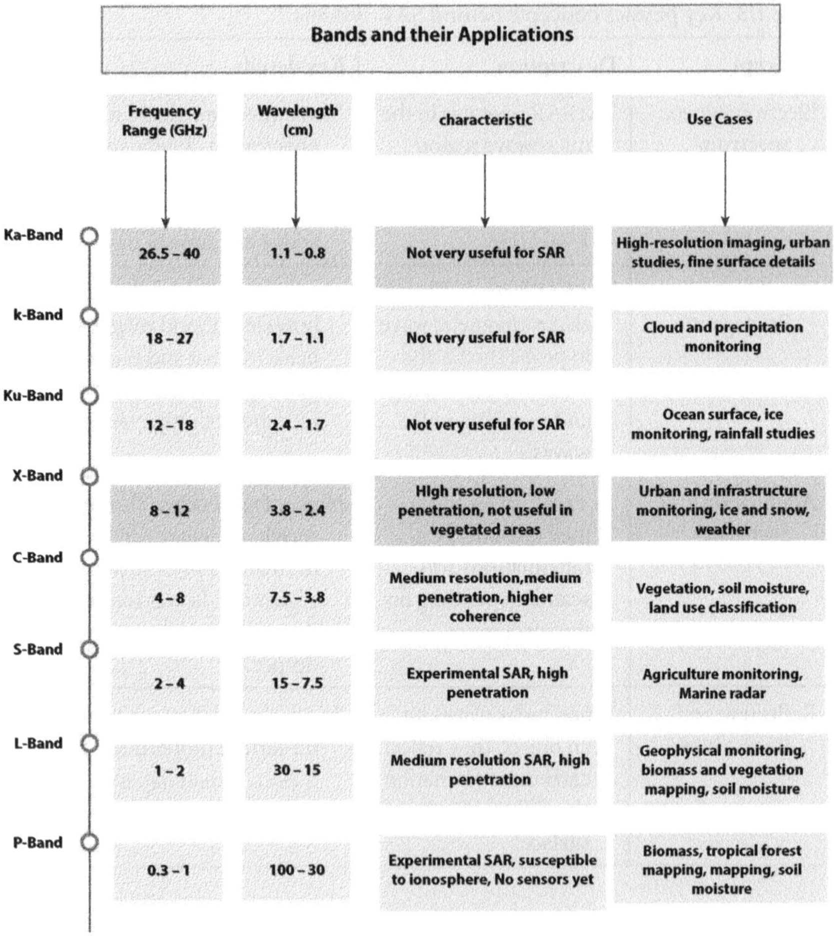

Figure 1.4 The SAR frequency bands within the electromagnetic spectrum.

Reflection and Backscatter: When radar waves are incident upon an object or surface, they can be absorbed, refracted, or reflected. The reflected waves come back to the radar and contain information on the object's characteristics, including surface roughness, shape, and material composition. The backscatter strength is a function of several parameters, including dielectric properties of the material, surface roughness, and incidence angle.

Higher Frequency Bands (Ka, K, Ku): These possess lower wavelengths, allowing improved-resolution SAR imaging. They best serve applications such as high-resolution reconnaissance and high-resolution surface mapping, although rain or atmospheric attenuation can degrade their performance.

Mid-Frequency Bands (X, C): These are the most frequent bands utilized in SAR systems since they offer a compromise between range and resolution. They are applied extensively in Earth observation, surveillance, and mapping.
Lower Frequency Bands (S, L, P): These have longer wavelengths, enabling more penetration through obstructions such as clouds, foliage, and even soil. Although they offer lower resolution, they are best suited for use where extensive coverage is required, such as long-range radar or surveillance of large areas like oceans or forests.

1.2.2 Doppler Effect and Range Resolution

The Doppler effect and range resolution are two important phenomena that enable SAR to generate high-resolution images. These are the main concepts behind the special capability of SAR to obtain fine azimuth and range resolution.

Doppler Effect: The Doppler effect takes place when there is relative motion between the target and the radar platform. When the radar travels along its flight path, it sends out waves that are reflected from the target. The frequency of the returned signal varies with the relative velocity of the radar with respect to the target. If the radar is approaching the target, the frequency of the wave that is reflected rises (positive Doppler shift), and if the radar is receding, the frequency drops (negative Doppler shift). This effect is critical for estimating the location and velocity of targets, playing a key role in SAR image fine azimuth resolution.
Positive Doppler Shift: When the radar platform is moving toward the target, the frequency of the returned wave increases. This is referred to as a positive Doppler shift.
Negative Doppler Shift: When the radar platform is moving away from the target, the frequency of the wave reflected off the target decreases, resulting in a negative Doppler shift.

The Doppler frequency shift can be mathematically described using the Doppler shift formula:

$$f_D = \frac{2v}{\lambda} \tag{1.6}$$

Where f is the Doppler shift frequency, v is the relative velocity between the radar and the target, λ is the wavelength of the radar signal.

In SAR, Doppler shift is employed to estimate the relative velocity between the target and radar, which aids in resolving the target positions with high precision. This frequency change is paramount in SAR imaging due to the system's ability to differentiate targets on the flight track [15]. Through the Doppler shift, the SAR can determine the speed of surface targets and precisely position them in the azimuth direction (the direction in the flight track), hence enhancing the overall image resolution.

Range Resolution: Range resolution is the ability of the radar to resolve two objects at different ranges from the sensor along the same line of sight. Range resolution is a function of pulse width (width of the radar signal transmitted) and the radar system frequency. Narrow pulse widths and higher frequencies provide improved range resolution, allowing the radar to discern small objects or details on the Earth's surface.

The range resolution (ΔR) is calculated by the bandwidth (B) of the radar signal and is expressed as:

$$\Delta R = \frac{C}{2B} \tag{1.7}$$

Where, ΔR is the range resolution, c is the speed of light in vacuum, and B is the radar signal's bandwidth.

A greater bandwidth gives greater range resolution, so the radar can more easily tell between closely spaced targets in range.

Azimuth Resolution: It defines how well the radar can separate two targets in cross-track direction (perpendicular to the flight course). The SAR achieves high azimuth resolution by taking advantage of the Doppler shift, which changes with the motion of the radar with respect to the target. SAR can thus generate a large antenna aperture, substantially improving its azimuth resolution.

The azimuth resolution (Δθ) is approximated by:

$$\Delta\theta = \frac{\lambda}{2L} \tag{1.8}$$

Where Δθ is the azimuth resolution, λ is the radar wavelength, and L is the synthetic aperture length (i.e., the range over which the radar travels during pulse transmission).

The combination of the synthetic aperture and Doppler shifts enables SAR systems to have very high azimuth resolution, even for targets in motion.

Radar Equation: The Radar equation is applied in the description of the power received at the radar by a target and is vital to ascertain the ability of the radar to detect. It is described as:

$$P_r = \frac{P_t G_\sigma^2 \lambda^2}{(4\pi)^3 R^4} \tag{1.9}$$

Where Pr is the received power, Pt is the transmitted power, G is the gain of the radar antenna, σ is the radar cross-section of the target, λ is the wavelength of the radar, and R is the range to the target.

This equation facilitates SAR systems' computation of a signal's received strength from a target at any given range; it is relevant in detecting and imaging targets whose ranges and radar cross-sections may vary. The SAR systems obtain range resolution through the method of chirped pulses. Chirped pulses are pulses of varying frequency. This means the radar can zero in on the target more precisely and get more accurate measurements of how long it takes for the signal to bounce back [16]. The smaller the time measurement, the smaller the distance the system can measure between objects.

1.2.3 Radar Signal and Wave Interaction with Surfaces

The behavior of radar signals against varying surfaces is at the core of comprehending how SAR systems create images. Physical characteristics, including surface roughness, composition, and dielectric properties, control such behavior.

Radar Backscatter: When a radar wave hits an object, some of the wave is bounced back to the sensor as backscatter. The strength of the backscatter varies with the type of surface and the angle of incidence of the radar waves. Smooth surfaces, such as flat metal or calm water, reflect most of the energy away from the sensor, resulting in weak backscatter. Coarser surfaces, like plant life, rocky terrain, or cities, scatter the waves in multiple directions and tend to give rise to more intense backscatter signals.

Surface Roughness and Scattering Mechanisms: Surface roughness is an important determinant of radar waves' scattering. When the surface features are greater than the radar wavelength, the waves will scatter in many directions, resulting in complicated backscatter patterns. Conversely, when the surface is smooth relative to the radar wavelength, the waves will be reflected, resulting in weaker backscatter [17]. The SAR systems can take advantage of these differences in backscatter to distinguish between various kinds of surfaces and ground features.

Dielectric Properties: Various materials have varying dielectric constants, which define the amount of radar energy that is reflected or transmitted upon interaction with them. For example, dry soil has a different dielectric constant than water or vegetation. SAR can be utilized to identify changes in the dielectric properties of the surface and is useful in giving information on the type of material (for example, detecting water bodies, forests, or urban regions).

Polarimetric SAR (PolSAR): Polarimetric SAR utilizes the polarization of radar waves to generate extra information regarding surface properties. Radar waves may be polarized in various ways, and the backscatter signal is generally characterized as horizontal or vertical polarization. The PolSAR systems send and receive waves with distinct polarization states (horizontal-horizontally or vertically-vertically) to better describe surface roughness, vegetation, and even humidity content [18].

1.2.4 Fundamentals of Synthetic Aperture Radar Image Formation

The SAR image formation takes raw radar data and transforms it into a coherent picture of the Earth's surface. It is a process of combining signals acquired at various times and locations, resolving the high-resolution images using advanced signal processing technology [19].

Signal Collection and Raw Data Acquisition: When the radar platform travels, it gathers radar signals reflected off the surface. These signals are converted into digital form and stored as raw data, containing details of the time delay (range) and the Doppler shift (azimuth). This raw data is not interpretable as an image.

Range and Azimuth Compression: To create an image, the radar system must process the unprocessed data to enhance range and azimuth resolution. Range compression is done through signal processing methods, such as matched filtering, which enables a refinement of the radar's capability to discern objects at varying ranges. Azimuth compression is done by exploiting the Doppler shift resulting from the movement of the radar platform [20]. Through the combination of information from various positions, the SAR emulates a much larger antenna (aperture) than physically exists.

Table 1.4 An overview of the SAR image formation process.

Step	Description	Key techniques used	Purpose
Signal Collection and Raw Data Acquisition	The radar platform moves, collecting reflected radar signals from the Earth's surface. These signals are digitized and recorded.	Radar platform motion and reflection processing, Doppler shift detection, and time delay measurement.	Gather raw data with information on range (time delay) and azimuth (Doppler shift) to be processed into an image.
Range and Azimuth Compression	Process the raw data to improve resolution in both range and azimuth directions.	Matched filtering (range compression), Doppler shift (azimuth compression), and data synthesis from different radar positions.	Improve the radar's ability to distinguish objects by enhancing resolution and simulating a larger antenna aperture.
Image Reconstruction Algorithms	Use algorithms to reconstruct the image, adjusting for platform motion and data distortions.	Range-Doppler, chirp scaling, and motion compensation.	Convert processed data into a high-resolution image, correcting distortions and accurately positioning targets.
Geometric Correction and Calibration	Align the image with the Earth's coordinate system and calibrate the data for radar sensitivity and atmospheric effects.	Geometric correction, calibration for radar sensitivity, and atmospheric influence.	Ensure the final image accurately maps features to real-world locations and accounts for environmental factors.

Image Reconstruction Algorithms: Following range and azimuth compression, SAR applies algorithms such as range-Doppler or chirp scaling to image reconstruction. These algorithms correct for the motion of the radar and compensate for distortion in the data, such as the motion of the target's relative position due to platform velocity. The product is a high-resolution focused image in which the spatial position of each pixel relates to an exact point on the Earth's surface.
Geometric Correction and Calibration: After the image formation, it might undergo geometric correction so that it aligns with the Earth's coordinate frame. This allows objects in the image to get correctly mapped to their true location. The image might get calibration done for eliminating sensitivity differences between the radar and accounting for the influence of the atmosphere, as well.

The creation of a SAR image is an intricate interaction of electromagnetic wave propagation, signal processing, and surface interactions. The underlying physics of SAR systems enables them to produce high-resolution images under varied conditions; it is an indispensable tool across a broad scope of applications from environmental monitoring to defense and surveillance [21]. This table (Table 1.4) emphasizes the key steps in forming SAR images, from the gathering of raw data to image calibration, stressing the signal processing methods to obtain high-resolution images.

1.3 Engineering of Synthetic Aperture Radar Systems

The engineering of SAR systems entails the design and integration of dedicated hardware, antennas, data acquisition modules, and sophisticated

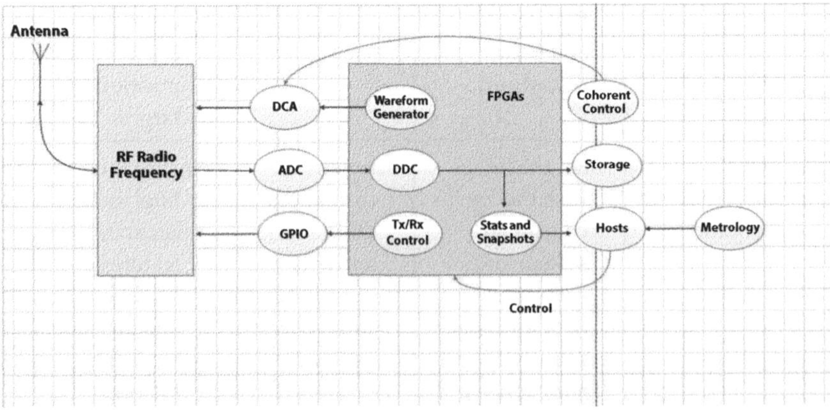

Figure 1.5 An overview of a standard SAR system.

signal processing algorithms (Figure 1.5). These systems collectively capture high-resolution imagery and transform raw radar data into useful geospatial information [22].

1.3.1 Radar Hardware Components

Hardware on the radar constitutes the core of a SAR system, comprising numerous important components upon which signal transmitting, receiving, and data obtaining depend. Some of these vital components are the radar transmitter, receiver, processing unit, and power supply unit, all designed to play their roles complementarily in achieving high-quality SAR images [23].

Transmitter: The transmitter produces microwave pulses that are radiated toward the Earth's surface. The transmitter most often employs solid-state technology, such as Gallium Arsenide (GaAs) or Gallium Nitride (GaN) to emit high-frequency, high-power signals. The frequency and pulse shape of the transmitter define the radar's resolution and penetration depth in various materials (e.g., soil, water, or plants).

Receiver: The receiver is tasked with detecting the radar signal reflected (backscattered) off the Earth's surface and converting it to a useful electronic signal. It must be very sensitive, since the signal reflected is usually much lower in power than the transmit signal. The receiver is tuned to detect the frequency shift due to the Doppler effect, which is essential for image formation and azimuth resolution.

Signal Processing Unit: The digital signal processor (DSP) or signal processing unit is used for filtering and processing the incoming raw radar signals into the system. The DSP executes critical operations, such as noise elimination, signal amplification, and data pre-processing for image reconstruction. The DSP is also in charge of data acquisition, saving the radar echoes received for further analysis and image creation.

Power Supply and Thermal Management: SAR systems need a reliable power supply to power the transmitter, receiver, and other electronic devices. Spaceborne SAR systems are usually powered by solar panels, with thermal management systems to avoid heating of sensitive components, especially in the space environment.

Interface and Communication Systems: Up-to-date SAR systems, particularly satellite ones, have complex data interfaces and communication. They consist of on-board processors to control data acquisition and delivery to the ground station. Real-time or near-real-time applications depend on efficient data transmission and compression, wherein huge amounts of SAR data are transmitted from satellites to the Earth.

1.3.2 Antenna Design and Beamforming

The antenna is an important factor in deciding the performance of a SAR system, especially image resolution and coverage. The radar antenna transmits the radar waves to the Earth and receives the reflected signals.

Antenna Types: SAR systems most often employ phased array antennas; these allow the radar beam to be electronically steered without physically moving the antenna. Such a feature is essential for a quick data acquisition over a large area. Phased array antennas utilize individual elements, which are scanned at varying phases to manipulate the direction and character of the radar beam [24]. Such antennas are especially valuable in spaceborne SAR systems, where mechanically steerable large antennas are not feasible.
Aperture Size and Resolution: The resolution of the system is controlled by the aperture size and geometry of the antenna. Greater apertures enable higher azimuth resolution since they can focus on smaller areas of the Earth's surface. The motion of the platform created the effective aperture in the SAR. However, the physical antenna size affects the performance of the entire system, particularly in the range direction.
Beamforming: Beamforming is a method employed to direct the radar beam. By rephasing signals radiated from the various antenna elements, the system can point the radar energy in a certain direction. This helps enhance signal-to-noise ratio (SNR), as well as improve the resolution in the target region [25]. In SAR systems, beamforming is applied to dynamically steer the radar beam and maximize data acquisition during the platform's motion.
Antennas for Specific Applications: Various antenna types are utilized based on the application. Low-frequency antennas (e.g., L-band or P-band), for instance, are the best for penetrating soil or vegetation, whereas high-frequency antennas are employed for fine-resolution surface feature imaging, like urban scenes or infrastructure. Utilizing dual-polarized antennas (horizontal and vertical polarization) also facilitates better target characterization and discrimination and is a requirement for advanced SAR applications such as PolSAR.

1.3.3 Data Acquisition and Signal Processing

Signal processing and data acquisition are at the heart of SAR's capability to formulate high-resolution images from raw radar signals. The radar's reflections are recorded over time by the data acquisition system, and advanced signal processing algorithms build up these reflections to form an image [26].

Pulse Compression: Pulse compression is utilized to obtain good range resolution in SAR. Pulse transmission is often very long compared to the required resolution; compression techniques for pulses are implemented to compact the pulse and improve the system's resolving capability against range-side lobed targets closely spaced from one another. Modulation in a chirp sequence is normally applied to produce pulses with varying linear frequency as time progresses.

Range and Azimuth Resolution: The two principal resolution types in SAR systems are range resolution (capable of distinguishing targets at various distances) and azimuth resolution (capable of distinguishing targets along the path of flight of the platform). Fine range resolution is obtained by reducing the pulse length, whereas azimuth resolution is obtained by simulating a large aperture through radar platform movement, as explained above.

Signal Processing Techniques: SAR signal processing requires sophisticated algorithms to transform unprocessed radar data into sharpened images.

a) **Matched Filtering:** Employed in range compression, this method enhances the signal's resolution and noise reduction.
b) **Fourier Transform:** A mathematical algorithm applied to transform the data from the time domain to the frequency domain to facilitate examination of Doppler shifts and other frequency-based features.
c) **Synthetic Aperture Processing:** Synthetic aperture processing employs the motion of the radar platform to produce a large aperture, enhancing azimuth resolution.
d) **Interferometric SAR (InSAR):** A method employing phase differences between two or more SAR images to quantify ground displacements, which is particularly beneficial for monitoring subsidence, earthquakes, and other geophysical processes.

Data Calibration and Noise Reduction: Once the raw data has been received, calibration and noise reduction algorithms are performed to enhance the quality of the SAR image. Calibration corrects for system biases, such as antenna performance variations, whereas noise reduction filters out random errors (e.g., thermal noise or interference) that can attenuate the quality of images.

Data Storage and Processing Speed: SAR systems produce large amounts of data, particularly in satellite-based systems, where there is a need for high-frequency sampling. Data storage and processing systems must process this data effectively, typically with high-performance computing (HPC) systems for real-time or near-real-time processing. Onboard processing is also implemented in some systems to minimize data volume before transmission to ground stations.

1.3.4 Advances in Synthetic Aperture Radar System Architecture

Advances in SAR system architecture have recently been prompted by technology improvements in computing power, sensor design, and miniaturization. These have translated into a marked improvement in both the performance and the range of applications of SAR systems [27].

Miniaturization and Compact SAR Systems: Advances in the miniaturization of SAR technology over recent years have enabled the application of such technology on smaller, multi-purpose platforms such as drones, CubeSats, and handheld devices. Miniaturization makes the cost-effective and effective application of SAR for both military and civilian use feasible. CubeSats, for example, are low-cost, small satellites that can support SAR payloads, which can be made more affordable to organizations that were not financially able to support expensive satellite systems previously. Miniature SAR systems, for example, within drones can reach remote or dangerous locations for disaster relief, environmental monitoring, and agriculture. Hand-held platforms, as with automobiles, can also utilize miniaturized SAR systems to collect real-time, on-the-ground data [28]. The push toward smaller yet more capable SAR units is pushing innovation, allowing more cost-effective and regular Earth observation in industries such as agriculture, urban planning, and infrastructure monitoring. Scientists are also optimizing these small systems' performance using better signal processing and AI to make them even more impactful in delivering real-time information. Miniaturization makes SAR technology more accessible, flexible, and useful for many types of applications.

Multi-frequency and Multi-polarization Systems: New SAR systems are under development to operate at multiple frequencies and polarizations, enhancing the amount of information that can be obtained. Multi-frequency SAR can detect a wider range of surface types, while multi-polarization provides better information on surface characteristics, such as vegetation type, moisture content, and roughness. These capabilities are particularly useful for environmental monitoring, agriculture, and geophysical surveys.

Interferometric SAR and Time Series Analysis: Advancements in InSAR technology permit the accurate quantification of displacement on the ground surface and time. Temporal analysis of SAR provides tracking of subsidence, movement due to volcanic eruption or movement, glacier motion, as well as worthwhile input data in geophysics and environmental research.

Table 1.5 Recent advancements in SAR system architecture.

Advancement area	Description	Impact
Miniaturization and Compact SAR Systems	Smaller, lighter components allow SAR deployment on drones, CubeSats, and mobile units.	Enables flexible, cost-effective deployment across civilian and military applications, expanding accessibility.
Multi-frequency and Multi-polarization Systems	New systems operate across multiple frequencies and polarizations, gathering more data.	Provides detailed insights into surface types, vegetation, moisture, and roughness, benefiting environmental monitoring and agriculture.
Interferometric SAR and Time Series Analysis	Advances in InSAR technology allow for accurate measurements of surface displacements over time.	Useful for monitoring ground movements, such as subsidence, volcanic activity, and glacier movement in geophysical and environmental research.
Artificial Intelligence and Machine Learning	Techniques in AI and ML automate image interpretation, improve detection, and enhance analysis.	Speeds up data analysis and enhances accuracy in applications like disaster response, urban planning, and defense.
Synthetic Aperture Radar Constellations	Development of radar constellations of satellites or UAVs working together to provide more coverage.	Increases revisit frequency, enables real-time monitoring, and improves data availability for critical regions.

Artificial Intelligence (AI) and Machine Learning (ML): Employing ML AI algorithms in SAR systems is revolutionizing SAR data processing. Both AI and ML approaches can be applied to interpret in an automated mode, enhance object detection, and maximize change detection accuracy. Artificial intelligence applications can be developed to identify SAR image patterns with faster and higher accuracy results in applications like disaster response, city planning, and defense.

Synthetic Aperture Radar Constellations: The potential for future SAR systems lies in developing constellations, or sets of satellites or UAVs, carrying SAR sensors that collaborate to make more frequent and more comprehensive measurements of the Earth's surface. The constellations can provide more frequent revisits, augment the data availability, and allow for real-time monitoring of crucial areas, including zones of catastrophe or war.

The SAR system design is an advanced, multi-disciplinary problem that needs the highest-level radar hardware, cutting-edge antenna design, and signal-processing algorithms. The insistent advancement of technology in system design, miniaturization, and numerical methods is pushing the frontiers of SAR development to be deployed in an increasingly broad array of fields and applications [29]. This table (Table 1.5) presents the most significant developments in SAR system architecture and their impact on improving the performance, accessibility, and flexibility of SAR technology.

1.4 Geospatial Analysis with Synthetic Aperture Radar Data

Synthetic aperture radar technology has become a valuable resource for geospatial analysis with high-resolution, all-weather, and day/night imaging capabilities for monitoring and observing the Earth's surface. Precise processing and interpretation of SAR data are required to harness its full potential in various geospatial applications such as mapping, surveillance, and disaster management [30]. It addresses the basic components of SAR geospatial analysis, including georeferencing, calibration, interpretation, and advanced techniques like change detection, interferometry, and polarimetry.

1.4.1 Georeferencing and Coordinate Systems

Georeferencing aligns SAR images into a coordinate system such that the accurate projection of the image data onto the real location is possible. This is the important component of integrating SAR data with other geospatial data, such as GIS layers, satellite images, or topographic maps [31].

Coordinate Systems: SAR data is often acquired in a radar coordinate system, depending on the radar platform position (e.g., airplane or satellite) when recorded. Georeferencing involves projecting this radar coordinate system into a geographic coordinate system (e.g., geodetic coordinates, UTM (Universal Transverse Mercator), or latitude/longitude). To perform this, the radar orientation, location, and sensor parameters, and the Earth model of an ellipsoid must be known.
Coordinate Transformation: The SAR data georeferencing normally involves radar geometry correction, which considers the sensor look angle, platform motion, and orbit track. For example, slant range must be converted to ground range, the actual distance over the Earth's surface. In addition, geodetic transformation is used to transform the SAR image from radar coordinates to a more conventional geographic coordinate system.
Ground Control Points (GCPs): To enhance georeferencing accuracy, ground control points may be employed. These are recognized points on the Earth's surface with known geographic coordinates, located in the SAR image, as well as in related ground truth information (e.g., GPS). The GCPs help remove errors induced by the movement of the platform or by other inaccuracies in the radar data.
Digital Elevation Models: To compensate for topographic distortions and position objects in their proper 3D locations, digital elevation models are routinely used. They provide elevation data for the Earth's surface so that radar images can be corrected to portray the true surface topography, especially in mountainous or hilly regions.

1.4.2 Synthetic Aperture Radar Image Calibration and Interpretation

The calibration and interpretation of SAR images are crucial steps in determining the usability and accuracy of the data for geospatial analysis. Calibration ensures that the values of backscatter on the SAR image are accurately interpreted in real surface properties, whereas interpretation focuses on extracting meaningful information from the SAR imagery [32].

Radiometric Calibration: Radiometric calibration compensates for the system's inherent bias and environmental factors that may impact the radar signal, including differences in sensor performance, atmospheric conditions, or surface backscattering characteristics. Calibration ensures that backscatter measurements are consistent and comparable over time or between locations.

Geometric Calibration: Geometric calibration ensures the SAR image is correctly aligned geographically and accurate in space. This corrects the distortions that result from sensor motion and look angle and is especially vital for SAR sensors gathering data on moving platforms (e.g., aircraft or satellites). Without geometric calibration, spatial locations of items within the image might be incorrectly represented, with inaccurate conclusions made in geospatial analysis.

Interpretation of SAR Data: Interpreting SAR data means examining the backscatter intensity and patterns to derive information regarding the Earth's surface.

a) **Surface Type Identification:** Separating among different surface types, like water bodies, urban, forest, or agricultural lands, using the backscatter signal.
b) **Surface Roughness Analysis:** Quantifying the smoothness or roughness of a surface. For example, smoother surfaces (such as water) generate less backscatter, whereas rough surfaces (such as forests or mountains) generate more scattering and produce greater backscatter.
c) **Vegetation and Water Analysis:** Varied surfaces like vegetation and water features result in unique backscatter signatures, which can be utilized to classify and observe these targets.

Image Enhancement Techniques: High-level image processing methods, including speckle filtering, contrast enhancement, and edge detection, may enhance the quality of SAR images. These methods reduce noise (speckle) that is inherent in radar images and emphasize important features for easier interpretation, particularly in intricate areas with a high degree of surface variation [33].

1.4.3 Advanced Data Analysis Techniques: Change Detection, Interferometry, and Polarimetry

Synthetic aperture radar data analysis with advanced techniques makes it possible to retrieve more comprehensive and subtle information from SAR images, including environment changes, surface deformations, and material properties.

Change Detection: Change detection with SAR requires the comparison of two or more SAR images over the same area at different points in time. This method is capable of disclosing changes in the terrain,

a) **Urban development:** Identifying new roads or buildings.
b) **Agricultural changes:** Crop growth or deforestation identification.
c) **Disaster monitoring:** Measuring damage resulting from earthquakes, floods, or landslides.

The greatest challenge with change detection is the need to compensate for noise and other variations in the images that are irrelevant. Image differencing, coherence mapping, or time series analysis techniques are employed to delineate important surface changes over time [34].

SAR Interferometry: Interferometric SAR (InSAR) is an excellent method for measuring surface displacement at the level of a millimeter. InSAR can be used to measure very small vertical or horizontal motions of the Earth's surface by comparing the phase difference between two or more SAR images, including:

a) **Land subsidence:** Sinkhole or urban subsidence monitoring.
b) **Earthquake displacements:** Quantifying ground deformation resulting from tectonic processes.
c) **Glacial motion:** Tracking the movement of ice sheets or glaciers.

InSAR is based on the coherence concept, the measure of how similar two SAR images are to each other. As long as there has not been substantial movement or alteration in the area between two SAR acquisitions, the coherence will be low, indicating surface movement. DInSAR can be employed to observe temporal changes in movement, e.g., monitoring a subsidence in progress [35].

Synthetic aperture radar polarimetry: PolSAR is the processing of the polarization of the radar signal, which yields further information on the Earth's surface. Transmitting and receiving with various polarized radar signals (e.g., horizontal or vertical) enables PolSAR to better characterize surface properties, including:

a) **Vegetation type and density:** Polarimetric data can discriminate between vegetation types (e.g., forest versus agricultural land) and estimate biomass.

b) **Surface roughness:** PolSAR may give indications regarding the surface roughness, i.e., identifying smooth versus rough surfaces.
c) **Water bodies:** PolSAR assists in the identification of water bodies by examining how the radar signal responds to the water surface.

Table 1.6 Overview of advanced SAR data analysis techniques: change detection, InSAR, and PolSAR.

Technique	Applications	Key considerations	Statistical data
Change Detection	**Urban Development:** New roads/ buildings. **Agricultural Changes:** Crop growth/ deforestation. **Disaster Monitoring:** Earthquake/ flood damage.	Requires noise reduction techniques. Uses image differencing, coherence mapping, or time series analysis.	**Urban Growth:** 20% increase in urban area over 5 years (Shanghai). **Deforestation:** 0.8% forest loss annually (Amazon). **Disaster:** 5–10% ground displacement (after earthquakes).
Interferometry synthetic aperture radar	**Land Subsidence:** Sinkholes/urban subsidence. **Earthquake Displacements:** Ground deformation. **Glacial Motion:** Ice sheet movement.	Measures millimeter-level displacement. Requires good coherence between images.	**Subsidence:** 30 cm/ year (Mexico City). **Earthquakes:** 2.5 meters displacement (Haiti 2010). **Glaciers:** 2–5 cm/ year movement (Antarctica).
Polarimetry synthetic aperture radar	**Vegetation:** Discriminates vegetation types and density. **Surface Roughness:** Smooth vs. rough surfaces. **Water Bodies:** Identifies water.	Enhances surface classification using multiple polarizations.	**Vegetation:** 90% biomass vs. 10–20% biomass (dense forest vs. sparse land). **Roughness:** -10 dB (rough) vs. -30 dB (smooth). **Water:** 97% detection accuracy.

The application of several polarizations in PolSAR greatly enhances surface classification and the detection of fine variations in surface properties, which makes it useful for environmental monitoring, land use/land cover classification, and resource management [36]. This table (Table 1.6) gives a brief description of three sophisticated SAR data processing methods: Change detection, InSAR, and PolSAR. It indicates their uses, important aspects, and statistical data examples, reflecting their strengths in observing environmental and structural alteration.

1.4.4 Application Case Studies: Mapping, Surveillance, and Disaster Management

Synthetic aperture radar technology has been utilized in industries, providing vital information for mapping, surveillance, and disaster relief. Some examples of real-world case studies that illustrate SAR's value and versatility are given below.

Mapping and Land Use Change: SAR has been utilized extensively for mapping and monitoring long-term land use and land cover change. For instance, SAR data have been utilized in the Amazon rainforest to track deforestation. Through analyzing SAR images across various years, scientists can detect areas where the forest cover has been removed due to agriculture, logging, or infrastructure construction [37]. The application of change detection methods assists in monitoring deforestation patterns and guides policy interventions in sustainable land use management.
Surveillance and Security: In surveillance, SAR is also responsible for border monitoring and security, illegal activity detection, and situational awareness in battlefield environments. For example, SAR is utilized in maritime surveillance to track shipping routes and identify illegal fishing or unreported ships. SAR solutions with high-resolution imaging can identify various types of ships even during bad weather, offering real-time intelligence for security forces.
Disaster Management and Emergency Response: SAR is extremely useful in disaster management, particularly after natural disasters like earthquakes, floods, or hurricanes. SAR systems can respond quickly by generating timely, high-resolution imagery of the affected region. For instance, after the 2010 Haiti earthquake, SAR was employed to map surface displacements and evaluate damage to infrastructure. By monitoring ground movements using InSAR, officials would be able to gauge the intensity of the earthquake and determine where emergency assistance was most needed [38].

Synthetic aperture radar is applied in flood monitoring to identify the water level changes and flood extent by comparing post-event and pre-event SAR images. This aids in disaster relief efforts by identifying flooded areas and allocating resources to affected communities.

The SAR imagery provides high-accuracy and dynamic geospatial analysis, which is of enormous benefit in interpreting surface conditions, temporal changes, and environmental dynamics. The incorporation of SAR into cutting-edge techniques such as change detection, interferometry, and polarimetry allows for improvements in the accuracy of mapping, monitoring, and decision-making over various applications [39].

1.5 Challenges and Future Directions

Although SAR technology has greatly transformed remote sensing and geospatial analysis, various challenges must be resolved to realize its potential. The subject is rapidly transforming with developments in signal processing, sensor technology, and computational methodologies that hold out the promise to overcome these hindrances.

1.5.1 Signal Processing Challenges in Synthetic Aperture Radar

Synthetic aperture radar technology is built on signal processing at its core since it facilitates the processing of radar echoes into recognizable images. With notable improvements notwithstanding, there remain some challenges of SAR signal processing that impact SAR data quality and efficiency [40].

Speckle Noise: Speckle noise is a salt-and-pepper-like granular pattern of noise caused by the interference of coherent radar waves that are scattered from different surface features. It is among the most intractable problems in SAR imagery. Speckle diminishes the quality of images and may render small features or slight changes in the terrain difficult to discern. More sophisticated denoising methods, including wavelet filtering or statistical modeling, are needed to remove speckle, but obtaining the best results without sacrificing detail is still a problem. Current approaches to speckle noise in SAR images aim at sophisticated methods that minimize noise while maintaining significant details. Deep learning techniques, particularly convolutional neural networks, are very efficient at learning to isolate speckle from genuine image features, resulting in cleaner images. Other methods include wavelet transformations, which partition the image into various frequency components to eliminate noise, and filtering by non-local means, which matches image patches to eliminate speckle. Anisotropic diffusion preserves edges while minimizing noise, and PolSAR imagery enhances noise elimination by employing several polarizations of the radar. Multi-look processing, which pools several SAR images together and then averages them, also minimizes speckle by eliminating random noise [41]. These new techniques, most of which utilize machine learning and sophisticated filtering, are enhancing SAR image quality and making the technology more useful for a wide range of applications.

Motion and Geometric Distortions: SAR data are prone to geometric distortions resulting from the relative motion of the radar sensor with respect to the target surface. Such distortions are significant in airborne SAR platforms or spaceborne platforms with non-ideal orbits. Correctly processing such distortions is computationally intensive and demands accurate motion models as well as calibration methods. Furthermore, mispositioning and a mis-set attitude of the platform can cause severe distortions and compromise image quality.

High-Resolution Imaging and Data Volume: SAR images at high resolution create tremendous amounts of data, presenting data storage, transmission, and processing problems. Finding a compromise between image resolution and data volume is an ever-existing problem, particularly in real-time applications. Approaches such as compressed sensing or data compression algorithms are considered in solving the problem, but ensuring that high-resolution imaging is possible and efficient is problematic.

Multi-dimensional Data Handling: SAR systems tend to gather multi-dimensional data, such as time-series, polarization, and interferometric data. Processing such multi-dimensional datasets demands advanced algorithms that can efficiently handle large, complex volumes of data while maintaining spatial and temporal correlations. Conventional techniques cannot extract useful information efficiently from such datasets, causing image processing and analysis delays.

Real-time Processing: For most applications, particularly in defense or emergencies, real-time processing of SAR data is essential. Processing data on board or in cloud platforms for quicker analysis is a major challenge. Effective algorithms and hardware to compress, process, and transmit data in real-time are still being developed.

1.5.2 Technological Innovations: From Traditional to Quantum Synthetic Aperture Radar

The future of SAR technology is defined by several groundbreaking innovations, ranging from breakthroughs in conventional radar technology to the emergence of quantum radar technology. These advances will widen the applications of SAR systems and overcome some of the present challenges.

Advances in Antenna and Transmitter Design: Conventional SAR systems use heavy, bulky antennas and sophisticated radar waveforms, but new technologies in phased-array and reconfigurable antennas are transforming SAR system design. Phased-array antennas, which electronically steer radar beams, enable more compact, lightweight, and agile systems. Metamaterials are also being investigated for their potential to improve antenna efficiency and size, reducing the overall spaceborne and airborne SAR payload footprint and weight.

Digital Beamforming: Digital beamforming technology is a paradigm shift in SAR system flexibility and performance. It offers the synthesis of multiple radar beams with higher precision and lower power consumption. It provides dynamic control of the direction of the radar beam without mechanical antenna movement, making the system more efficient and with higher resolution. It is especially advantageous in spaceborne SAR systems, where weight and power must be minimized.

Synthetic Aperture Radar in the Quantum Age: Quantum radar, a latest development in radar technology, holds the potential to transform SAR systems. Leaning on quantum properties of light, such as quantum entanglement, quantum SAR systems can offer never-before-imagined resolution, increased sensitivity, and the ability to identify weak signals even in the presence of noise. This proves useful in hostile environments where standard radar waves would be overwhelmed by interference or noise. Quantum radar also has the potential for improved security and anti-jamming capability, which could make SAR systems less susceptible to electronic warfare [42].

Artificial Intelligence in SAR Signal Processing: AI and ML techniques are finding their way in SAR systems, from the acquisition to the analysis phase. Deep learning, and specifically convolutional neural networks, are being utilized for autonomous target recognition, image classification, and change detection. The AI-based SAR systems can process vast volumes of information almost in real-time, enabling the identification of patterns and anomalies that would be dissipated under normal methods [43]. Furthermore, AI can optimize image processing, denoising, and resolution enhancement, reducing processing time while enhancing accuracy.

Integration with Other Sensors: Another innovation is integrating SAR with other remote sensing methods, including optical, light detection and ranging (LiDAR), and thermal infrared measurements. Multi-sensor fusion also increases SAR capability to better delineate surface particulars. For instance, integration with optical imagery can exceed cloud cover or low-visibility constraints, and SAR is an even better tool for environmental monitoring, urban planning, and disaster management.
Miniaturization and CubeSats: Miniaturization is becoming more available for SAR technology. Low-cost, small-scale systems using miniature satellites, or CubeSats, that carry onboard SAR systems are being launched in high numbers. With these, there can be frequent monitoring and global coverage on a large scale, offering democratized access to high-resolution SAR imagery. The systems can be implemented in constellations to offer improved revisit times with more reporting on changing conditions.

1.5.3 Emerging Applications and the Future of Synthetic Aperture Radar Technology

As SAR technology continues to develop, its applications are being widened to a wide range of industries, from environmental monitoring to defense, agriculture, and urban planning.

Disaster and Emergency Response: SAR's ability to generate high-resolution information in any conditions and during day or night makes it an invaluable instrument in disaster monitoring and emergency response. For instance, the SAR is increasingly used for flood mapping, landslide monitoring, and post-earthquake loss estimation. As SAR technology continues to improve to become more accurate and, in a position, to provide near-real-time data, it will be at the forefront of predicting disasters, emergency response, and recovery.
Agriculture and Land Use Monitoring: The SAR is utilized in monitoring crop health, irrigation levels, and the moisture levels of soil in agricultural land. It is also possible with this technology to differentiate crop type and drought/pest-caused stress conditions. This periodical monitoring in farmland assists policymakers, as well as farmers, by conveying effective details concerning farm outputs, land degradation, and sustainability-based cultivation methods.
Urban Monitoring and Infrastructure Management: The potential of SAR for monitoring urban growth, infrastructure construction, and environmental hazards is vast. Urban areas are more susceptible to subsidence, flooding, and structural change, and SAR can monitor minute deformations of buildings, bridges, and roads. This capability is critical

for infrastructure management, urban planning, and ensuring the climate resilience of cities and other environmental hazards.

Ocean and Maritime Surveillance: The SAR is also valuable for ocean monitoring, providing information on sea surface height, wave patterns, and shipping traffic. Its day-night, all-weather capability renders it ideal for shipping route surveillance, illegal fishing detection, and oil spill observation. The SAR is also employed for iceberg detection and polar region monitoring, giving valuable data for climate science.

Climate Change and Environmental Monitoring: The SAR is critical in observing climate-change-induced changes in the Earth's surface, including glacier withdrawal, land subsidence due to the melting of permafrost, and deforestation. Its capacity to track vegetation cover, soil humidity, and carbon sequestration is critical in determining the effects of anthropogenic activities and natural events on the environment. With increasingly visible impacts from climate change, SAR will take on a primary function of supplying actionable information for researchers and policymakers.

Security and Defense Applications: The SAR is still an indispensable tool for defense and national security, especially in border monitoring, target tracking, and military surveillance. The SAR's technological advancements, especially with the combination of AI and quantum radar, will improve defense systems' security and detection levels. The SAR will also remain integral to disaster relief missions in areas of conflict, offering geospatial intelligence even in denied or contested environments.

The prospects for SAR technology are excellent, with continuous innovations in signal processing, hardware development, and computational methods continuing to push the frontiers of what can be achieved. Challenges, including speckle noise, volume of data, and real-time processing, persist, but advances in the fields of AI, quantum radar, and miniaturization will ensure these constraints are addressed [44, 45]. New uses in disaster relief, earth observation, agriculture, and military defense will propel the future development of SAR systems, solidifying their status as a necessity for global surveillance, security, and decision-making for decades to come.

1.6 Conclusion

Synthetic aperture radar technology has become a leading, potent, and multipurpose Earth observing tool, giving high-resolution imagery independent of weather and the time of day. From the underlying principles established in radar physics to the interface with sophisticated signal processing algorithms, the SAR has proved to be worth its weight in gold in the broad spectrum of scientific, commercial, and defense applications. Looking ahead, the SAR's contribution to Earth observation will grow as advancements in technology, data analysis, and application areas keep advancing the field. This chapter summarizes the main findings of the previous chapters and investigates the essential role the SAR will have in future Earth observation systems.

1.6.1 Summary of Key Insights

Synthetic aperture radar uses high-frequency radio waves to acquire fine-scale images of the Earth's surface by measuring the reflected signal. The SAR is distinct from optical imaging because it can function regardless of weather or day/night, and is suitable for global monitoring and ecological tracking. The SAR is based on principles such as the wave propagation of electromagnetic waves and the Doppler effect, with growing accuracy being presented due to radar hardware and signal processing advancements. Digital beamforming and phased-array antennas are just a few examples of technologies that have enabled SAR to be implemented across multiple platforms, from satellites to unmanned air vehicles. The capability to supply precise, georeferenced data makes SAR beneficial for such uses as land use management, calamity assessment, and infrastructure monitoring. Its challenges, such as speckle noise, high data volume, and real-time computation, need continuous research studies. Emerging technologies like AI, machine learning, and quantum radar offer solutions to these and are enhancing SAR efficiency and accessibility. The SAR is venturing into new areas like agriculture, climate monitoring, and national defense, and its ability to monitor environmental change makes it a powerful tool in solving global issues like climate change and food security.

1.6.2 The Role of Synthetic Aperture Radar in Future Earth Observation Systems

While Earth observing systems evolve in the coming decades, SAR will lead the way toward remote sensing's future. Some of the most important factors that indicate the significance of SAR in the next generation of Earth observation systems are:

Comprehensive Global Monitoring: As climate change, natural disasters, and environmental degradation continue to pose problems, there arises a greater need for persistent and dependable monitoring globally. SAR's capability of day/night, all-weather imaging is essential in ensuring that continuous surveillance of the Earth's surface is sustained. SAR systems will play a central role in tracking global environmental alterations such as forest cover loss, ice sheet dynamics, land subsidence, and coastal erosion.

Enhanced Disaster Risk Management: The ability of SAR to measure surface deformation (e.g., using InSAR) and detect environmental changes will be critical in disaster risk management. The combination of SAR with other Earth observation data, including optical imagery and weather data, will enable decision-makers to receive timely and accurate information during natural disasters. Whether it is flood mapping, earthquake damage, or wildfire detection, SAR will remain an important asset in limiting the effects of disasters and enhancing the effectiveness of emergency response operations.

Smarter and More Efficient Data Integration: Future Earth observation networks will be based on multi-sensor data fusion, integrating SAR data with other remote sensing modalities, e.g., optical, thermal, and LiDAR data. Such an integrated system will produce richer information on land cover, urban sprawl, water resources, and environmental well-being. For instance, the fusion of SAR and optical data can improve land use classification and vegetation health monitoring, whereas the fusion of SAR and LiDAR can improve topographic modeling in mountainous regions.

Expanding the Role of SAR in Space-Based Platforms: With the advent of small satellites and CubeSats, spaceborne SAR will be more economical, making it possible to monitor regional and global phenomena frequently and in detail. Constellations of small SAR satellites will enhance revisit times, with near-real-time data for essential applications like crop monitoring, urban growth, and disaster relief. The advancements in miniaturization and quantum radar will also enable new forms of spaceborne SAR systems with better sensitivity and resolution.

Artificial Intelligence-Based SAR Data Analytics: Machine learning and AI are changing the nature of SAR processing and interpretation by making it user-friendlier and more efficient. The AI-powered target detection, change detection, and image segmentation automated processes streamline SAR systems by analyzing faster and accurately without the need for expert operation. Deep learning and neural network progress in the last few years have enhanced SAR capacity for object detection as well as distinguishing environmental modifications under adverse situations. These AI-powered features enable real-time processing of data, making SAR systems more effective in applications like agriculture, defense, disaster management, and environmental monitoring. Artificial intelligence also helps in system calibration and the correction of errors, making the overall accuracy of SAR data greater. Therefore, AI-based SAR analytics provide more actionable intelligence in a timely fashion, facilitating improved decision-making across industries.

Enabling Sustainable Development Goals: SAR technology's ability to track and fulfill the United Nations' Sustainable Development Goals cannot be denied. A picture of how deforestation and land degradation, monitoring water resources, and agricultural sustainability have been depicted as examples is presented. By tracking urban growth, climate change, and detecting environmental threats, the SAR will directly contribute to making the world adopt sustainable and resilient infrastructure, as well as improve natural resources management [46].

Lastly, SAR technology is going to remain the backbone of Earth observation systems due to its unique ability to acquire high-quality, day-night, and all-weather imagery. Its application across different sectors, ranging from climate monitoring to disaster management, security, and agriculture, makes it evident that it plays a key role in solving some of the world's most complex challenges. As SAR continues to develop further, driven by the development of AI, quantum radar, and miniaturized sensors, its role in future Earth observation systems will become ever more central, delivering comprehensive, reliable, and timely information to guide decision-making and sustainable development on a worldwide basis.

AI disclosure statement: The author acknowledges the use of AI-based tools for minor language editing or grammatical correction. However, all intellectual contributions, analyses, and conclusions presented in this chapter are entirely the author's own.

References

1. Hein, A., Introduction, in: *Signals and Communication Technology*, pp. 1–15, 2004, doi: 10.1007/978-3-662-09457-0_1.
2. Pavlikov, V., Zhyla, S., Pozdniakov, P., Kolesnikov, D., Cherepnin, H., Malashta, P., Basic Geometric Relations, Parameters and Characteristics of Theory for Phantoming of SAR Images, in: *17th International Conference on Advanced Trends in Radioelectronics, Telecommunications and Computer Engineering (TCSET)*, Lviv, Ukraine, pp. 312–316, 2024, doi: 10.1109/TCSET64720.2024.10755612.

3. Hein, A., SAR basics, in: *Signals and Communication Technology*, pp. 17–97, 2004, doi: 10.1007/978-3-662-09457-0_2.
4. Yamaguchi, Y., Introduction, pp. 1–15, CRC Press eBooks, Boca Raton, 2020, doi: 10.1201/9781003049753-1.
5. Gough, P. and Hayes, M., Synthetic Aperture Sonar: The Past, The Present and The Future, in: *Sonar Signal Processing*, 2004, doi: 10.25144/18014.
6. Caprais, P., Towards A Forward-Looking Synthetic Aperture Sonar, in: *Sonar Signal Processing*, 1998, doi: 10.25144/19042.
7. Kumar, S., Siqueira, P., Govil, H., Agrawal, S. (Eds.), *Spaceborne Synthetic Aperture Radar Remote Sensing: Techniques and Applications*, CRC Press, 2023, https://doi.org/10.1201/9781003204466.
8. Bamler, R., Principles of Synthetic Aperture Radar. *Surv. Geophys.*, 21, 2/3, 147–157, 2000, Doi:10.1023/a:1006790026612.
9. Nicolas, J. and Adragna, F., The Principles of Synthetic Aperture Radar, in: *Processing of Synthetic Aperture Radar Images*, Wiley Online library, pp. 25–55, Portico, 2008, doi: 10.1002/9780470611111.ch2.
10. Macdonald, G.J., Scharien, R.K., Duncan, K., Farrell, S.L., Rezania, P., Tavri, A., Arctic Sea Ice topography information from RADARSAT Constellation Mission (RCM) Synthetic Aperture Radar (SAR) Backscatter. *Geophys. Res. Lett.*, 51, 4, Feb. 2024, doi: 10.1029/2023gl107261.
11. Rysz, M., Tsokas, A., Dipple, K.M., Fair, K.L., Pardalos, P.M., Synthetic Aperture Radar (SAR) data Applications, in: *Springer Optimization and Its Applications*, 2022, doi: 10.1007/978-3-031-21225-3.
12. Yamaguchi, Y., *Polarimetric SAR Imaging*, CRC Press eBooks, 2020, https://doi.org/10.1201/9781003049753.
13. Kazar, B.M. and Celik, M., Theory behind the SAR Model, in: *SpringerBriefs in computer science*, pp. 7–17, 2012, doi: 10.1007/978-1-4614-1842-9_2.
14. Krishnapuram, B., Sichina, J., Carin, L., Physics-based detection of targets in SAR imagery using support vector machines. *IEEE Sens. J.*, 3, 2, 147–157, Apr. 2003, doi: 10.1109/jsen.2002.805552.
15. Munoz-Ferreras, J.M. and Perez-Martinez, F., On the Doppler Spreading Effect for the Range-Instantaneous-Doppler Technique in Inverse Synthetic Aperture Radar Imagery. *IEEE Geosci. Remote Sens. Lett.*, 7, 1, 180–184, Oct. 200, doi: 10.1109/lgrs.2009.2030372.
16. Wang, Y., Inverse synthetic aperture radar imaging of manoeuvring target based on range-instantaneous-Doppler and range-instantaneous-chirp-rate algorithms. *IET Radar Sonar Navig.*, 6, 9, 921–928, Dec. 2012, doi: 10.1049/iet-rsn.2012.0091.
17. Meng, L., Yan, C., Yan, X.-H., Unlocking the power of synthetic aperture radar for geosciences. *Eos*, 105, Oct. 2024, doi: 10.1029/2024eo245030.
18. Chua, M.Y., SAR image formation, pp. 231–260, CRC Press eBooks, 2023, doi: 10.1201/9781003282693-8.

19. Rahayu, E.S., Hidayat, R., Ariananda, D.D., Iswandi, Sequential-Compressive-Range Azimuth Estimation in Radar Signal Processing, in: *13th International Conference on Information Technology and Electrical Engineering (ICITEE)*, pp. 189–195, 2021, Doi: 10.119/icitee53064.2021.9611822.
20. Rahayu, E.S., Ariananda, D.D., Hidayat, R., Single Snapshot-Spatial compressive beamforming for azimuth estimation and backscatter reconstruction. *2020 3rd International Seminar on Research of Information Technology and Intelligent Systems (ISRITI)*, pp. 245–250, Dec. 2020, doi: 10.1109/isriti51436.2020.9315503.
21. Cruz, H., Véstias, M., Monteiro, J., Neto, H., Duarte, R.P., A review of Synthetic-Aperture Radar Image Formation Algorithms and Implementations: A Computational Perspective. *Remote Sens.*, 14, 5, 1258, Mar. 2022, doi: 10.3390/rs14051258.
22. Singh, P., Shankar, A., Diwakar, M., Review on nontraditional perspectives of synthetic aperture radar image despeckling. *J. Electron. Imaging*, 32, 02, Nov. 2022, doi: 10.1117/1.jei.32.2.021609.
23. Saponara, S., Hardware accelerator IP cores for real time Radar and camera-based ADAS. *J. Real-Time Image Process.*, 16, 5, 1493–1510, Dec. 2016, doi: 10.1007/s11554-016-0657-0.
24. Sahoo, G.S. and Ghosh, A., Antenna array design for higher spectral effeiciency for hybrid beamforming 5G MIMO wireless communication system. *2022 IEEE Wireless Antenna and Microwave Symposium (WAMS)*, Jun. 2022, pp. 1–5, doi: 10.1109/wams54719.2022.9848089.
25. Noordin, N.H. and Z. Khalidin, Z., Beamforming algorithms for adaptive array antenna. *2nd International Conference on Electronic Design (ICED)*, pp. 5–9, 2014, Doi:10.1109/iced.2014.7015761.
26. Sartipi, M. and Fletcher, R., Energy-Efficient data acquisition in wireless sensor networks using compressed sensing. *Data Compression Conference*, Mar. 2011, pp. 223–232, doi: 10.1109/dcc.2011.29.
27. Villano, M., Moreira, A., Krieger, G., Staggered-SAR for high-resolution wide-swath imaging, in: *IET International Conference on Radar Systems (Radar 2012)*, pp. 23–23, 2012, doi: 10.1049/cp.2012.1600.
28. Alghamdi, A., Alshammari, A., Chang, L., Iqbal, A., Mabrouk, I.B., Miniaturized Implantable Antenna with Ultra-Wide Bandwidth Characteristics for Leadless Pacemakers, in: *16th European Conference on Antennas and Propagation (EuCAP)*, Mar. 2024, pp. 1–5, doi: 10.23919/eucap60739.2024.10501669.
29. De Almeida, F.Q., Younis, M., Krieger, G., Moreira, A., On a Dual PRI Pulse Sequence Mode for High-Resolution Wide-Swath SAR Imaging, in: *IEEE International Geoscience and Remote Sensing Symposium IGARSS*, pp. 2967–2970, 2021, doi: 10.1109/igarss47720.2021.9554666.
30. Kamusoko, C., Towards explainable AI and Data-Centric approaches for geospatial data analysis, pp. 228–234, CRC Press eBooks, 2024, doi: 10.1201/9781003398257-13.

31. Khoi, D.D. and Munthali, K.G., Multispectral classification of remote sensing data for geospatial analysis, pp. 13–28, Springer eBooks, 2012, doi: 10.1007/978-4-431-54000-7_2.
32. Imperatore, P., SAR Imaging Distortions Induced by Topography: A compact analytical formulation for radiometric calibration. *Remote Sens.*, 13, 16, 3318, Aug. 2021, doi: 10.3390/rs13163318.
33. Simms, E.L., Image Acquisition Specifications into Effect, pp. 111–122, CRC Press eBooks, 2019, doi: 10.1201/9780429264771-6.
34. Tsokas, A., Rysz, M., Pardalos, P.M., Dipple, K., SAR data applications in earth observation: An overview. *Expert Syst. Appl.*, 205, 117342, May 2022, doi: 10.1016/j.eswa.2022.117342.
35. Sica, F., Cozzolino, D., Verdoliva, L., Poggi, G., The Offset-Compensated nonlocal filtering of interferometric phase. *Remote Sens.*, 10, 9, 1359, Aug. 2018, doi: 10.3390/rs10091359.
36. López-Martínez, C. and Pottier, E., Basic Principles of SAR Polarimetry, in: *Remote Sensing and Digital Image Processing*, pp. 1–58, 2021, doi: 10.1007/978-3-030-56504-6_1.
37. Abdikan, S., Sanli, F.B., Ustuner, M., Calò, F., Land Cover Mapping Using Sentinel-1 Sar Data, in: *the International Archives of the Photogrammetry*, Jun. 2016, vol. XLI-B7, pp. 757–761, doi: 10.5194/isprs-archives-xli-b7-757-2016.
38. Kose, S., Koytak, E., Hascicek, Y.S., An overview on the use of satellite communications for disaster management and emergency response. *Int. J. Emergency Manage.*, 8, 4, 350, Jan. 2012, doi: 10.1504/ijem.2012.051647.
39. Dillon, B., Understanding the emergency response, Oxford University Press eBooks, 2014, doi: 10.1093/law/9780198712909.003.0004.
40. Wang, W.-Q., MIMO SAR imaging: Potential and challenges. *IEEE Aerosp. Electron. Syst. Mag.*, 28, 8, 18–23, Aug. 2013, doi: 10.1109/maes.2013.6575407.
41. Chan, D., Gambini, J., Frery, A.C., Speckle Noise Reduction in Sar Images Using Information Theory, in: *the International Archives of the Photogrammetry*, Nov. 2020, vol. XLII-3/W12-2020, pp. 141–146, doi: 10.5194/isprs-archives-xlii-3-w12-2020-141-2020.
42. Majji, S.R., Chalumuri, A., Kune, R., BS, M., Quantum Processing in Fusion of SAR and Optical Images for Deep Learning: A Data-Centric Approach, 73743–73757, 2022, doi:10.36227/techrxiv. 19753891.v1.
43. Zhou, L., Yu, H., Lan, Y., Xing, M., Artificial intelligence in Interferometric Synthetic Aperture Radar Phase Unwrapping: A review. *IEEE Geosci. Remote Sens. Mag.*, 9, 2, 10–28, Apr. 2021, doi: 10.1109/mgrs.2021.3065811.

44. Moreira, A., Krieger, G., Younis, M., Zink, M., Future spaceborne SAR technologies and mission concepts. *IGARSS 2022 - 2022 IEEE International Geoscience and Remote Sensing Symposium*, Jul. 2023, pp. 576–578, doi: 10.1109/igarss52108.2023.10282219.
45. Douglas, M., Kuster, N., Pokovic, K., Past, present, and future of SAR evaluations, in: *URSI Asia-Pacific Radio Science Conference (URSI AP-RASC)*, pp. 426–428, 2016, doi:10.1109/ursiap-rasc.2016.7601391.
46. Liang, H., Xu, W., Ding, X., Zhang, L., Wu, S., Urban Sensing with Spaceborne Interferometric Synthetic Aperture Radar, in: *The Urban Book Series*, pp. 345–365, 2021, doi: 10.1007/978-981-15-8983-6_21.

2

Extracting Reservoir Surface Features and Land Cover Changes from Optical and Synthetic Aperture Radar Imagery: Tumariya Reservoir Case Study

Asha Thapliyal[1], Sanjeev Kimothi[2]* and Himani Bisht[3]

[1]*Department of Water Resource, Uttarakhand Space Application Centre, Dehradun, Uttarakhand, India*
[2]*Department of Physics, Graphic Era (Deemed to be University), Dehradun, Uttarakhand, India*
[3]*Centre of Excellence for NRDMS, Uttarakhand Remote Sensing & GIS, Soban Singh University, Campus Almora, Uttarakhand, India*

Abstract

Water bodies are vital for sustaining biodiversity, agriculture, and human livelihoods. Accurate monitoring of these ecosystems is critical in the context of climate variability, increasing human pressure, and land-use change. Traditional techniques for monitoring water bodies and aquatic vegetation often face challenges due to cloud cover, limited revisit times, and high costs. In contrast, satellite-based remote sensing techniques, such as vegetation, water indices (normalized difference vegetation index, normalized difference water index, and normalized difference turbidity index), and synthetic aperture radar (SAR), offer robust solutions for consistent monitoring across time and space. This study aims to evaluate the spread of water and aquatic vegetation dynamics in the Tumariya reservoirs over 10 years using composite indices and SAR data. The focus is on generating and assessing the effectiveness of normalized difference vegetation index, normalized difference water index, and normalized difference turbidity index composite images for mapping water bodies and aquatic vegetation, and employing SAR data (Sentinel-1, years 2016 and 2024) for detailed water surface analysis in the Tumariya Reservoir.

***Keywords*:** Synthetic aperture radar (SAR), NDVI, NDWI, NDTI

**Corresponding author*: drskimothi@gmail.com

Pooja Joshi, Ashutosh Bhatt, Neelam Sharma and Kapil Joshi (eds.) Synthetic Aperture Radar for Disaster Management: Bridging Science and Solutions, (47–58) © 2026 Scrivener Publishing LLC

2.1 Introduction

Wetlands are one of the most productive and biologically diverse ecosystems [1–3]. As a result, inventorying and monitoring wetlands and their neighboring uplands is critical, since the eighteenth century, up to 87% of wetlands have been lost globally, and severely degraded in Asia and high- and mid-latitude regions [4, 5]. Now, remote sensing has made it much easier to research wetlands and other environmental resources in recent decades [6, 7]. Satellite remote sensing techniques equipped with highly advanced sensing capabilities have been effectively used in monitoring wetlands around the world [8–22]. These findings stressed that the wetlands are changing rapidly, which impacts the water balance and aquatic ecosystems. Worldwide, it is observed that these results are due to massive land system changes in urbanization, thus their impacts can be seen directly or indirectly, resulting in a decline in regional hydrology [23–28]. As a result, efficient wetlands monitoring and evaluation are now essential to make precautionary preservation of these regions. Wetland preservation, planning, and sustainable management can be effectively facilitated by geospatial technology, especially advanced satellite-based remote sensing [29].

Hyperspectral remotely sensed data have been widely used for mapping and monitoring of various structural components of wetlands, such as water spread, aquatic vegetation, and water quality, among others [30–33]. Remote sensing has emerged as an important tool for mapping and monitoring wetland habitat types or structural components. Several studies have been conducted globally to inventory wetlands; grassland conversion is also concentrated near wetlands [34–36]. An algorithm was developed to automatically extract wetland features using a hierarchical decision tree approach, incorporating surface water spread, wetlands, turbidity, and wet vegetation, including aquatic species for both pre- and post-monsoon periods [15, 37]. False-color image display, which includes modified normalized difference water index (MNDWI), normalized difference pond index (NDPI), and normalized difference vegetation index (NDVI), will be instrumental in generating the required knowledge.

Water bodies are crucial for biodiversity, agriculture, and human livelihoods. Monitoring them is essential amid climate change, human activity, and land-use shifts. Traditional methods face limitations like cloud cover, low revisit frequency, and high costs. Satellite-based remote sensing, including vegetation and water indices, including NDVI, normalized difference water index (NDWI), and normalized difference turbidity index (NDTI), and synthetic aperture radar (SAR), provides a consistent, cost-effective alternative. This study assesses water spread and aquatic vegetation dynamics in the Tumariya Reservoir over 10 years, using composite NDVI, NDWI, and NDTI images, and SAR data (Sentinel-1, 2016 and 2024) for detailed water surface analysis.

- Monitor water spread and aquatic vegetation in the Tumariya Reservoir over 10 years using NDVI, NDWI, and NDTI indices.
- Evaluate the application of SAR (Sentinel-1) data from 2016 and 2024 for water spread analysis in the Tumariya Reservoir.

2.2 Study Area

Tumariya's ecosystem includes marshlands, agricultural areas, riverine forests, tall grasslands, and aquatic vegetation. The wetland supports a complex food web of birds, amphibians, mollusks, annelids, fish, and mammals. The Tumariya Reservoir, located at 29°18′N and 78°55′E, is an artificial wetland approximately 15 km from Kashipur, Uttarakhand, India. It receives water primarily from the Dhela and Tumaria rivers, and lies at the ecological junction of West Gangetic moist mixed deciduous forests and dry Siwalik Sal forests. The area features a subtropical monsoon climate with hot summers (up to 42°C), cold winters (down to 1°C), and annual rainfall of 1500–1600 mm, mostly concentrated in July and August. However, anthropogenic threats such as overgrazing, fuelwood collection, and urban expansion are degrading the habitat quality and leading to wetland shrinkage.

2.3 Materials and Methods

2.3.1 Remote Sensing Dataset Used

2.3.1.1 Landsat Data

Landsat 8 images for 2013 and 2020 were obtained from the open-source platform of USGS for multitemporal analysis.

2.3.1.2 Synthetic Aperture Radar Data

Sentinel-1 SAR data from 2016 and 2024 were used to analyze water spread dynamics in the Tumariya Reservoir. The primary dataset used in this study was Sentinel-1 ground range detected products, acquired through the European Space Agency's Copernicus Open Access Hub (https://earthexplorer.usgs.gov/). The analysis focused on data from VV polarization, which is effective in differentiating water bodies due to its sensitivity to smooth surfaces, such as calm water, which typically exhibits low radar backscatter. The SAR images were collected in October 2016 and 2024, providing sufficient temporal resolution to assess seasonal fluctuations and short-term changes in water surface coverage.

2.3.2 Auxiliary Data

Shuttle radar topography mission digital elevation model (SRTM DEM) (30m) for terrain correction, Sentinel-2 imagery for visual validation, and limited *in-situ* water level data from local authorities.

2.4 Generation of Spectral Indices

The spectral indices used to identify various features are estimated through different band combinations; the demonstration of the multi-band index is shown in Table 2.1. Figure 2.1 and Figure 2.2 illustrate the derived indices NDWI, NDVI, MNDWI, NDPI, and NDTI using Landsat 8 data for 2013 and 2020, respectively, along with the FCC covering the reservoir area.

- The NDVI is used to investigate and monitor vegetation status through the studied datasets.
- The NDWI is used to detect surface water.
- The NDTI is obtained for monitoring turbidity and aquatic vegetation over the study area.

Table 2.1 A demonstration of the multi-band index, formula, and its utility.

Multi-band index	Equation	Remark
Normalized Difference Water Index	NDWI = (Green-NIR)/(Green+ NIR)	Water has a positive value.
Normalized Difference Vegetation Index	NDVI = (NIR-R)/(NIR+R)	Water has a positive value.
Modified Normalized Difference Water Index	MNDWI = (Green-SWIR_1)/(Green+ SWIR_1)	Water has a positive value.
Normalized Difference Turbidity Index	NDTI = (Red - Green)/(Red + Green)	Vegetation can be identified.

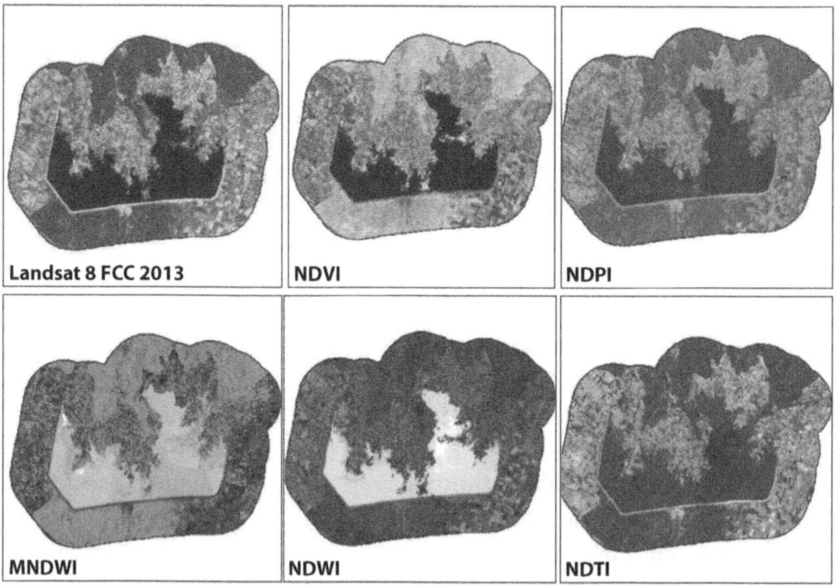

Figure 2.1 The NDWI, NDVI, MNDWI, NDPI, and NDTI using Landsat 8 data of 2013, along with the FCC covering the reservoir area.

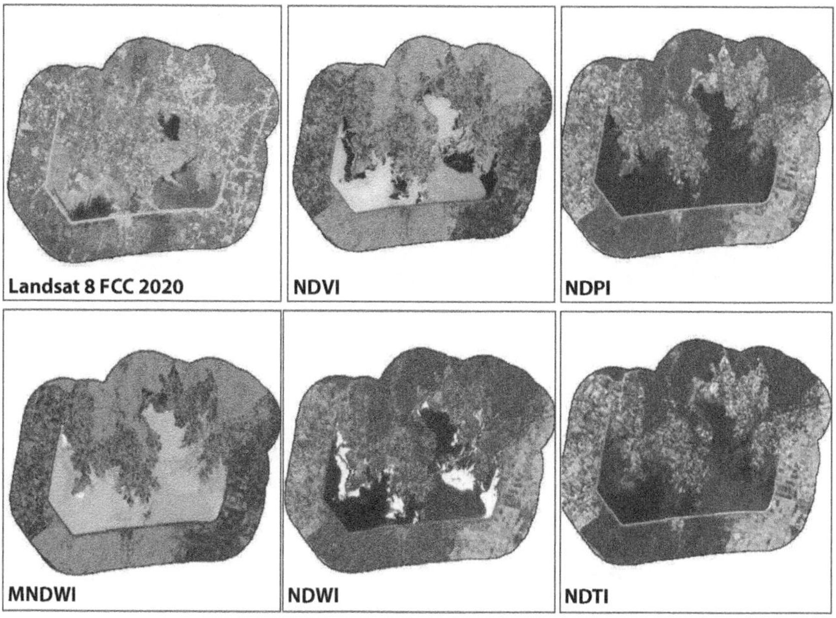

Figure 2.2 The NDWI, NDVI, MNDWI, NDPI, and NDTI using Landsat 8 data of 2020, along with the FCC covering the reservoir area.

2.4.1 Synthetic Aperture Radar-Based Water Spread

Figures 2.3 and 2.4 illustrate the Sentinel-1 datasets from 2016 and 2024 for water body delineation and water body signature extraction, respectively.

Monitoring Methodology: The Sentinel-1 data for 2016 and 2024 were processed using the SNAP software as follows:

2.4.1.1 Pre-Processing of Sentinel-1 Data

- **Orbit File Application**: Applied precise orbital data to enhance geolocation accuracy.
- **Thermal Noise Removal**: Removed background noise from SAR signals.
- **Radiometric Calibration**: Converted raw SAR data to back-scatter coefficients (σ) in decibels (dB).
- **Speckle Filtering**: Applied the Refined Lee filter to reduce SAR speckle noise.
- **Terrain Correction**: Used the range-Doppler method with SRTM DEM for orthorectification.

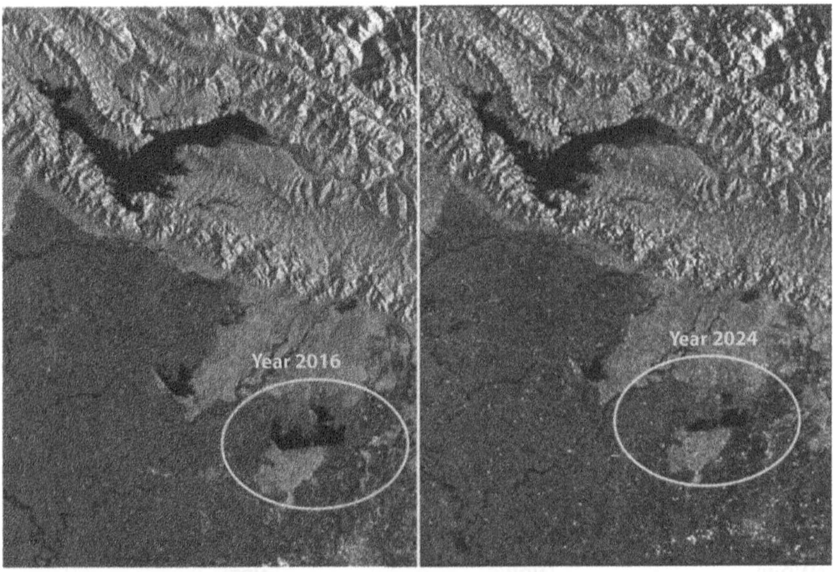

Figure 2.3 Sentinel-1 data from 2016 and 2024 used for water body delineation.

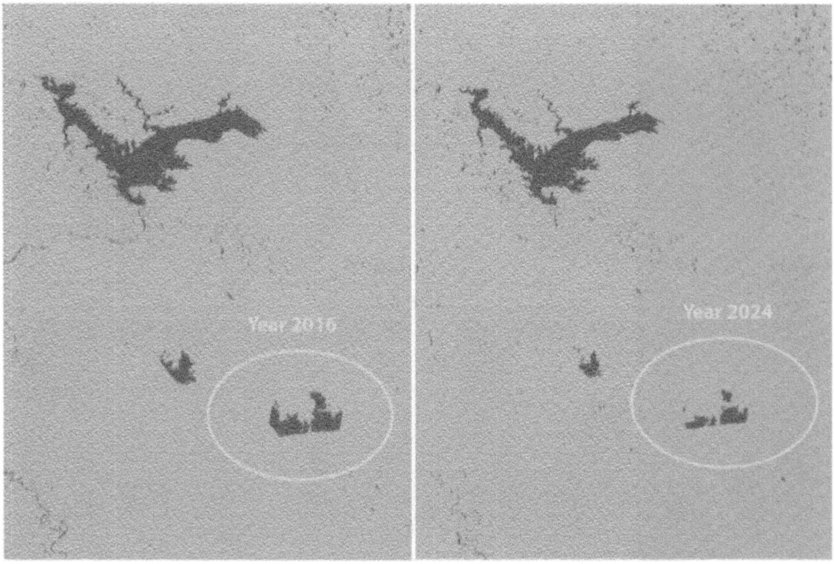

Figure 2.4 Sentinel-1 data in 2016 and 2024 used for water body extraction.

2.4.2 Water Surface Extraction

The following steps were carried out to extract water bodies from SAR data.

- **Thresholding**: Water pixels identified using backscatter thresholds below -13 dB in VV polarization.
- **Band Math and Binary Water Masks**: Generated using SNAP tools.
- **Morphological Operations**: Dilation and erosion to refine water masks.

2.5 Wetland and Land Use Land Cover Mapping

Figure 2.5 illustrates the detailed land use and land cover classification for 2013 and 2020. A supervised classification approach at a 1:50,000 scale was used with a Level III classification scheme. Wetland classes included water bodies, aquatic vegetation, grasses, swamps, and marshes. Non-wetland classes included cropland, settlements, mixed riverine forest, and scrub. Accuracy assessment and transition change analysis were conducted using GIS tools. The land use and land cover analysis between 2013 and

Figure 2.5 Land use and land cover classification for 2013 and 2020.

2020 revealed notable changes in the Tumariya Reservoir region. The reservoir area and forest cover remained unchanged over the period, with both maintaining stable extents of 20.95 km^2 and 19.12 km^2, respectively. However, significant shifts were observed in agricultural and urban land use. The agricultural area decreased from 16.15 km^2 in 2013 to 11.62 km^2 in 2020, indicating a reduction in cultivated land. Conversely, the urban area expanded from 12.80 km^2 to 17.33 km^2, reflecting increased development and urbanization in the region. These trends suggest a shift from agricultural land use toward urban infrastructure, which may have implications for water demand, land planning, and environmental sustainability in the surrounding areas.

2.6 Conclusion

This study demonstrates the effectiveness of Sentinel-1 SAR data for accurately extracting and monitoring surface water bodies, particularly within the Tumariya Reservoir. Using ESA SNAP software and VV polarization backscatter analysis, we successfully delineated the water surface and tracked its temporal variations, overcoming limitations posed by cloud cover and daylight dependency typical of optical imagery. The SAR-based thresholding techniques proved reliable for generating water masks, which are critical for water resource planning and flood risk assessment. In parallel, optical data was utilized to assess land use and land cover changes between 2013 and 2020. While the reservoir and forest areas remained unchanged, significant reductions in agricultural land and expansions in urban areas were observed. These findings indicate growing urbanization and a shift in land use that may increase pressure on local water resources. Together, the integration of SAR for water monitoring and optical data for land analysis provides a comprehensive understanding of environmental changes in the region, supporting better-informed management strategies.

Acknowledgments

The authors acknowledge the European Space Agency's Copernicus Open Access Hub (https://earthexplorer.usgs.gov/) for providing free access to the used datasets.

AI disclosure statement: The author acknowledges the use of AI-based tools for minor language editing or grammatical correction. However, all intellectual contributions, analyses, and conclusions presented in this chapter are entirely the author's own.

References

1. Ghermandi, A., van den Bergh, J.C., Brander, L.M., de Groot, H.L., Nunes, P.A., The economic value of wetland conservation and creation: A meta-analysis. *Ecol. Econ.*, 69, 2, 261–269, 2008.
2. Keddy, P.A., Fraser, L.H., Solomeshch, A.I., Junk, W.J., Campbell, D.R., Arroyo, M.T.K., Alho, C.J.R., Wet and wonderful: The world's largest wetlands are conservation priorities. *BioScience*, 59, 1, 39–51, 2009.
3. Amani, M., Salehi, B., Mahdavi, S., Brisco, B., Huang, W., Wetland classification using multi-source SAR data and a random forest classifier. *Remote Sens.*, 11, 5, 538, 2019.
4. Asselen, S., Verburg, P.H., Vermaat, J.E., Janse, J.H., Drivers of wetland conversion: A global meta-analysis. *PLoS One*, 8, 11, e81292, 2013.
5. Mao, D., Wang, Z., Tian, Y., Wu, J., Wetland loss in China and its causes. *Sci. Total Environ.*, 774, 145645, 2021.
6. Murphy, C.E. and Thomas, M.H., Remote sensing of freshwater wetland habitats using optical data. *Remote Sens. Rev.*, 26, 25–42, 2007.

7. Heumann, B.W., Satellite remote sensing of wetlands: A review. *GISci. Remote Sens.*, 48, 4, 360–389, 2011.
8. Best, R.G. and Moore, D.M., Landsat image classification of wetlands in the Green Bay area. *Photogramm. Eng. Remote Sens.*, 45, 715–722, 1979.
9. Hutton, R.F. and Dincer, T., Use of Landsat data for wetland monitoring in Africa. *Remote Sens. Environ.*, 8, 333–342, 1979.
10. Nayak, S.R. and Sahi, G.S., Wetland mapping using Landsat data in India. *J. Indian Soc. Remote Sens.*, 13, 45–50, 1985.
11. Anon, Wetlands of India: Conservation area series, Zoological Survey of India, 1993.
12. Hess, L.L., Melack, J.M., Simonett, D.S., Radar detection of flooding beneath the forest canopy: A review. *Int. J. Remote Sens.*, 11, 1313–1325, 1990.
13. Hess, L.L., Melack, J.M., Filoso, S., Wang, Y., Delineation of inundated area and vegetation along the Amazon floodplain using SIR-C imagery. *IEEE Trans. Geosci. Remote Sens.*, 33, 896–904, 1995.
14. Kasischke, E.S., Bourgeau-Chavez, L.L., Smith, K.B., Wetland mapping in the boreal forest using radar imagery. *Remote Sens. Environ.*, 85, 248–263, 2003.
15. Wang, J., Xiao, X., Zou, Z., Mapping paddy rice using Landsat and MODIS imagery with decision tree algorithms. *Remote Sens. Environ.*, 140, 1–13, 2014.
16. Campbell, D., Fraser, L.H., Lemay, V., Remote sensing and GIS analysis of wetland loss in Western Canada. *Wetlands Ecol. Manage.*, 25, 125–139, 2017.
17. Campbell, D., Lemay, V., Fraser, L.H., Spatial patterns of wetland change in the Canadian Prairie Pothole Region. *Ecol. Indic.*, 85, 319–328, 2018.
18. Mukherjee, K., Hazra, S., Dey, S., Monitoring spatiotemporal changes of Indian wetlands using Landsat data. *Remote Sens. Appl.: Soc. Environ.*, 10, 36–46, 2018.
19. Jafari, R., Bahrami, S., Sadeghi, S.H., Wetland dynamics assessment using satellite imagery. *Wetlands Ecol. Manage.*, 29, 541–556, 2021.
20. Fekri, M., Amini, J., Varshosaz, M., Monitoring of wetland changes using SAR data and. *J. Appl. Remote Sens.*, 15, 2, 026504, 2021.
21. Qiande, L., Ji, M., Zhang, H., Change detection of wetland environments using remote sensing. *Ecol. Indic.*, 123, 107303, 2021.
22. Irfan, M., Mahmood, Q., Khan, Z., Wetland mapping using multi-temporal satellite data. *Environ. Monit. Assess.*, 192, 342, 2020.
23. Mitsch, W.J., Managing the world's wetlands—Preserving and enhancing their ecological functions. *Int. Ver. Theor. Angew. Limnol. Verh.*, 26, 1, 139–147, 1996.
24. Nitin, K., Prasad, D., Mahapatra, S., Urban sprawl and wetland degradation in Guwahati. *Int. J. Remote Sens. Gis*, 3, 1–9, 2014.
25. Davidson, N.C., Fluet-Chouinard, E., Finlayson, C.M., Global extent and distribution of wetlands: Trends and issues. *Mar. Freshwater Res.*, 69, 620–627, 2018.

26. Pramod, K., Singh, S., Sharma, P.K., Study of wetland loss and degradation in Doon Valley, India. *Environ. Earth Sci.*, 64, 1041–1049, 2011.
27. Ahmad, T., Kumar, P., Aadil, R., Assessment of wetland dynamics using geospatial techniques in Wular lake, Kashmir Himalaya. *Environ. Earth Sci.*, 79, 357, 2020.
28. Bassi, N., Kumar, M.D., Sharma, A., Pardha-Saradhi, P., Status of wetlands in India: A review of extent, ecosystem benefits, threats and management strategies. *J. Hydrol.: Reg. Stud.*, 2, 1–19, 2014.
29. Nath, A., Ghosh, S., Sinha, A., GIS and remote sensing for wetland management. *Gis Dev.*, 4, 11, 34–38, 2000.
30. Bolch, T., Menounos, B., Wheate, R., Remote sensing of wetland hydrology: A case study of glacial meltwater-fed wetlands. *Remote Sens. Environ.*, 255, 112302, 2021.
31. Muhammad, S., Ahmed, R., Wagan, B., Mapping and monitoring wetland ecosystem using geospatial technologies. *Environ. Earth Sci.*, 78, 114, 2019.
32. Al-Shaibah, B., Wetland mapping and monitoring using remote sensing techniques: A review. *J. Environ. Manage.*, 294, 112948, 2021.
33. Gaziolu, M., Wetland monitoring using Landsat and MODIS images in Turkey. *Wetlands*, 38, 1001–1014, 2018.
34. Christopher, K.W., Land use change near wetland regions in the Prairie Pothole Region. *BioScience*, 63, 4, 301–311, 2013.
35. Kumar, S., Remote sensing-based monitoring of wetland vegetation in India. *J. Environ. Biol.*, 30, 871–876, 2009.
36. Matta, M., Kumar, A., Sharma, S., Monitoring wetland change using remote sensing in Harike wetland, India. *Environ. Monit. Assess.*, 189, 118, 2017.
37. Subramanian, S.K. and Saxena, R., Development of a decision tree-based classification algorithm for wetland feature extraction. *Indian J. Remote Sens.*, 39, 267–275, 2011.

Part 2

TECHNIQUES FOR DISASTER MONITORING AND ASSESSMENT

3

Synthetic Aperture Radar Polarimetry for Disaster Analysis

Vijaya Bhandari[1*], Jaspreet Singh[1] and Arun Balodi[2]

[1]Department of Electronics and Communication Engineering, BTKIT, Dwarahat, Almora, Uttarakhand, India
[2]Department of Electronics and Communication Engineering, Dayananda Sagar University, Bengaluru, Karnataka, India

Abstract
Synthetic aperture radar (SAR) polarimetry has reshaped disaster analysis through its precise and reliable framework, which detects the Earth's surface movements with unprecedented accuracy. This chapter examines SAR polarimetry's significant disaster management and mitigation capabilities, showing its vital value for worldwide disaster management platforms. It delivers detailed situational insight into disaster sites by combining its ability to see beyond clouds with its all-weather adaptability, while giving high-definition cartographic data. The chapter investigates advanced techniques anchored in polarimetric decomposition and coherence-based analysis combined with machine learning algorithms to improve our capacity to respond to floods alongside earthquakes, landslides, and other catastrophic occurrences. Multiple case studies accompany this section with real-world examples, demonstrating the revolutionary results of SAR polarimetry with multisensor systems and real-time data analytics integration. Through practical implementation and scientific innovation, this work presents a forward-thinking approach to what SAR polarimetry accomplishes in redefining disaster resilience and adaptability while establishing itself as the fundamental component of contemporary disaster analysis.

Keywords: SAR polarimetry, disaster analysis, remote sensing, machine learning, risk assessment

**Corresponding author*: viji27.gbpec@gmail.com

Pooja Joshi, Ashutosh Bhatt, Neelam Sharma and Kapil Joshi (eds.) Synthetic Aperture Radar for Disaster Management: Bridging Science and Solutions, (61–88) © 2026 Scrivener Publishing LLC

3.1 Introduction

The Growing Imperative for Advanced Disaster Analysis

Earthquakes, tsunamis, and hurricanes are natural disasters that cause significant death tolls and infrastructure damage. They have become more frequent due to climate change and urbanization, and thus, sophisticated monitoring is required. Remote sensing supports disaster management, but optical techniques are hindered by darkness and cloud cover. These are overcome *via* synthetic aperture radar (SAR), which makes use of day-and-night, all-weather photos, which are therefore critical for disaster evaluation. Technological improvements such as SAR polarimetry (PolSAR) improve surface change detection, enhancing damage assessment accuracy [1].

3.2 The Evolution of Synthetic Aperture Radar Technology: From Imaging to Polarimetry

Innovations in SAR technology have increased disaster analysis through refined damage detection accuracy. Initial SAR utilized single-polarization imagery, constraining terrain and structural damage identification. Dual- and full-polarimetric SAR innovations expanded damage mapping by examining scattering mechanisms. The PolSAR identifies surface materials and helps in earthquake and tsunami damage assessment by detecting destroyed buildings and infrastructure damage [2]. These advances highlight SAR polarimetry's increasing significance in disaster observation.

3.3 Synthetic Aperture Radar Polarimetry as a Game-Changer in Disaster Management

Synthetic aperture radar polarimetry has proven to revolutionize disaster relief in providing detailed, real-time insights into affected regions. Unlike conventional SAR imaging, which relies on intensity variations, PolSAR utilizes different polarization channels to extract information on surface roughness, moisture content, and structural changes. This capability is particularly useful for identifying collapsed buildings, detecting landslides, and monitoring post-disaster recovery efforts [3].

Integrating PolSAR with advanced data processing techniques, such as machine learning (ML) and change detection algorithms, has further improved disaster analysis. For example, multi-temporal SAR data has been successfully used to assess earthquake damage, as demonstrated in studies of the 2016 Kumamoto earthquake and the 2011 Tohoku earthquake. By leveraging PolSAR data, disaster response teams can make informed decisions regarding emergency aid distribution, infrastructure rebuilding, and long-term resilience planning.

3.4 Fundamentals of Synthetic Aperture Radar Polarimetry

3.4.1 Core Principles of Synthetic Aperture Radar

A modern mapping method called synthetic aperture radar (SAR) allows for high-quality images of the planet's surface captured at any time of day or in various weather conditions. Unlike optical imaging, the SAR operates in the microwave spectrum, penetrating clouds and providing consistent monitoring capabilities. The SAR relies on the motion of the radar platform to simulate an extended antenna aperture, thereby improving spatial resolution beyond the limitations of physical antenna size [4]. The capability of the SAR to capture detailed surface characteristics has made it instrumental in applications such as disaster assessment, land cover classification, and environmental monitoring [5].

3.5 Polarization and Its Critical Role in Synthetic Aperture Radar Imaging

Polarization is a fundamental aspect of SAR imaging that significantly enhances target characterization by measuring the orientation of electromagnetic waves. Polarimetric SAR utilizes different polarization states—horizontal (H) and vertical (V)—to analyze scattering mechanisms from various surfaces. The ability to distinguish between surface types, vegetation structures, and urban areas is greatly enhanced by using entirely polarimetric data. The scattering properties of targets are mathematically represented by the scattering matrix, which plays a key role in decomposition techniques used for target classification. The analysis of these matrices has been further refined using methodologies such as the Wishart distribution and spectral clustering.

3.6 Polarimetric Synthetic Aperture Radar Modes: Single, Dual, and Full Polarization

The PolSAR systems operate in three primary modes:

1. **Single Polarization (HH or VV):** This type of camera captures data in only one polarization mode, limiting its ability to differentiate target characteristics.
2. **Dual Polarization (HH+HV or VV+VH):** This technique provides additional scattering information by capturing two polarization combinations, enhancing vegetation and surface characterization.
3. **Full Polarization (HH, HV, VH, VV):** Offers the most comprehensive information by capturing all polarization states, allowing for advanced decomposition techniques and improved classification accuracy [6].

Advanced techniques such as the intensity-driven adaptive neighborhood method and fuzzy clustering have been developed to enhance the extraction of polarimetric information.

3.7 Comparative Advantages of Optical and Passive Radar Systems

Compared to optical imaging and passive radar systems, SAR provides several key advantages:

- **All-weather capability:** Unlike optical sensors, SAR is unaffected by atmospheric elements, including rainfall, fog, and clouds.
- **Day-and-night imaging:** SAR operates in the microwave spectrum, enabling continuous monitoring regardless of sunlight availability [6].
- **Penetration capability:** Certain SAR waves (like the L-band) may pass through plants and soil, revealing subsurface structures.
- **Improved disaster response:** Rapid damage assessment using SAR has been demonstrated in post-earthquake scenarios, such as the 2016 Kumamoto earthquake.

3.8 Techniques in Synthetic Aperture Radar Polarimetry

3.8.1 Polarimetric Decomposition Methods: Freeman-Durden, Cloude-Pottier, and Yamaguchi Models

Different scattering components may be broken down using the polarimetric decomposition method. Buildings in various stages are represented by distinct scattered elements in a PolSAR image taken after an earthquake. For example, double-bounce dispersion elements are the most prevalent with a standing structure perpendicular to the SAR flight path; hence, it is possible to utilize the polarimetric decomposition technique directly to identify the structure in the hazard zone. However, being oriented structures, the standing and fallen buildings in the earthquake zone primarily consist of volume-scattering elements [3].

3.9 Analysis of Coherence and Phase Information for Damage Assessment

Damage detection based on SAR is generally categorized into incoherent and coherent change detection methods [7].

3.9.1 Incoherent Change Detection

Incoherent techniques use the backscattering loudness of SAR pictures taken before and after an incident to use amplitude contrast to identify structural changes. To increase accuracy, a Log-ratio study and loudness-normalized differences were shown [8, 9]. Other automated methods are amplitude correlation, discriminant analysis, and time-series-based difference analysis of correlation using advanced land observing satellite (ALOS)/phased array type L-band synthetic aperture radar (PALSAR) images for less-damaged structures. Double-bounce changes in COSMO-SkyMed images were also utilized by [10] to identify destroyed buildings following the L'Aquila earthquake.

3.9.2 Coherent Change Detection

Consistent methods examine the connection between SAR pictures using interferometry to estimate phase decorrelation with more accurate measures of structural damage [11]. Coherent change detection is used in earthquake damage mapping, flood monitoring [12], analysis of soil liquefaction [13], and monitoring of volcanic activity. Research indicates that damage caused by earthquakes is related to coherence loss, with increased damage leading to greater phase decorrelation [14].

Advancing coherent change detection applications, researchers have explored coherence time series analysis for monitoring damage evolution over time. Markov random field classifiers were applied with multi-temporal SAR imagery to assess post-earthquake damage in Bam, Iran [15]. A temporal decorrelation model was developed to distinguish areas of abnormal phase decorrelation beyond natural environmental variations [16].

3.10 Feature Extraction Using Advanced Algorithms

3.10.1 Polarimetric Characteristics of a Stricken Area

3.10.1.1 H-α-Wishart Unsupervised Classification Algorithm

Based on H/α target decomposition [63], Lee *et al.* [64] introduced the robust unsupervised classification technique, H-α-Wishart. The scattering processes are described by α, and their unpredictability is measured by the entropy H. Terrain type designation is aided by the physical scattering properties of the H/α plane's zones. Marino, Cloude, and Woodhouse [65] and Jager *et al.* [17] claim that the Wishart classifier highly depends on the backscattering intensity. Lakes, barren ground, and roads are examples of targets with weak backscattering that are readily categorized into strange scatter groups. The Z6 and Z9 portions of the H/α plane, which correspond with these lessons, are considered non-structures, while the remaining pixels are seen as either rubbish or whole structures. To avoid establishing a threshold, we use the weak backscattering components and disregard the overall accuracy of the classification, even if the Wishart allocation no longer applies to the urban region.

3.10.2 Yamaguchi Model

More appropriate for urban settings, the Yamaguchi decomposition [6] models the covariance or coherence matrix $\langle T \rangle$ Considering outcomes of four separate dispersion procedures: helix bouncing $(_{helix})$, volume scattering $(_{vol})$, surface scattering $(_{odd})$, and double-bounce scattering $(_{dbl})$:

$$\langle T \rangle = P_s \langle T_{odd} \rangle + P_d \langle T_{dbl} \rangle + P_v \langle T_{vol} \rangle + P_h \langle T_{helix} \rangle \tag{3.1}$$

The averaging operator over a sample estimates each component's scattering power, as explained in [6]. Strong double-bounce scattering happens when buildings create a "dihedral reflector" with the ground. Destruction breaks this effect, leaving rubble oriented randomly and prevailing volume scattering if buildings collapse orthogonally to the radar line of sight. Oblique buildings also generate heavy cross-pol backscattering and are thus difficult to separate from vegetation in classical polarization analysis [17]. Polarimetric orientation angle compensation techniques [18] increase double scattering power, but encounter difficulties with big orientation angles and complex disaster scenarios involving irregularly oriented buildings.

3.10.2.1 *Normalized Circular-Pol Correlation Coefficient*

For reflection-asymmetric structures, one useful characteristic index is the normalized circular-pol correlation index (NCCC), which may be written as:

$$NCCC = \left| \frac{\rho_{RRLL}}{\rho(0)} \right| \tag{3.2}$$

Where

$$\rho_{RRLL} \frac{\langle S_{RR} S^*{}_{LL} \rangle}{\sqrt{\langle S_{RR} S^*{}_{RR} \rangle} \sqrt{S_{LL} S^*{}_{LL}}} \tag{3.3}$$

is the circle-shaped polarimetric basis's coefficient of correlation between left-left (LL) and right-right (RR) polarimetric channels, whereas * denotes the conjugate operator. The following is the conversion switching between the linear and circular bases:

$$S_{RR} = iS_{hv} + \frac{1}{2}(S_{hh} - S_{vv}) \tag{3.4}$$

$$S_{LL} = iS_{hv} - \frac{1}{2}(S_{hh} - S_{vv}) \tag{3.5}$$

where the horizontal (h)–vertical (v) polarimetric dispersion vector of the basic components is denoted by $S_{hh}S_{hv}$ and S_{vv}

$$\rho(0) = \frac{< 4\left|S_{hv}\right|^2 - \left|S_{hh} - S_{vv}\right|^2 >}{< 4\left|S_{hv}\right|^2 + \left|S_{hh} - S_{vv}\right|^2 >} \tag{3.6}$$

3.10.2.2 *Texture Feature Grey-Level Co-Occurrence Matrix*

Among HR distant sensing's most obvious aspects, imaging is texture. A popular method for extracting texture features is the grey-level co-occurrence matrix (GLCM) [66]. It explains the statistical correlations of the second order. Between two pixels and tabulates the frequency with which many grey-level combinations co-occur in a picture segment. The co-occurrence matrix's "homogeneity" (Hom) characteristic was selected:

$$Hom = \sum_i \sum_j \frac{P(i,j)}{1+(i-j)^2} \tag{3.7}$$

3.11 Automated Change Detection with Artificial Intelligence and Machine Learning Techniques

Remote sensing (RS) is crucial in identifying and tracking Earth's surface changes through multi-temporal imagery. Optical RS sensors are extensively utilized in change detection (CD) for earthquake hazard mapping, crop monitoring, vegetation analysis, urbanization, and tracking snow cover [19]. Change detection compares time-series images to determine land cover changes, employing supervised or unsupervised techniques with knowledge of the scene beforehand.

Though SAR images have been less used for CD in the past because of speckle noise, their all-weather and light independence make them valuable for uninterrupted surveillance. Deep learning methods, including convolutional neural networks and recurrent neural networks, along with AI, have dramatically improved SAR-based CD [20].

Merging SAR polarimetry with AI and ML facilitates automatic CD through feature extraction, object-based analysis, and self-supervised learning. Deep clustering, contrastive learning, and probabilistic anomaly detection methods enhance noise compensation and improve damage assessment. The technologies allow for the faster and more precise identification of structural damages due to natural disasters, reinforcing early warning systems and post-disaster relief activities.

3.12 Applications in Disaster Analysis

3.12.1 Flood Monitoring and Mapping

Remote sensing technologies can offer fast and economical data for mapping and tracking flooded areas, making them particularly important during the response phase [21]. This is possible in almost real-time nowadays because of the many satellites and constellations in orbit around the planet run by both public and commercial organizations. Compared to optical SAR, satellite sensors are currently especially useful in this situation because of their ability to image in all conditions and at all times [22].

Appropriate processing methods are essential for analyzing the knowledge provided by cloud SAR data to build trustworthy technologies and services for satellite-based flood monitoring and detection. Studies on the spatiotemporal description of flood events, extracting flooded regions, and calculating water levels from SAR pictures are particularly active and difficult.

3.12.2 Earthquake Damage Assessment

Timely and effective damage assessment is essential for emergency response, saving lives, and alleviating humanitarian crises [23]. Limited data on the severity and magnitude of damage inflicted by natural disasters, especially in rural landscapes, is available. Satellite imagery is used extensively for quick and inexpensive damage assessment, facilitating the production of hazard maps [5]. Earthquakes are among the most frequent and unpredictable disasters [24], with cities being more prone to them due to high populations and poor building codes. Although most research maps urban damage [25], the impact of rural disasters is poorly researched, mainly reporting building damage. Rural regions, with small towns, agriculture, and the natural environment [26], have varied damage patterns compared to cities. The UN approximates that most people reside in rural regions, which will likely grow. Yet, "rural" is a different concept by nation and by year.

Rural areas have varied land cover types, ranging from forests and natural areas to agricultural land [27]. Disasters often impact these places because urban areas frequently conform to more stringent building regulations or avoid high-risk territories. In hilly rural areas, earthquakes may be followed by secondary events such as landslides [28]. Moreover, rural houses with local brick, wood, and clay materials are constructed with low seismic resistance and, therefore, have high vulnerability [29]. For example, 85% of rural residences were destroyed during the Yushu's 2010 earthquake (magnitude 7.1) and Wenchuan's 2008 earthquake (magnitude 8.0), which reached 90%.

3.12.3 Landslide Detection and Monitoring

3.12.3.1 Detection

Landslide detection involves measuring their frequency, distribution, and intensity to support post-emergency tasks such as rebuilding and evaluat ing dams. Multispectral imaging is used to detect landslides, while improvements in processing power, cloud computing, and high-resolution sensors have enabled better detection accuracy. Landslides triggered by rapid rainfalls are mapped with classification techniques over very high-resolution panchromatic and HR multispectral satellite images [30].

Automated mapping methods involve pixel-by-pixel and object-based analysis. Both DInSAR and MTInSAR are InSAR methods that estimate slow-moving landslides and ground displacement, although vegetation and rapid changes can affect accuracy. L- and wavelengths improve displacement mapping in vegetated areas, and SAR amplitude-based methods identify fast landslides in difficult terrains. Raar and multispectral satellite imaging provide cheap, broad-area monitoring for landslide monitoring. Using sensors that cross X-band, L-band, and C-band, in addition to visible and infrared spectral intervals, is more accurate. Both DInSAR and MTInSAR measure slow-moving landslides, interferograms, and multispectral change detection work better in cases of quick displacements.

Ground-based observations, such as interferometry and light detection and ranging (LiDAR), enable high-resolution local analysis; LiDAR enables more accurate mapping of landslides by enabling 3D reconstruction and pinpointing terrain indicators like cracks and debris deposits. Comparing subsequent point clouds can help calculate the landslide volume and identify the rockfall source zone [31].

3.12.3.2 Monitoring

Monitoring landslide surface displacement over time is crucial for tracking its evolution. Multispectral imagery, though initially limited, evolved with very high-resolution satellites (IKONOS, QuickBird) and Sentinel-2, making monitoring more effective. Image correlation, a common technique for sub-pixel displacement detection, is used for several satellite data (Pléiades, Landsat-8, SPOT) and allows the kinematics of landslide movement, including progressive and retrogressive movements, to be followed.

Most image correlation methods are appropriate for landslides traveling meter-to-decameter per year, whereas specialized methods identify slower movement (centimeters to millimeters per day) with sub-pixel precision. Egregating several satellite sources (Sentinel-2, Landsat-8, Landsat-7) strengthens time series reliability by eliminating uncertainties arising from seasonal effects and background noise [32].

Satellite SAR interferometry yields high-resolution displacement time series, facilitating landslide characterization, zonation, seasonal trends, and remedial work monitoring. Satellite missions such as Sentinel-1 and TerraSAR-X have high-frequency revisits, facilitating regional-scale monitoring. Interferon entry is dependent on stable targets, such as buildings and artificial reflectors, since vegetation and bare soil tend to return weak signals.

Artificial targets [33], such as corner reflectors, either active or passive, can be used to evaluate the accuracy of InSAR measurements and investigate deformation tracking in agriculture and bare soil environments.

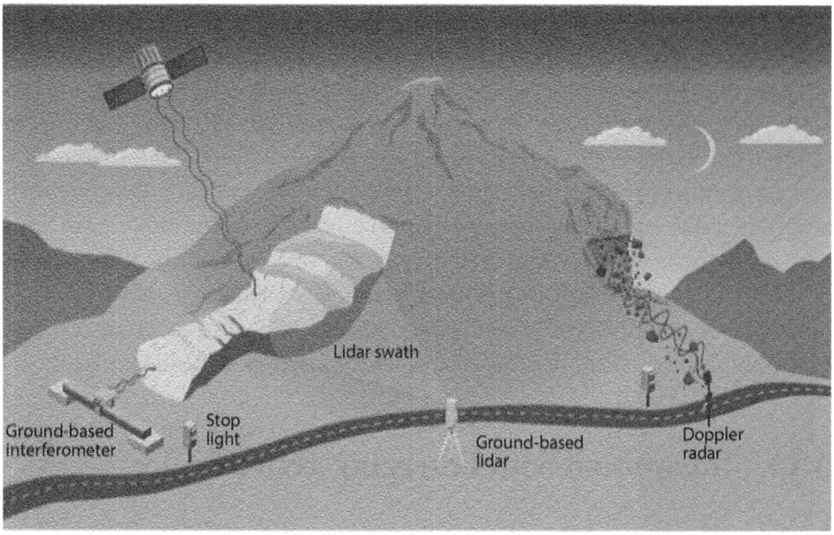

Figure 3.1 Site monitoring with remote-sensing techniques. Elite and ground-based interferometry, LiDAR, can monitor slower landslides (on the left side of the hill). Doppler radar is used for detecting swift movements, such as rockfalls (on the right side of the hill). The traffic lights next to the ground-based interferometer and Doppler radar indicate the use of near-real-time measurements to provide early warnings.

Ground-based radar interferometry can detect Everyday movement rates of a few meters because of its near-real-time collection frequency, even though interferometric landslide monitoring is effective at slower landslide rates. Thus, this method can monitor most landslides during a potentially catastrophic collapse, including huge rockslides, unstable volcano flanks, earthflows, and landslides. Doppler radar may also be used in conjunction with early warning systems, such as blocking roadways at risk of being destroyed (Figure 3.1).

Multitemporal LiDAR surveys, similar to radar-based surveillance, monitor landslide processes with enhanced data acquisition from automated laser scanners. Monitoring, however, is still restricted due to the necessity for automated data processing, particularly for early warnings. Satellite methods allow for regional ground deformation monitoring, and *in situ* sensors offer high-frequency data for interpreting deformation and sensing early warning indications of collapse [34].

3.12.4 Urban Disaster Resilience

Urban disaster resilience encompasses building up infrastructure and recovery capacity. One important aspect is infrastructure mapping, which determines weaknesses and optimizes the response to emergencies. Proper mapping allows for advanced risk assessment and preparedness against disasters [35]. Reconstruction after disasters follows the "Building Back Better" principle, strengthening infrastructure with intelligent technologies such as digital twins and predictive modeling using AI-based approaches [36]. Long-term disaster avoidance is supported through sustainable approaches like energy-efficient structures and climate-resilient materials.

Like green infrastructure, nature-based solutions enhance flood control and reduce the urban heat island phenomenon [37]. Community participation strengthens response and recovery efforts, whereas coordination among governments, NGOs, and city planners guarantees resilience in urban policies. Through the combined use of technology, sustainability, and governance, cities can minimize the effects of disasters and create safer, more resilient places.

3.13 Integration with Emerging Technologies

3.13.1 Harnessing Machine Learning and Artificial Intelligence for Polarimetric Data Interpretation

Using satellite imagery, land use and land cover were categorized. For decades, researchers have been using unique scattering signatures.

Optical sensors depend on spectral data but suffer from mixed spectral responses and limited performance in adverse weather or low light [38]. The SAR avoids these disadvantages with day-and-night, all-weather monitoring, but single-polarization SAR limits target interpretation [39]. Introducing PolSAR enhanced classification accuracy through decomposition models [40]. More recently, advances in AI and ML enhanced PolSAR data interpretation by feature extraction, classification, and anomaly detection automation. Deep learning architectures, such as convolutional neural networks and transformers, perform well in differentiating land cover classes. In contrast, the AI-fusion of PolSAR and interferometric SAR enhances classification accuracy. Both AI and ML are crucial for automated land-use and land-cover mapping.

3.14 Multisensor Fusion: Combining Synthetic Aperture Radar with Optical, Light Detection and Ranging, and Ground-Based Systems

Integrating various remote sensing data sources during the past several years has become a major point of emphasis in enhancing land cover mapping, environmental monitoring, and geospatial analysis. Combining SAR with round-based, visual, and LiDAR technologies is one area of multisensor fusion that has demonstrated significant improvements in classification accuracy and overcome the limitations of individual sensors. The SAR provides information about the phase and amplitude of the most often used remote sensing technique, microwave contact with the Earth's surface. In contrast to sensor optics, the SAR does not depend on solar illumination or weather; thus, it remains very effective even in conditions of clouds, fog, or darkness [41]. Yet, the peculiar geometric and spectral characteristics of the SAR and the presence of speckle noise tend to make SAR imagery difficult to interpret. Scientists have increasingly explored data fusion methods, combining SAR with optical imagery, LiDAR, and ground observations to address these shortcomings. Optics sensors provide spectrum data with excellent resolution. While ground observations provide localized validation and supplementary contextual information, LiDAR offers accurate elevation and structural information. Research has demonstrated that integrating optical and SAR data improves the precision of classification, especially in cloud-sensitive areas where optical data is limited.

Furthermore, integrating the SAR with LiDAR enhances the interpretation of urban and forest areas by using surface reflectance and structural information in 3D. Work [42] showed that multi-sensor fusion provided moderate to high accuracy gain relative to single-sensor methods, especially for land cover classification. Likewise, SIR-B SA data measured Florida's land cover using multisensory fusion with Landsat TM [43]. While they found that only a marginal advantage was provided by the fused data over TM by itself, this was far more than that given by SAR. In general, multi-sensor data fusion enables a richer representation of the Earth's outermost layer by leveraging the advantages of disparate data sources to enhance classification accuracy, feature extraction, and environmental monitoring.

3.15 Real-Time Data Processing for Faster Disaster Response

In disaster management, real-time data processing is essential in analyzing remote sensing information, like LiDAR, optical pictures, and SAR, to evaluate damage and identify changes in real-time following a disaster. Both ML and AI have significantly increased the speed and accuracy of disaster response by automating the processing of enormous datasets. Artificial intelligence-based change detection methods provide quick detection of impacted areas, minimizing manual interpretation time and enhancing emergency responders' situational awareness. High-performance computing and edge computing solutions also support real-time processing by offloading computational workloads onto multiple nodes, making data analysis possible even in distant or disaster areas. These developments enable quicker decisions, enhance resource allocation, and implement more efficient disaster mitigation measures.

3.16 Cloud-Based Platforms for Global Accessibility and Scalability

Cloud platforms have revolutionized disaster response by offering scalable infrastructure to store, process, and analyze geospatial data. Microsoft Azure, Amazon Web Services, and Google Earth Engine are platforms that enable real-time access to satellite imagery, SAR data, and other key datasets, facilitating global collaboration for disaster response teams. These platforms obviate the requirements for large-scale local storage and computing capacities, thus making sophisticated geospatial analytics available in resource-scarce regions. Additionally, cloud-based platforms are interoperable with the Internet of Things and geographic information systems networks, enabling seamless data integration to support end-to-end disaster monitoring. Through cloud computing, disaster response initiatives are streamlined, minimizing delays in decision-making and improving aggregate resilience against natural disasters.

3.17 Case Studies

3.17.1 Flood Mapping in Southeast Asia: Real-Time Decision Support for Humanitarian Aid

One of the world's most susceptible regions to flooding is the Southeast Asian region. Frequent and costly floods have occurred in Southeast Asia's emerging nations, where a sizable population lives on or near the floodplain [44]. According to regional research [45] and forecasts issued *via* the United Nations Framework Convention on Climate Change [46], floods and heavy rains happen more often. A typical monsoon in Cambodia causes Tonlé Sap Lake's surface water area to increase by more than 500% (Figure 3.1; [67]). When the period of precipitation comes to an end, water levels at Tonlé Sap may range from 50–80 km^3 to as little as 1–2 km^3 at low times during drought [47]. Sea level rise increased steadily, and droughts have occurred in the region for 30 to 50 years [26]. Devastating floods struck the basin of the Lower Kong River in 2000, 2011, 2013, and 2016. posing a danger to economic livelihoods, causing significant infrastructure damage, and resulting in multiple fatalities. Rapid, automated, and adaptable technologies are thus required to monitor floods in these susceptible locations.

3.17.1.1 Flood Detection

For Southeast Asia, which comprises Thailand, Vietnam, Cambodia, and Laos, NASA Goddard researchers created software for flood detection (Ahamed *et al.*, [68]). The software minimizes cloud cover and contrasts new data to previous conditions by analyzing 4-day, 8-day, and daily composites using NASA LANCE servers' near-real-time MODIS data. Every day from 10:30 a.m. to 1:30 p.m., observations are taken by the MODIS cameras on the Aquarius and Terra satellites [48].

Tucker [69] developed the normalized difference vegetation index (NDVI), which tracks changes in plant life and applies to flood detection. Figure 3.2 illustrates the seasonality of the Mekong River, whereby monsoon-season floods from May to October are indicated by the NDVI

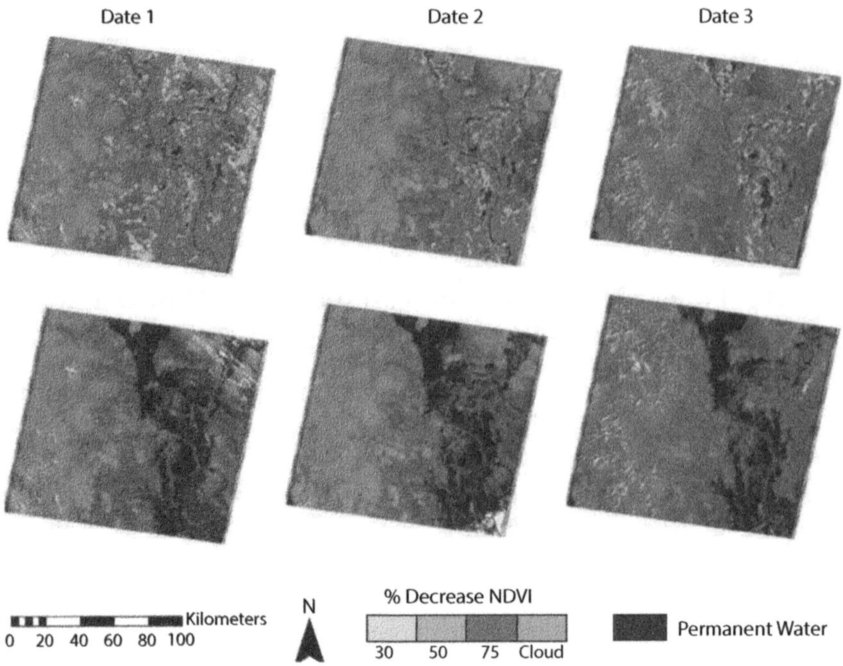

Figure 3.2 Landsat scenes (bottom, Path/Row 126/52) show good agreement for flooded area detection with the corresponding MODIS NDVI change product (top) in the Phnom Penh region of Cambodia. Date 1 shows the MODIS NDVI change product from an 8-day composite acquired 8/29 – 9/5/2006 and Landsat 5 TM acquired 9/3/2006. Date 2 shows the MODIS NDVI change product from 8-day composite acquired 10/16/2006 – 10/23/2006 and Landsat 5 TM acquired 10/21/2006. Date 3 shows the MODIS NDVI change product from an 8-day composite acquired 11/1/2006 – 11/8/2006 and Landsat 5 TM acquired 11/6/2006. Credit: C. Doyle.

(green) and discharge (blue) [70]; NDVI is an important indicator for detecting floods because it is reduced when flooding hides the vegetation [49].

To determine the extent of flooding, NASA LANCE servers provide near-real-time multispectral MODIS images. The following formula is used to get the NDVI:

$$NDVI_{NRT} = \frac{NIR - \text{Red}}{NIR + \text{Red}} \tag{3.8}$$

The position in near-infrared band 1 of the near-infrared spectrum is the $NDVI_{NIR}$, and the red band is the red band (band 2). The 32-day January $NDVI$ was averaged between 2003 and 2011. Historical $NDVI_{Dry}$ preprocessed and contrasted variables to $NDVI_{NRT}$. The percentage difference between the recent and historical circumstances is used to identify changes on a per-pixel basis.

$$\Delta NDVI = \frac{NDVI_{Flood} - NDVI_{Dry}}{NDVI_{Dry}} \times 100 \tag{3.9}$$

Pixels are categorized as "flooded" when their ΔNDVI is less than -75 and as "non-flooded" when their ΔNDVI is more than -75%. Areas with steep slopes and permanent air are hidden. Assessments of product correctness are continuing, and this approach is being improved (Ahamed *et al.*, [68] in prep). A GeoServer (http://geoserver.org) stores and serves the ΔNDVI pictures online. The flood length model was applied retroactively to floods between August and December 2011. Its accuracy was evaluated against Landsat images (Figure 3.3). The MODIS imagery-derived flood extent shows excellent agreement with SAR data (Radarsat 2, Envisat ASAR, TerraSAR-X) and higher-resolution Landsat images (Figure 3.4) acquired by the UN Institute for Instruction and Research Operational Satellite Applications program at this time.

(a) (b)

Figure 3.3 Aerial photographs of the Minamihama district (Region 1); (a) before (September 1975) and (b) after the disaster (March 12, 2011).

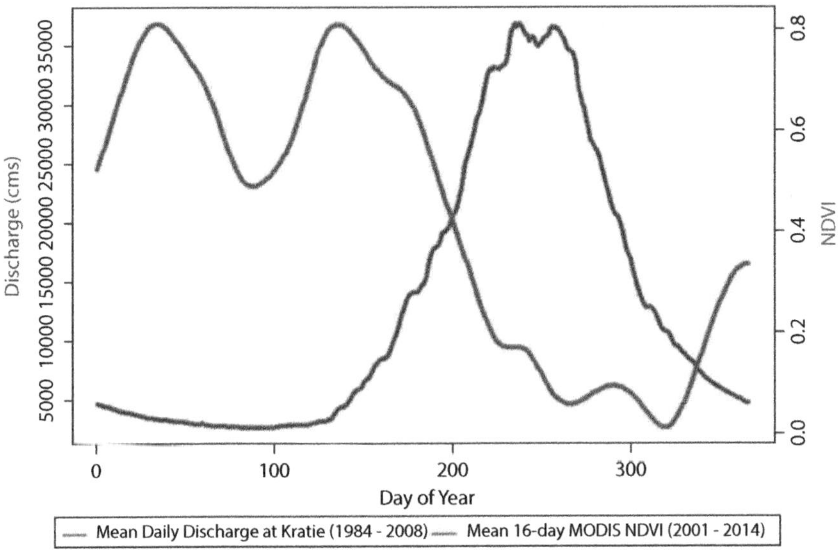

Figure 3.4 Illustrates the variability in the regional hydrologic cycle by showing NDVI (green) and discharge (blue).

3.17.2 Earthquake Damage in Urban Environment Comments: Polarimetric Insights from the 2011 Tōhoku Earthquake

Japan's eastern coast was shaken by a strong magnitude 9.0 earthquake on March 11, 2011, which generated a tsunami over 10 meters in height and ravaged beaches. Damaged areas were identified by emergency inspections using optical and SAR sensors. Apart from a tsunami inundation map, orthorectified aerial photomaps were created by Japan's Geospatial Information Authority [50].

Applying interferometry to examine ALOS/P LSAR data, crustal displacement (approximately 6 cm in Ishinomaki) could be observed, but not damaged buildings [51].

Damaged cities were identified using PRISM high-resolution imagery (2.5 m), whereas areas of tsunami impact were found using panchromatic remote sensing and ALOS stereo mapping [52]. Notwithstanding the challenges of complex X-band radar reflections in urban areas, TerraSAR-X efficiently surveyed flooded areas.

This research employs Radar backscattering in impacted metropolitan areas was investigated using PALSAR full-polarimetric data collected both before and following the tragedy. It finds characteristics useful for detecting destroyed buildings. Polarimetric interferometric synthetic aperture radar (PiSAR) full-polarimetric information gathered shortly thereafter the calamity is utilized to confirm the results.

3.17.3 Polarimetric Insights from the 2011 Tōhoku Earthquake

Using PALSAR (single-polarization mode), the Japan Aerospace Exploration Agency made several urgent comments about the catastrophe after the incident. On April 8, the first full-polarimetric measure after the catastrophe was carried out in the ascending orbit with an off-nadir angle of 21.5°. The Oshika Peninsula, Onagawa Town, and Ishinomaki City were inspected. The same observation mode was used for two earlier PALSAR observations: April 2, 2009, and November 21, 2010. The almost identical outcomes of the two datasets, which were collected before the catastrophe, demonstrate the PALSAR sensor's stability. The findings from the data collected on November 21, 2010, are reported in this letter.

Five days after the full-polarimetric PALSAR investigation on April 13, PiSAR L-band measurements were conducted over disaster areas, including Ishinomaki City. Aerial photographs indicated that the tsunami flooding was three meters inland. Three badly damaged areas were delineated: Minamihama (Region 1), Nakamura and Shintate (Region 2), and Kamihama (Region 3), as shown in Figure 3.5. Aerial photographs of Area 1 before and after the accident are shown in Figure 3.5. On July 13, 2011, a field inspection confirmed the near-destruction of homes and structures.

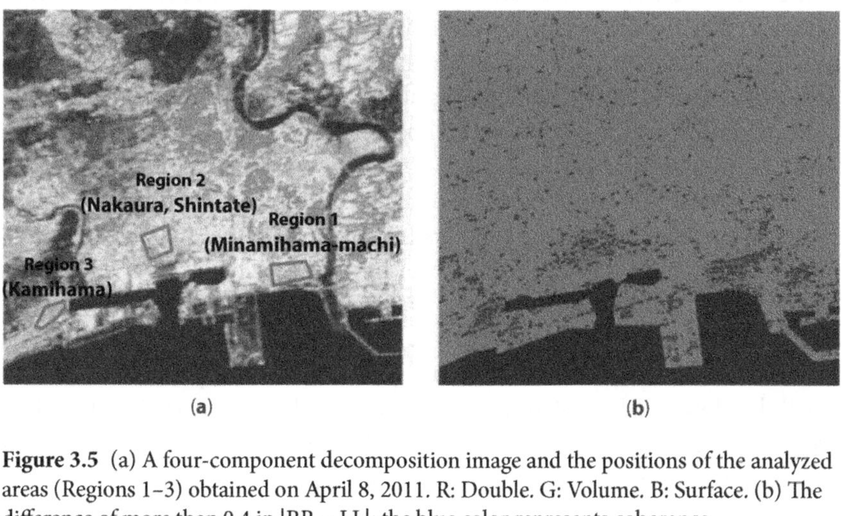

Figure 3.5 (a) A four-component decomposition image and the positions of the analyzed areas (Regions 1–3) obtained on April 8, 2011. R: Double. G: Volume. B: Surface. (b) The difference of more than 0.4 in |RR – LL|; the blue color represents coherence.

3.17.4 Landslide Detection in Himalayan Regions: Enhancing Early Warning Systems

Landslides are a significant risk in the Himalayas, but inventories usually are not timestamped, restricting dynamic hazard modeling [53]. Earth observation technology aids in landslide mapping, but optical data need sunny skies, resulting in delays [54]. Sentinel-1 SAR data has enhanced spatiotemporal mapping, providing daily timestamps for 30% of landslides with 80% accuracy [55].

Rain intensity-duration thresholds are essential for landslide early warnings, yet low data availability and low temporal resolution complicate predictions [56]. A debris flow model coupled with WRF in a prototype system models precipitation for hazard estimation, providing a viable alternative, though with concerns about accuracy [57].

Himalayan highways, which are crucial for connectivity, are often interrupted by landslides. A Bayesian model study of NH 7 (Rishikesh–Joshimath) revealed that road sections that have widened in the last decade have twice the risk of landslides, highlighting the importance of improved geohazard management and road planning [58].

3.17.5 Urban Resilience Monitoring: Post-Hurricane Reconstruction in Coastal Cities

Large and small island urban towns confront particular challenges in recovery efforts because they are encircled by water and cut off from the support that cities on the coast and interior "mainland" may get. Such special situations for catastrophe resilience are seen in the urban areas of Caribbean Island republics and countries. Puerto Rico's position puts it in several hurricanes' and other storms' paths, which devastate houses, infrastructure, and ecosystems in nature. A lack of essential technical and financial assets further slows down recovery. We chose Hurricanes Irma and Maria to represent the catastrophic catastrophes and Puerto Rico as the example study location because of the storms' substantial and long-lasting impacts on the island, resulting in months-long water shortages and power outages over the island [59]. Puerto Rico had 2.87 million residents as of 2020, losing almost 600,000 due to emigration after Hurricanes Maria and Irma. One of the larger Caribbean islands is Puerto Rico, which is (around 8870 sq kilometers) compared to its nearby small developing island states, with 94% of its inhabitants residing in urban areas (the biggest city, San Juan, has 347,000 citizens). It also has a higher level of economic development (median GDP per capita of US$32K) [60].

We are driven by two main inquiries about the resilience of metropolitan areas to disasters in island states or countries: (1) How do residents of islands react to hurricanes and quakes in Puerto Rico and its Caribbean neighbors? (Puerto Rico, Haiti, etc.)? Do they occur frequently? (2) What are the differences in recovery processes between these urban populations on the island and American cities? We propose that the Caribbean's island states have limited internal resources for disaster recovery because of their lengthy history of colonialism and present socioeconomic issues. They deal with the ongoing dearth of essential services by adjusting and expanding on regional inventiveness. The interconnection of social systems and physical infrastructure at the regional level leads to variations in community resilience, which we analyze by looking at Puerto Rico's recovery after two consecutive hurricanes (Irma and Maria). A quantitative assessment of urban catastrophe resilience was conducted using indicators related to commercial activity recovery and the key services' restoration process.

3.18 Challenges and Future Directions

3.18.1 Technical Constraints: Resolution, Coverage, and Volume of Data

Resolution Limitations: Satellites and aerial imagery with greater resolution are essential for extensive damage mapping, but tend to be associated with compromises in processing time or expense [61].

Coverage Problem: Most satellite-based monitoring systems face cloud cover, signal interference, or a lack of coverage in disaster-prone areas.

Data Overload: The enormous volumes of SAR, optical, and geospatial data necessitate sophisticated processing methods and computational resources to efficiently derive useful insights.

3.18.2 Breaking Down Operational Barriers: Cost, Accessibility, and Training Requirements

Cost Considerations: Although satellite data is more readily available, high-resolution imagery and real-time analytics are expensive, which constrains broad applications.

Accessibility Challenges: Several developing areas lack the infrastructure and human resources to apply advanced remote sensing tools in disaster response.

Training and Skill Gaps: Specialized training is needed to enable decision-makers and emergency responders to use SAR- and AI-based analytics efficiently.

3.18.3 Improving the Spatial and Temporal Resolution of Synthetic Aperture Radar Systems

Higher Frequency Observations: Reducing satellite revisit times would allow monitoring catastrophes in near real-time, especially for quickly changing hazards such as floods and landslides [62].

Multisensor Integration: Combining SAR with optical LiDAR and UAV-based imagery can improve precision and dependability in disaster assessments.

Miniaturized and Affordable SAR Satellites: The creation of small, affordable SAR satellites can democratize access to high-resolution disaster information.

3.18.4 Global Disaster Management Network Vision with Artificial Intelligence-Augmented Automation

Damage Detection Powered by AI: Machine learning algorithms can speed up damage classification, minimizing dependence on manual interpretation.

Predictive Analytics for Early Warning Systems: AI can leverage historical and real-time data to enhance disaster forecasting and preparedness.

Global Collaborative Networks: Global data-sharing programs, such as OpenSAR and Sentinel Hub, can support coordinated disaster response worldwide.

3.19 Conclusion

The role of SAR polarimetry in transforming disaster analysis is undisputable, providing unmatched capabilities in real-time observation, risk evaluation, and disaster prevention. With its all-weather capability, high spatial resolution, and sophisticated polarimetric decomposition method, SAR polarimetry presents an accurate, data-based response to disaster events. The integration of ML, AI, and m-lit-sensor fusion has amplified its potential to include automated damage assessment and predictive analysis for several calamity categories, including earthquakes, floods, and landslides. As SAR polarimetry advances, it will be critical to bridge scientific progress, technological advancements, and policy frameworks to improve global disaster resilience. Data accessibility, processing limitations, and operational expenses must be overcome to maximize its potential in practical applications. The future of disaster management is developing AI-based, networked worldwide systems that utilize SAR polarimetry and other remote sensing tools to create more proactive, responsive, and effective disaster response systems. Cooperative innovation among governments, scientific communities, and technology developers is essential to realize this vision. Investment in high-resolution SAR systems, cutting-edge computing infrastructure, and capacity-building programs will keep SAR polarimetry at the cutting edge of contemporary disaster analysis and management. This research highlights the need for interdisciplinary collaboration to unlock the full potential of SAR polarimetry, building more resilient communities and sustainable disaster preparedness measures worldwide.

AI disclosure statement: The author acknowledges the use of AI-based tools for minor language editing or grammatical correction. However, all intellectual contributions, analyses, and conclusions presented in this chapter are entirely the author's own.

References

1. Pulvirenti, L., Chini, M., Pierdicca, N., Boni, G., Use of SAR data for detecting floodwater in urban and agricultural areas: The role of the interferometric coherence. *IEEE Trans. Geosci. Remote Sens.*, 54, 3, 1532–1544, 2015.
2. Chen, S.-W. and Sato, M., Tsunami damage investigation of built-up areas using multitemporal spaceborne full polarimetric SAR images. *IEEE Trans. Geosci. Remote Sens.*, 51, 4, 1985–1997, 2012.
3. Zhai, W., Wang, X., Bi, Y., Liu, J., Zhu, G., Du, J., Extraction of earthquake damage information and mapping of buildings from single post-earthquake polarimetric synthetic aperture radar image based on polarimetric decomposition and texture features. *Sens. Mater.*, 34, 12, 4451–4462, 2022.
4. Zebker, H.A. and Villasenor, J., Decorrelation in interferometric radar echoes. *IEEE Trans. Geosci. Remote Sens.*, 30, 5, 950–959, 1992.

5. Plank, S., Rapid damage assessment using multi-temporal SAR—A comprehensive review and outlook to Sentinel-1. *Remote Sens.*, 6, 6, 4870–4906, 2014.
6. Yamaguchi, Y., Moriyama, T., Ishido, M., Yamada, H., Four-component scattering model for polarimetric SAR image decomposition. *IEEE Trans. Geosci. Remote Sens.*, 43, 8, 1699–1706, 2005.
7. Preiss, M., Gray, D.A., Stacy, N.J.S., Detecting scene changes using synthetic aperture Radar interferometry. *IEEE Trans. Geosci. Remote Sens.*, 44, 8, 2041–2054, 2006, doi: 10.1109/TGRS.2006.872910.
8. Bovolo, F. and Bruzzone, L., A detail-preserving scale-driven approach to change detection in multitemporal SAR images. *IEEE Trans. Geosci. Remote Sens.*, 43, 12, 2963–2972, 2005, doi: 10.1109/TGRS.2005.857987.
9. Takeuchi, S., Suga, Y., Yonezawa, C., Chen, A.J., Detection of urban disaster using InSAR. A case study for the 1999 Great Taiwan Earthquake, in: *IGARSS 2000. IEEE 2000 International Geoscience and Remote Sensing Symposium. Taking the Pulse of the Planet: The Role of Remote Sensing in Managing the Environment. Proceedings (Cat. No. 00CH37120)*, IEEE, pp. 339–341, 2000.
10. Guida, R., Iodice, A., Riccio, D., Monitoring of collapsed built-up areas with high-resolution SAR images, in: *2010 IEEE International Geoscience and Remote Sensing Symposium*, pp. 2422–2425, 2010, doi: 10.1109/IGARSS.2010.5650083.
11. Yonezawa, C. and Takeuchi, S., Decorrelation of SAR data by urban damages caused by the 1995 Hyogoken-nanbu earthquake. *Int. J. Remote Sens.*, 22, 8, 1585–1600, 2001.
12. Li, Y., Martinis, S., Wieland, M., Urban flood mapping with an active self-learning convolutional neural network based on TerraSAR-X intensity and interferometric coherence. *ISPRS J. Photogramm. Remote Sens.*, 152, 178–191, 2019, doi: https://doi.org/10.1016/j.isprsjprs.2019.04.014.
13. ElGharbawi, T. and Tamura, M., Estimating deformation due to soil liquefaction in Urayasu city, Japan using permanent scatterers. *ISPRS J. Photogramm. Remote Sens.*, 109, 152–164, 2015, doi: https://doi.org/10.1016/j.isprsjprs.2015.09.002.
14. Dong, L. and Shan, J., A comprehensive review of earthquake-induced building damage detection with remote sensing techniques. *ISPRS J. Photogramm. Remote Sens.*, 84, 85–99, 2013, doi: https://doi.org/10.1016/j.isprsjprs.2013.06.011.
15. Gamba, P., Dell'Acqua, F., Trianni, G., Rapid Damage Detection in the Bam Area Using Multitemporal SAR and Exploiting Ancillary Data. *IEEE Trans. Geosci. Remote Sens.*, 45, 6, 1582–1589, 2007, doi: 10.1109/TGRS.2006.885392.
16. Lu, C.-H., Chuang, R.Y., Chiang, P.-C., Yen, J.-Y., Ching, K.-E., Chen, Y.-G., Use of Multi-Temporal Sar Satellite Images to Detect Hazards to Infrastructure Induced by Postseismic Surface Deformation: A Case Study of 2016 Meinong Earthquake in Southwestern Taiwan, Available SSRN 4435103.

17. Jager, M., Neumann, M., Guillaso, S., Reigber, A., A self-initializing PolInSAR classifier using interferometric phase differences. *IEEE Trans. Geosci. Remote Sens.*, 45, 11, 3503–3518, 2007.
18. Kimura, H., Radar polarization orientation shifts in built-up areas. *IEEE Geosci. Remote Sens. Lett.*, 5, 2, 217–221, 2008.
19. Wieland, M., Pittore, M., Parolai, S., Zschau, J., Exposure estimation from multi-resolution optical satellite imagery for seismic risk assessment. *ISPRS Int. J. Geo-Inf.*, 1, 1, 69–88, 2012.
20. Gao, F., Dong, J., Li, B., Xu, Q., Xie, C., Change detection from synthetic aperture radar images based on neighbourhood-based ratio and extreme learning machine. *J. Appl. Remote Sens.*, 10, 4, 46019, 2016.
21. Amitrano, D., Di Martino, G., Iodice, A., Riccio, D., Ruello, G., Unsupervised rapid flood mapping using Sentinel-1 GRD SAR images. *IEEE Trans. Geosci. Remote Sens.*, 56, 6, 3290–3299, 2018.
22. Moreira, A., Prats-Iraola, P., Younis, M., Krieger, G., Hajnsek, I., Papathanassiou, K.P., A tutorial on synthetic aperture radar. *IEEE Geosci. Remote Sens. Mag.*, 1, 1, 6–43, 2013.
23. Havivi, S., Schvartzman, I., Maman, S., Otman, S.R., Blumberg, D.G., Combining TerraSAR-X and Landsat images for emergency response in urban environments. *Remote Sens.*, 10, 5, 802, 2018.
24. Dell'Acqua, F. and Gamba, P., Remote sensing and earthquake damage assessment: Experiences, limits, and perspectives. *Proc. IEEE*, 100, 10, 2876–2890, 2012.
25. Cooner, A.J., Shao, Y., Ca bell, J.B., Detection of urban damage using remote sensing and machine learning algorithms: Revisiting the 2010 Haiti earthquake. *Remote Sens.*, 8, 10, 868, 2016.
26. IPCC, 2014. Climate Change 2014: Impacts, Adaptation, and Vulnerability. Part A: Global and Sectoral Aspects. Contribution of Working Group II to the Fifth Assessment Report of the Intergovernmental Panel on Climate Change [Field, C.B., V.R. Barros, D.J. Dokken, K.J. Mach, M.D. Mastrandrea, T.E. Bilir, M. Chatterjee, K.L. Ebi, Y.O. Estrada, R.C. Genova, B. Girma, E.S. Kissel, A.N. Levy, S. MacCracken, P.R. Mastrandrea, and L.L. White (eds.)]. Cambridge University Press, Cambridge, United Kingdom and New York, NY, USA.
27. Pizzoli, E. and Gong, X., How to best classify rural and urban, in: *ponencia presentada en la Fourth International Conference on Agriculture Statistics*, pp. 22–24, 2007.
28. Hewitt, K., *Regions of risk: A geographical introduction to disasters*, Routledge, London, 2014.
29. Jingya, L., Lei, W., Kailing, W., Kaihong, G., Jian, H., Earthq ake-resistant performance investigation for rural buildings in Zhongxiang area, China. *Good. Geodyn.*, 4, 1, 55–60, 2013.
30. Brunsden, D., Mass movement; the research frontier and beyond: A geomorphological approach. *Geomorphology*, 7, 1–3, 85–128, 1993.

31. Frodella, W., *et al.*, Synergic use of satellite and ground-based remote sensing methods for monitoring the San Leo rock cliff (Northern Italy). *Geomorphology*, 264, 80–94, 2016.
32. Delacourt, C., Allemand, P., Casson, B., Vadon, H., Velocity field of the 'La Clapière' landslide measured by the correlation of aerial and QuickBird satellite images. *Geophys. Res. Lett.*, 31, 15, 2004.
33. Garthwaite, M.C., On the design of radar corner reflectors for deformation monitoring in multi-frequency InSAR. *Remote Sens.*, 9, 7, 648, 2017.
34. Kromer, R.A., *et al.*, Automated terrestrial laser scanning with near-real-time change detection–monitoring of the Séchilienne landslide. *Earth Surf. Dyn.*, 5, 2, 293–310, 2017.
35. Hallegatte, S., Rentschler, J., Wal h, B., Building back better: achieving resilience through stronger, faster, and more inclusive post-disaster reconstruction, 2018.
36. dos Santos, H.A., da Silva Santana, E., Bueno, R.E., Bonilla, S.H., Technologies helping smart cities to build resilience: Focus on COVID-19, in: *Advances in Production Management Systems. Artificial Intelligence for Sustainable and Resilient Production Systems: IFIP WG 5.7 International Conference, APMS 2021, Nantes, France, September 5–9, 2021, Proceedings, Part IV*, Springer, pp. 714–723, 2021.
37. Kalantari, Z., Gadnert, A., Santos Ferreira, C.S., *EGU Gen. Assembly Conf. Abstr.*, 5963, 2020.
38. Deng, C. and Wu, C., BCI: A biophysical composition index for remote sensing of urban environments. *Remote Sens. Environ.*, 127, 247–259, 2012.
39. Ferro-Famil, L. and Pottier, E., Synthetic aperture radar imaging, in: *Microwave Remote Sensing of Land Surface*, Baghdadi, N., Zribi, M. (Eds.), *Baghdadi, N., Zribi, M. (Eds.)*, pp. 1–65, Elsevier, Amsterdam, 2016.
40. Yamaguchi, Y., Yajima, Y., Yamada, H., A four-component decomposition of POLSAR images based on the coherency matrix. *IEEE Geosci. Remote Sens. Lett.*, 3, 3, 292–296, 2006.
41. Bigdeli, B. and Pahlavani, P., High-resolution multisensor fusion of SAR, optical and LiDAR data based on crisp vs. fuzzy and feature vs. decision ensemble systems. *Int. J. Appl. Earth Obs. Geoinf.*, 52, 126–136, 2016.
42. Solberg, A.H.S., Jain, A.K., Taxt, T., Multisource classification of remotely sensed data: fusion of Landsat TM and SAR images. *IEEE Trans. Geosci. Remote Sens.*, 32, 4, 768–778, 1994.
43. Lozano-Garcia, D.F. and Hoffer, R.M., Synergistic effects of combined Landsat-TM and SIR-B data for forest resources assessment. *Int. J. Remote Sens.*, 14, 14, 2677–2694, 1993.
44. Strömberg, D., Natural disasters, economic development, and humanitarian aid. *J. Econ. Perspect.*, 21, 3, 199–222, 2007.
45. Västilä, K., Kummu, M., Sangmanee, C., Chinvanno, S., Modelling climate change impacts on the flood pulse in the Lower Mekong floodplains. *J. Water Clim. Chang.*, 1, 1, 67–86, 2010.

46. Solomon, S., Qin, D., Manning, M., Chen, Z., Marquis, M., Averyt, K.B., Tignor, M., Miller, H.L. (Eds.), *Climate Change 2007: The Physical Science Basis. Contribution of Working Group I to the Fourth Assessment Report of the Intergovernmental Panel on Climate Change*, Cambridge University Press, Cambridge, United Kingdom and New York, NY, USA, 2007.
47. Mekong River Commission, *Social Impact Monitoring and Vulnerability Assessment: Report on a Regional Pilot Study for the Mekong Corridor*, no. 30, Mekong River Commission, Vientiane, Lao PDR, 2010.
48. Nigro, J., Slayback, D., Policelli, F., Br Kentridge, G.R., NASA/DFO MODIS near real-time (NRT) global flood mapping product evaluation of flood and permanent water detection. *Eval. Greenbelt MD*, 27, 2014.
49. Brakenridge, R. and Anderson, E., MODIS-based flood detection, mapping and measurement: the potential for operational hydrological applications, in: *Transboundary Floods: Reducing Risks through Flood Management*, pp. 1–12, Springer, Dordrecht, 2006.
50. Miura, T. and Nagai, S., Landslide detection with Himawari-8 geostationary satellite data: A case study of a torrential rain event in Kyushu, Japan. *Remote Sens.*, 12, 11, 1734, 2020.
51. Watanabe, M., Motohka, T., Miyagi, Y., Yonezawa, C., Shimada, M., Analysis of urban areas affected by 2011 off the Pacific Coast of Tohoku earthquake and tsunami with L-band SAR full-polarimetric mode. *IEEE Geosci. Remote Sens. Lett.*, 9, 3, 472–476, 2012.
52. Takahashi, M., *et al.*, Disaster monitoring for Japan Earthquake with satellites by JAXA. *J. Jpn. Soc. Photogramm. Remote Sens.*, 50, 4, 198–205, 2011.
53. Jo es, J.N., Oulton, S.J., Bennett, G.L., Stokes, M., Whitworth, M.R.Z., Temporal variations in landslide distributions following extreme events: Implications for landslide susceptibility modelling. *J. Geophys. Res. Earth Surf.*, 126, 7, e2021JF006067, 2021.
54. Williams, J.G., *et al.*, Satellite-based emergency mapping using optical imagery: experience and reflections from the 2015 Nepal earthquakes. *Nat. Hazards Earth Syst. Sci.*, 18, 1, 185–205, 2018.
55. Burrows, K., Marc, O., Remy, D., Using Sentinel-1 radar amplitude time series to constrain the timings of individual landslides: a step towards understanding the controls on monsoon-triggered landslide. *Nat. Hazards Earth Syst. Sci.*, 22, 8, 2637–2653, 2022.
56. Gariano, S.L., Melillo, M., Brunetti, M.T., Kumar, S., Mathiyalagan, R., Peruccacci, S., Challenges in defining frequentist rainfall thresholds to be implemented in a landslide early warning system in India, in: *Progress in Landslide Research and Technology*, vol. *1*, Issue 1, pp. 409–416, Springer, Cham, 2022.
57. Dixit, S., Siva Subramanian, S., Srivastava, P., Yunus, A.P., Martha, T.R., Sen, S., Numerical-model-derived intensity–duration thresholds for early

warning of rainfall-induced debris flows in a Himalayan catchment. *Nat. Hazards Earth Syst. Sci.*, 24, 2, 465–480, 2024.

58. Mey, J., Guntu, R.K., Plakias, A., Silva de Almeida, I., Schwanghart, W., More than one landslide per road kilometre–surveying and modelling mass movements along the Rishikesh–Joshimath (NH-7) highway, Uttarakhand, India. *Nat. Hazards Earth Syst. Sci.*, 24, 9, 3207–3223, 2024.
59. Aldrich, D.P., *Building resilience: Social capital in post-disaster recovery*, University of Chicago Press, Chicago, IL, 2012.
60. Yabe, T., Rao, P.S.C., Ukkusuri, S.V., Regional differences in resilience of social and physical systems: A case study of Puerto Rico after Hurricane Maria. *Environ. Plan. B Urban Anal. City Sci.*, 48, 5, 1042–1057, 2021.
61. Parra, L., *et al.*, Advancements in On-Board Processing of Synthetic Aperture Radar (SAR) Data: Enhancing Efficiency and Real-Time Capabilities. *IEEE J. Sel. Top. Appl. Earth Obs. Remote Sens.*, 1–21, Jan. 2024, doi: 10.1109/JSTARS.2024.3406155.
62. Nagaiah, K., Kalaivani, K., Palamalai, R., Suresh, K., Sethuraman, V., Karuppiah, V., A Logical Remote Sensing Based Disaster Management and Alert System Using AI-Assisted Internet of Things Technology. *Remote Sens. Earth Syst. Sci.*, 7, 4, 457–471, 2024.
63. Cloude, S.R. and Pottier, E., An entropy based classification scheme for land applications of polarimetric SAR. *IEEE Trans. Geosci. Remote Sens.*, 35, 1, 68–78, Jan. 1997.
64. Lee, J.S., Du, Q., Schuler, D.L., An unsupervised classification technique for polarimetric SAR images using a decomposition of the coherency matrix. *IEEE Trans. Geosci. Remote Sens.*, 42, 4, 844–853, Apr. 2004.
65. Marino, A., Cloude, S.R., Woodhouse, I.H., A new polarimetric change detector in radar imagery. *IEEE Trans. Geosci. Remote Sens.*, 50, 11, 4302–4318, Nov. 2012.
66. Haralick, R.M., Shanmugam, K., Dinstein, I., Textural features for image classification. *IEEE Trans. Syst., Man, Cybern.*, SMC-3, 6, 610–621, Nov. 1973.
67. Mekong River Commission., Annual Flood Report 2005. Mekong River Commission, Vientiane, Lao PDR, 2005.
68. Ahamed, A., Bolten, J.D., Doyle, C.S., Fayne, J., A MODIS-based automated flood monitoring system for Southeast Asia. *Int. J. Appl. Earth Observ. Geoinf.*, 61, 104–117, 2017, DOI: 10.1016/j.jag.2017.05.006.
69. Tucker, C.J., Red and photographic infrared linear combinations for monitoring vegetation. *Remote Sens. Environ.*, 8, 2, 127–150, 1979.
70. Mekong River Commission (MRC)., State of the Basin Report 2010. Vientiane, Lao PDR, 2010.

4

Synthetic Aperture Radar Interferometry

Suryakant Annasaheb Shinde

Department of Applied Science (Physics), MGM University, Chhatrapati Sambhajinagar, India

Abstract

Nowadays, the interferometric synthetic aperture radar (InSAR) is regarded as a potent instrument for Earth observation. It helps in extremely precise surveillance of tiny surface movements. The most important thing is that it works at night and even in any weather conditions. In this chapter, the basic concepts of InSAR are described. It goes over how to record SAR signals, analyze phase differences, and use this technique to create digital elevation models (DEMs). The history of the development of InSAR is discussed, starting with its initial experimental use and up to its use in modern satellite missions. These include Sentinel-1, TanDEM-X, ERS, SRTM, and the upcoming NISAR mission. Numerous practical uses of InSAR are also investigated. They are topography mapping, glacier monitoring, volcano observation, urban structure checking, and earthquake change tracking. Some advanced InSAR methods are also mentioned in this chapter. These include multi-temporal InSAR (MT-InSAR), polarimetric InSAR (PolInSAR), persistent scatterer interferometry (PSI), and differential InSAR (DInSAR). Both methods have advantages when it comes to long-term monitoring of environmental and natural hazards. Some of InSAR's limitations are also discussed. These are issues like atmospheric disturbances and loss of coherence with time. InSAR technology is expected to evolve in the future with the help of artificial intelligence. The next generation of higher-resolution radar satellites is also going to be critical. This chapter is aimed at giving an overview of InSAR. It also gives a prospective understanding that would be of assistance to researchers, practitioners, and students in the geoscience and remote sensing professions.

***Keywords*:** Interferometric SAR (InSAR), surface deformation monitoring, digital elevation model (DEM), differential InSAR (DInSAR)

Email: sshinde@mgmu.ac.in

Pooja Joshi, Ashutosh Bhatt, Neelam Sharma and Kapil Joshi (eds.) Synthetic Aperture Radar for Disaster Management: Bridging Science and Solutions, (89–116) © 2026 Scrivener Publishing LLC

4.1 Introduction

Interferometric synthetic aperture radar (InSAR) has become an important technology in Earth observation. It can detect surface changes as small as a few millimeters. It is a remote sensing method that captures small changes on the Earth's surface, which also helps in generating high-resolution digital elevation models (DEMs) [1, 2]. It works by comparing the phase differences between two or more SAR images. These pictures are captured at a slightly different angle or at various times. The shift of the phase occurs due to variations in the distance between the satellite and the ground. This allows detection of terrain height or surface movement.

The strength of the InSAR lies in its use of coherent SAR signals. Because of this, even tiny deformations can be measured accurately. Today, InSAR is widely used for monitoring land surfaces, infrastructure, and natural disasters. Mathematically, the interferometric phase (ϕ) between two SAR acquisitions is given by:

$$\phi = \frac{4\pi}{\lambda} \cdot \Delta R \tag{4.1}$$

Where: ϕ = interferometric phase difference
λ= radar wavelength
ΔR = difference in range distance between two SAR observations

By comparing the phase differences between SAR image pairs, InSAR can measure surface elevation and ground movement with very high precision. Under ideal conditions, the accuracy can reach sub-centimeter levels [1]. This method has transformed Earth observation. It allows real-time and high-resolution surface observations, and it is independent of weather and daylight. One of the most significant outcomes of InSAR is the generation of accurate DEMs. Missions like the Shuttle Radar Topography Mission (SRTM) [3] and TanDEM-X [4] have made this possible. The SAR data at several viewing angles are gathered by these missions to determine the height of the terrain.

The InSAR is very sensitive. It can detect even millimetric changes in the Earth's surface. These changes can be caused by earthquakes, volcanoes, or subsidence of the land because of loss of groundwater, or glaciers. Due to these capabilities of repeatedly and in small detail monitoring such phenomena, InSAR has become a necessity in geophysical research. Besides natural processes, InSAR is currently applied in infrastructure monitoring.

It is used to evaluate bridges, pipelines, dams, and other important structures. It is possible to know the early indicators of deformation, which assist in disaster prevention and risk management.

In such a way, InSAR can be used not only in environmental studies but also in civil engineering. It is applied in areas such as decision-making in natural hazard assessment and urban planning. It also plays a very important role in observing disasters such as earthquakes, landslides, and floods. It aids in realizing how the ground moves during such events and gives information that is useful in enhancing preparedness and mitigation. Among the key advantages of InSAR compared to optical methods, InSAR is all-weather. It does not require sunlight and can be used at night or even during cloud cover. This ensures consistent and reliable monitoring.

The essence of the method is in phase shift analysis of repeated SAR acquisitions. Originally, it was primarily employed to make static elevation models. However, nowadays, it is extensively applied in monitoring dynamic processes. These are earthquake studies, glacier studies, urban growth, and environmental alterations. Advanced InSAR methods such as differential InSAR (DInSAR), persistent scatterer interferometry (PSI), small baseline subset (SBAS), and polarimetric InSAR (PolInSAR) have greatly expanded their utility. These techniques, when supplemented with artificial intelligence (AI) and contemporary computing, provide an effective means of real-time monitoring of Earth and geohazard research [3].

4.2 Historical Evolution of Synthetic Aperture Radar Interferometry

The development of InSAR marks a major step forward in remote sensing and Earth observation. It is a mix of advancements in radar technology, coherent signal processing, and an increased need for precise terrain or deformation mapping. The InSAR started as a theoretical idea tested in laboratories. It has gradually been transformed into a very potent system of global monitoring. This journey has been supported by advanced satellite missions. It has moved on to become a major component of massive geospatial analysis after it was initially a specialized experimental tool. The technological growth of InSAR can be divided into five main stages. Each phase shows an important milestone in its progress and operational development. The field of SAR Interferometry has changed significantly over the last few decades. As shown in Figure 4.1, early research in the 1960s laid the foundation for the later development of InSAR systems in the 1980s.

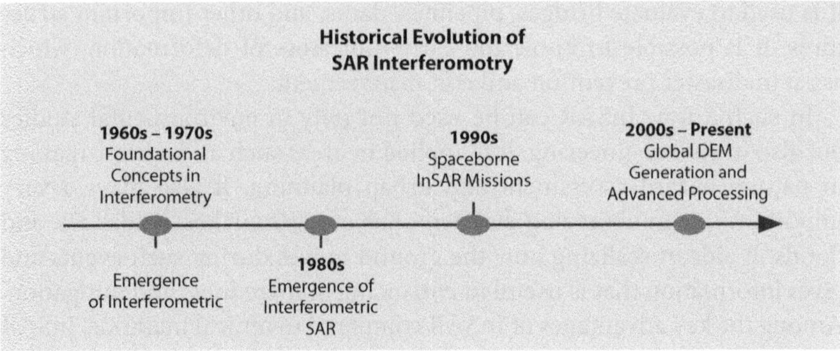

Figure 4.1 Chronological development of synthetic aperture radar interferometry.

In the following decades, several dedicated InSAR satellites were launched. These missions made it possible to monitor ground deformation and map terrain on a large scale.

4.2.1 Foundational Developments (1960s–1970s)

The basic ideas behind radar interferometry were developed in the 1960s. These were inspired by earlier work in radio interferometry and microwave imaging. At that time, scientists began studying how radar waves could be used for measuring phase differences. In 1965, successful laboratory tests showed that radar signals could indeed be used for interferometric measurements. This was the first practical proof of the radar interferometry concept. By the late 1970s, major progress had been made in airborne radar systems. Most of this work was driven by defense and aerospace research. Although the main aim was not terrain mapping at that time, these studies introduced the concept of phase-based radar imaging. The techniques developed during this period gave important insights. Together, they laid the groundwork for the future development of InSAR.

4.2.2 Emergence of the Interferometric Synthetic Aperture Radar (1980s)

The 1980s marked the beginning of what is now called the modern era of InSAR.

Between 1982 and 1985, a breakthrough came from researchers at NASA's Jet Propulsion Laboratory. They successfully created the first airborne SAR interferograms. In these experiments, two antennas were placed at different positions on the same aircraft. This arrangement allowed the measurement of terrain elevation by single-pass interferometry. The main benefit was that it eliminated the necessity of several flights over the same territory. Correct elevation data would be attainable in one pass. By 1986, scientists started working on repeat-pass InSAR. This technique allowed the detection of changes in the Earth's surface over time. Such developments provided a new area of research on dynamic geophysical processes with radar. However, this also came with its own challenges that persist to this day in the application of InSAR. Issues such as temporal decorrelation, which is a result of surface changes between images, also appeared. Another issue was baseline sensitivity, which impacts measurement accuracy. Nevertheless, the studies conducted in the 1980s provided a solid ground. It has been a key step toward taking InSAR to full-scale satellite operations.

4.2.3 Advent of Spaceborne Interferometric Synthetic Aperture Radar (1990s)

The 1990s marked a fresh development in the history of InSAR. This decade saw the first dedicated Earth observation spaceborne SAR missions. A major milestone came in 1991, when the European Space Agency (ESA) launched ERS-1 [2]. It became the initial satellite to enable space-based repeat-pass interferometry. This led to frequent and widespread applications of InSAR. In 1993, data from ERS-1 [2] were utilized to create the first high-resolution DEM by utilizing InSAR techniques. The launch of ERS-2 [3] in 1995 improved these capabilities further. It allowed tandem interferometry with a 1-day revisit time. Such a setup increased temporal continuity and expanded the surface observation zone. Around the same time, NASA's SIR-C/X-SAR mission flew aboard the Space Shuttle in 1994. Its data were extremely valuable as dual-frequency SAR and were applied early in radar polarimetry and interferometry. Together, these missions proved the practical usefulness of InSAR. They contributed to the topographic mapping, tectonic research, and volcanic monitoring. The time has transformed InSAR into a powerful and reliable tool in earth science.

4.2.4 Global Digital Elevation Model Production and Advanced Time-Series Techniques (2000s)

With the start of the 21st century, InSAR research shifted focus toward large-scale DEM generation [1, 3, 4] and advanced techniques for tracking ground deformation. A major achievement during this time was SRTM [3], performed in 2000 aboard the Space Shuttle Endeavour. The mission utilized single-pass interferometry, comprising two C-band and X-band antennas [1, 3, 4]. This setup allowed near-global collection of elevation data. The resulting DEMs were at a spatial resolution of 30 meters, hence a major improvement in topographic mapping of the globe. In the meantime, new ways of time-series analysis were developed to observe continuous deformation. Two key methods have been the PSI [3] and the SBAS approach. The techniques enabled long-term monitoring of ground displacement in areas where temporal decorrelation occurs. Their introduction marked the beginning of a new phase in InSAR applications, focused on long-term geophysical monitoring.

Together, these advances demonstrated the growing capacity of InSAR to support both scientific research and practical monitoring applications on a global scale. These methods enabled long-term deformation tracking in urban and vegetated areas, significantly enhancing applications in geohazards, infrastructure monitoring, and land subsidence analysis.

4.2.5 Modern Interferometric Synthetic Aperture Radar Era and Next-Generation Missions (2010–Present)

The current era is defined by high-resolution, high-repeat-cycle SAR missions that deliver free, globally available data for operational InSAR applications. Germany's TerraSAR-X and TanDEM-X [4] missions (launched in 2007 and 2010, respectively) enabled bistatic, single-pass interferometry, culminating in the global TanDEM-X [4] DEM with ~12 m resolution [7] and high vertical accuracy. ESA's Sentinel-1 [2] A (2014) and Sentinel-1 [2] B (2016), part of the Copernicus program, revolutionized global InSAR access by providing systematic, freely available C-band SAR data with a revisit interval of 6–12 days. These missions have been extensively used in deformation monitoring, disaster assessment, urban infrastructure management, and glacier tracking. Looking ahead, the NISAR [3] mission, a collaboration between NASA and ISRO, is poised to be the first SAR mission to operate in dual frequencies (L- and S-band) with full polarimetric capability. Scheduled for launch in 2025, NISAR is expected to provide systematic InSAR coverage of Earth's land surface every 6–12 days, facilitating studies in tectonic deformation, agriculture, biomass, glacier movement, and landslide monitoring.

4.2.6 Overview of Key Spaceborne Synthetic Aperture Radar Missions Relevant to the Interferometric Synthetic Aperture Radar

Spaceborne SAR missions have played a significant role in expanding the capabilities of InSAR. These platforms provide reliable and repeatable radar observations. They are important for generating DEMs, monitoring ground deformation, or studying dynamic Earth processes. The modern era of InSAR started in the 1980s. Between 1982 and 1985, researchers at NASA's Jet Propulsion Laboratory achieved a breakthrough. They created the first airborne SAR interferograms, which showed that single-pass interferometry was feasible. This was done through the installation of two antennas at different locations on the same airborne platform. It allowed adequate measurements of the elevations without having to fly multiple times on the same territory.

By 1986, the focus shifted to repeat-pass InSAR. This method could unveil time variations on the planet's surface, and it was possible to trace geophysical events with the assistance of radar. However, this period also exposed several challenges. The temporal decorrelation (caused by surface variations between acquisitions) and the baseline sensitivity (causing the accuracy of measurements) were the two large issues. However, these limitations notwithstanding, the work of the 1980s has been a good foundation. It helped push InSAR toward satellite-based operations. These developments later supported time-series methods such as PSI, which is broadly utilized for tracking urban deformation, volcanic activity, or groundwater-induced subsidence. This phase of SAR systems on the air was concluded with a row of missions, and some of them, like SIR-C/X-SAR, officially stopped in 2012.

One major advancement in the 2000s was the launch of ALOS PALSAR by the Japan Aerospace Exploration Agency (JAXA) in 2006. Operating at L-band (1.27 GHz), JAXA offered deeper vegetation penetration than earlier missions. This made ALOS PALSAR particularly useful for InSAR studies in forested and tropical regions, where short-wavelength systems struggle due to signal scattering from vegetation. It was followed by ALOS-2 in 2014, improving resolution and observation frequency. TerraSAR-X/TanDEM-X [4] (DLR) Launch Years: 2007/2010 | Frequency Band: X-band (9.65 GHz). These German satellites flew in close tandem (approx. 500 m apart) to perform bistatic, single-pass InSAR. Their mission produced the TanDEM-X [4] global DEM and enabled precision applications in glaciology, tectonics, and urban deformation.

Sentinel-1 [2] A and 1B (ESA) Launch Years: 2014/2016 | Frequency Band: C-band Operating in Interferometric Wide Swath Mode (IW), Sentinel-1 provides regular, free, and high-resolution SAR data. With its 6- to 12-day repeat cycle, it is a backbone for global geohazard monitoring and operational InSAR services. NISAR (NASA–ISRO) Scheduled Launch: 2025 | F. Bands: L-band (1.25 GHz) & S-band (3.2 GHz). This mission will be the first to offer dual-frequency, fully polarimetric SAR imaging from space. NISAR is designed to support broad InSAR applications, such as crustal deformation, agricultural monitoring, biomass estimation, and disaster resilience planning. It is expected to deliver regular displacement measurements over Earth's landmass with a 6–12 day revisit rate.

4.2.7 Summary Table of Key Interferometric Synthetic Aperture Radar Missions

Several international satellite missions have significantly advanced SAR interferometry. Early platforms like ERS-1/2 and ENVISAT established repeat-pass capabilities, while newer systems such as TerraSAR-X/TanDEM-X enabled bistatic acquisitions. As summarized in Table 4.1, recent missions like Sentinel-1 and the upcoming NISAR offer enhanced coverage, resolution, and dual-frequency observations for diverse geophysical applications.

Table 4.1 Major Spaceborne SAR missions for interferometric applications.

Satellite	Agency	Year	Band	InSAR type	Key applications
ERS-1/ERS-2	ESA	1991/1995	C-band	Repeat-pass	DEM generation, deformation mapping
ENVISAT	ESA	2002	C-band	Repeat-pass	Urban subsidence, PSI, volcano monitoring
ALOS PALSAR	JAXA	2006	L-band	Repeat-pass	Tectonics, forests, and agricultural land motion
TerraSAR-X/ TanDEM-X	DLR	2007/2010	X-band	Single-pass (bistatic)	High-res DEM, glaciology, and infrastructure monitoring
Sentinel-1A/1B	ESA	2014/2016	C-band	Repeat-pass	Global InSAR, geohazards, and operational monitoring
NISAR (planned)	NASA/ ISRO	2025 (expected)	L + S band	Dual-pol InSAR	Crustal motion, biomass, agriculture, and landslides

4.3 Fundamental Principles and Working Mechanism of Synthetic Aperture Radars and Interferometric Synthetic Aperture Radars

Synthetic aperture radars operate as active microwave remote sensing systems, transmitting their own signal and measuring the reflected backscatter from the Earth's (or lunar) surface. Unlike optical sensors, the SAR can acquire high-resolution images independent of sunlight and through cloud cover, making it ideal for uninterrupted Earth and planetary monitoring across varied weather and illumination conditions.

The SAR achieves high spatial resolution by synthetically enlarging the antenna aperture through the motion of the platform (e.g., satellite or aircraft). The principle rests on coherent signal processing that integrates the echoes received over successive positions of the radar, thereby simulating a large antenna—hence the term "synthetic aperture."

4.3.1 Radar Signal Transmission and Reception

At each position along its flight path, the SAR system emits a short-duration pulse toward the target. The pulse has a bandwidth B, center frequency fc, and duration τ. The returned echo from the target is received and recorded as a function of time (range) and platform position (azimuth) [4, 5]. The transmitted signal can be represented as: [4, 5]

$$st(t)=A\cdot\exp\,[\,j\,(2\pi\,fct + \phi(t))]$$

where:

A- is the amplitude,

fc - is the carrier frequency,

$\phi(t)$ - represents the frequency modulation (e.g., linear chirp: $\phi(t) = \pi Kt2$,

$K = B/\tau$ is the chirp rate.

As shown in Figure 4.2 below, InSAR uses two radar acquisitions from different positions to measure phase differences arising from path length variations, which relate to surface elevation or displacement.

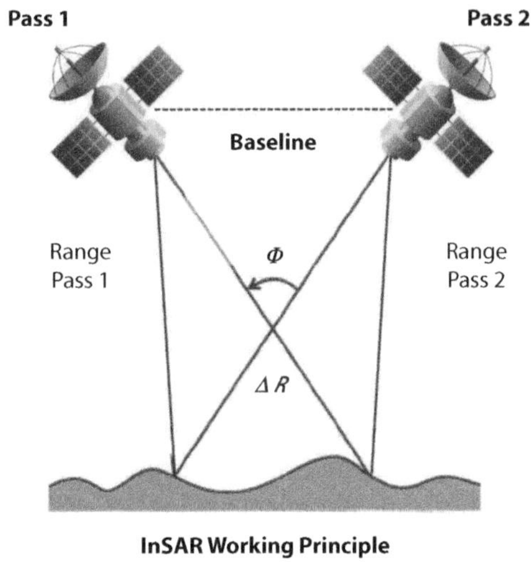

Figure 4.2 Basic principle of InSAR.

4.3.2 Range Resolution (δ_r)

Range resolution refers to a radar system's capability to distinguish between two targets located at different distances along the radar's line of sight. In the SAR, this resolution is primarily governed by the bandwidth of the transmitted signal and is mathematically expressed as: [4, 6]

$$\delta_r = \frac{C}{2B} \tag{4.2}$$

where:

C is the speed of light in a vacuum (3×108m/s),
B is the bandwidth of the transmitted radar pulse.

This expression indicates that the range resolution improves with an increase in bandwidth. A higher bandwidth compresses the transmitted pulse in time, allowing the radar to better resolve closely spaced targets in the slant-range direction. This characteristic is crucial for capturing fine-scale surface features in SAR imagery [4].

4.3.3 Azimuth Resolution (δa)

In real aperture radar, the azimuth resolution δa real is dependent on the physical antenna length L:

$$\delta_{a,real} = R\lambda / L \tag{4.3}$$

In contrast, SAR synthesizes an aperture length by moving the antenna and coherently combining received echoes. The synthetic aperture dramatically improves azimuth resolution:

$$\delta_{a,SAR} = \frac{L}{2} \tag{4.4}$$

Alternatively, in terms of wavelength λ and synthetic aperture length D_{sa}:

$$\delta_{a,SAR} = \frac{\lambda R}{2Dsa} \tag{4.5}$$

where R is the slant range to the target [4, 5].

4.3.4 Slant Range and Ground Range

The slant range R is the direct distance from the radar to the target, while the ground range is the projection of this distance on the Earth's surface. For flat terrain, the relation is:

$$R = \sqrt{h2 + x2} \tag{4.6}$$

where: h - is the platform altitude,
x - is the ground distance from nadir.

The relationship between slant range, ground range, and incidence angle, as depicted in Figure 4.3, plays a critical role in SAR image formation and interpretation.

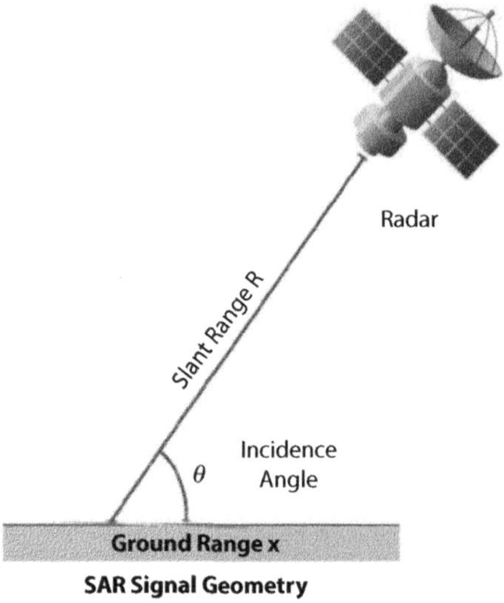

Figure 4.3 Fundamental geometry of SAR imaging.

The ground range resolution is:

$$\delta_g = c / 2B\cos(\theta) \tag{4.7}$$

where θ - is the incidence angle [4, 6].

4.3.5 Doppler Frequency and Synthetic Aperture Formation

As the radar moves, the relative velocity between the radar and the target induces a Doppler shift:

$$f_D = (2v / \lambda) \sin(\theta) \tag{4.8}$$

As illustrated in Figure 4.4, Doppler frequency shifts arising from relative motion are fundamental to SAR image formation, enabling accurate target localization in the azimuth direction. This Doppler frequency varies across the synthetic aperture and is exploited to determine the azimuth position of the target. By integrating the signal over time, SAR synthesizes a narrow beam in the azimuth direction, enhancing resolution far beyond that of real-aperture systems.

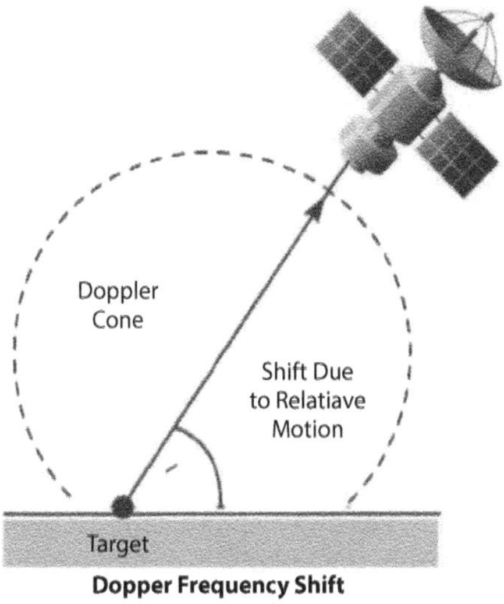

Figure 4.4 Doppler frequency shift in SAR imaging.

4.3.6 Signal Processing and Image Formation

The SAR image formation involves the use of matched filtering in both the range and azimuth directions. Range compression is done by convolving the received signal with a replica of the transmitted chirp. This step improves resolution in the slant range direction. In the azimuth direction, compression is achieved by focusing the Doppler history of the moving radar platform. This helps sharpen the image in the along-track direction. The final output is a complex-valued SAR image. In this image, each pixel contains phase information and amplitude [9].

$$I(x,y) = |A(x,y)| \cdot \exp\,[j\phi(x,y)] \tag{4.9}$$

These complex-valued SAR images form the foundation for many advanced analysis methods. The key among them are InSAR, PolSAR, and polarimetric decomposition methods, which are used for studying surface scattering behavior [4, 6].

4.3.7 Topographic Mapping: Digital Elevation Model Generation

One of the key applications of InSAR is the creation of DEMs. These models offer a three-dimensional view of the Earth's surface. To accomplish this, two SAR images of the same area are obtained at slightly different points. These positions are separated by a known distance, called the baseline. This spatial separation helps extract elevation data using the geometric relationship between the radar sensor and the ground. The core idea is based on analyzing the phase difference between the two SAR acquisitions. This phase difference, usually represented as ϕ, is directly related to the slant range difference (ΔR).

Mathematically, this relationship can be written as:

$$\phi = \frac{4\pi}{\lambda} \cdot \Delta R \tag{4.10}$$

Here, λ represents the radar wavelength, and ΔR denotes the difference in slant range distance from the radar sensor to the ground target between the two acquisition positions. This phase measurement allows the computation of elevation values for each pixel, given the known imaging geometry. The actual height h of a point on the ground can be derived from the measured phase using the following equation:

$$h = \frac{\lambda R \Delta\phi}{4\,\pi B_{\perp} \sin(\theta)} \tag{4.11}$$

In this equation, R represents the slant range, $\Delta\phi$ is the measured interferometric phase difference, $B_{\perp}$ is the perpendicular baseline (component of the baseline orthogonal to the radar's line of sight), and θ is the incidence angle. Under ideal conditions, the technique can provide elevation measurements with meter or even sub-meter vertical accuracy.

Prominent satellite missions such as the SRTM and TanDEM-X have successfully implemented this method. The SRTM, launched aboard the Space Shuttle Endeavour in 2000, used a single-pass, dual-antenna interferometric configuration to map nearly 80% of Earth's land surface with a resolution of 30 meters. More recently, the TanDEM-X mission—consisting of twin satellites TerraSAR-X and TanDEM-X flying in close formation—provided a global DEM at 12-meter resolution with unprecedented accuracy [7].

The workflow for DEM generation [1, 3, 4] using InSAR includes several key steps: acquiring a pair of SAR images, performing precise co-registration to align pixels, generating the interferogram, filtering to reduce noise, unwrapping the phase to remove ambiguities, and finally converting the unwrapped phase into elevation values through geocoding and projection into geographic coordinates. This application has become vital in topographic analysis, hydrological modeling, and as input for geospatial systems.

4.3.8 Deformation Monitoring: Earthquakes, Landslides, and Subsidence

The InSAR is a powerful technique for monitoring ground surface deformation over time, offering sub-centimeter precision across large spatial scales. When two SAR images of the same area are acquired from nearly identical positions but at different times, any change in the position of the ground surface between acquisitions will manifest as a phase shift in the resulting interferogram.

This temporal phase change, attributed to ground displacement along the radar's line of sight (LOS), is expressed as:

$$\Delta\phi_{defo} = \frac{4\pi}{\lambda} \cdot \Delta d \tag{4.12}$$

With $\Delta\phi$defo, the interferometric phase difference caused by the deformation, λ the radar wavelength, and Δd the LOS displacement. This renders InSAR extremely useful in such applications as analyzing tectonic movement, volcanic monitoring, ground subsidence, and landslide evaluation [10]. The technique is further improved by DInSAR. It improves the capability of sensing surface changes as the topographic phase component is filtered out of the interferogram. This is done using a reference DEM, such as the one generated from the SRTM mission. The remaining phase component thus corresponds only to surface displacement. The differential phase is computed as:

$$\Delta\phi_{diff} = \phi_{total} - \phi_{topo} \tag{4.13}$$

This technique enables highly accurate mapping of ground deformation between two time-separated SAR acquisitions. Applications include measuring co-seismic and post-seismic displacements, tracking inflation or deflation of volcanic domes, detecting landslides, and assessing settlement in urban and agricultural regions due to factors such as groundwater extraction or tunneling.

To observe long-term or slow-moving deformation trends, multi-temporal approaches are used. Persistent scatterer interferometry [3, 12] focuses on stable reflectors, such as buildings and rocks, whose radar signal remains coherent across multiple acquisitions. On the other hand, SBAS techniques combine multiple interferograms with small spatial and temporal baselines to mitigate decorrelation and atmospheric noise, allowing analysis even in less coherent areas such as vegetated terrain.

Real-world examples of InSAR deformation monitoring include the 1999 İzmit earthquake in Turkey, where co-seismic displacements were mapped with high accuracy; tracking subsidence in Mexico City [13] due to groundwater pumping; and monitoring landslide activity in the Himalayan and Andean Mountain ranges.

Visualization of deformation is typically done through LOS displacement maps, where colored fringes represent displacement values. Each full fringe typically corresponds to a change of half the radar wavelength. These maps provide a clear and easy-to-understand view of the magnitude and spatial extent of surface movement. They are extremely valuable for both disaster management and scientific research [14].

4.3.9 Glaciology and Ice Monitoring: Ice Sheet Movement and Permafrost Changes

The InSAR has become an important tool in glaciological research. It gives precise measurements of ice sheet dynamics, glacier flow velocities, and permafrost changes in polar and high-altitude areas. Conventional field-based glacier monitoring has a limited spatial area. They are also risky due to harsh environmental conditions. The InSAR overcomes these challenges. It has high resolution, a wide area, and all-weather capability, making it most suited to such remote places. Glaciers or ice sheets are constantly in motion. They move between a few centimeters to several meters a day.

By capturing SAR images at distinct times or analyzing their phase differences, InSAR can detect LOS motion with accuracy from centimeters to millimeters. The glacier movement-induced deformation is extracted based on the interferometric phase difference equation:

$$\Delta\phi_{\text{ice}} = \frac{4\pi}{\lambda} \cdot \Delta d \tag{4.14}$$

Here, $\Delta\phi_{(\text{ice})}$ denotes the interferometric phase shift caused by ice movement. λ is the radar wavelength, Δd refers to the displacement along the radar's LOS. These measurements help in mapping glacier velocity fields, locating crevasse zones, and identifying basal sliding or glacier surging behavior.

In addition, InSAR is extremely effective in observing permafrost degradation, especially in Arctic and sub-Arctic regions, where climate change is rapidly accelerating thaw processes. Seasonal freeze-thaw cycles lead to vertical ground movement. This is captured as phase variations in the InSAR interferograms. Through repeated observations, InSAR can monitor changes in active layer thickness, detect subsidence due to melting permafrost, and identify areas of terrain instability.

For example, data from missions like Sentinel-1 [2] and ALOS PALSAR have been widely used to monitor outlet glaciers in Greenland and Antarctica, track changes in Canadian permafrost zones, and assess mass loss trends in Himalayan glaciers.

4.3.10 Volcanic Activity Monitoring

The InSAR is considered one of the most powerful remote sensing techniques for volcano monitoring. It enables the early detection of volcanic deformation, including inflation, deflation, and caldera subsidence. Such deformations are indicators of magma chamber pressurization or drainage, which precede eruptions or reflect ongoing magmatic processes. The phase shift between two SAR images acquired before and after volcanic activity can indicate whether the ground has moved due to subterranean pressure changes. The basic phase-displacement relationship used in volcanic deformation studies is:

$$\Delta\phi_{\text{volcano}} = \frac{4\pi}{\lambda} \cdot \Delta d \tag{4.15}$$

where $\Delta\phi_{\text{volcano}}$ corresponds to the deformation-induced phase change, λ is the radar wavelength, and Δd represents the LOS surface displacement. These deformations can be vertical (e.g., ground uplift or subsidence) or horizontal, depending on the geometry of the magma intrusion.

Volcanic deformation is often modeled using simplified pressure source models, such as the Mogi model, which assumes a point pressure source at a certain depth. The InSAR-derived deformation fields are fitted with these models to estimate depth, volume change, and location of magma chambers [8].

Several case studies have demonstrated the efficiency of InSAR in volcano monitoring:

Mount Etna (Italy) [8]: Continuous monitoring of inflation or flank movement.

Kīlauea (Hawaii) [8]: Capturing co-eruptive deflation and post-eruptive inflation cycles.

An example of InSAR's effectiveness is seen at Nevado del Ruiz volcano in Colombia [8], where it helped identify precursory deformation patterns before an eruption. When volcano observatories and disaster management organizations are in inaccessible or remote locations, InSAR provides significant early warning capabilities. Using open-source SAR data, like from the Sentinel-1 mission [2], InSAR can generate near-real-time deformation maps. This goes a long way in the volcanic hazard analysis and improved preparedness toward any volcanic risk in volcanically active regions.

4.4 Case Studies

4.4.1 Co-Seismic Deformation from the 2015 Nepal Earthquake

The 2015 Gorkha earthquake (Mw 7.8) in Nepal triggered intense seismic activity across the central Himalayan area, with noticeable effects reaching northern India. In response, Sharma *et al.* [15] from IIRS–ISRO applied DInSAR methods to research co-seismic ground deformation by utilizing Envisat ASAR data. They used two SAR images, namely pre-earthquake and post-earthquake, to create interferograms or approximate LOS ground displacement.

Research employed interferometric velocity maps generated from multiple SAR missions, involving ERS-1/2, Envisat, and Sentinel-1, to evaluate long-term slope deformations in the research region. These displacement readings were coupled with a series of thematic layers, i.e., slope angle, lithology, land use, rainfall intensity, and drainage density, to develop a full picture of slope instability factors. To quantify the relative importance of every parameter, the analytical hierarchy process (AHP) has been utilized, allowing for the assignment of weights or ranks depending on expert judgment or the individual contribution of every factor to landslide occurrence. The combination of deformation data of SAR and geospatial variables resulted in the creation of a classified map of landslide susceptibility, which divided the area into four hazard levels: Very Low, Low, Moderate, and High. The product is a useful spatial value in the mitigation of risks or land-use planning in landslide-prone areas.

4.4.2 Landslide Monitoring: The Kotrupi Event

One of the slope failures that occurred in the Himalayan region was the Kotrupi landslide, which occurred in 2017, during the monsoon season in Himachal Pradesh. In response, Niraj *et al.* [16] from IIRS–ISRO adopted an integrated approach by utilizing DInSAR and multi-temporal InSAR (MTInSAR) to assess pre- and post-failure ground deformation. They utilized Sentinel-1 C-band SAR data in ascending mode and processed it using StaMPS (Stanford Method for Persistent Scatterers). The DInSAR analysis revealed co-event ground displacement of up to ±60 millimeters in the LOS direction. In the landslide, the MTInSAR process showed that slope creep continued after landslides at mean rates of about -45 mm per year. The DEM of difference analysis was conducted by utilizing pre- and post-event elevation models. It could show terrain variation within the range of +/-3 to 3.5m. Such InSAR-based results have been validated using high-resolution optical satellite images or ground measurements, thus providing more accurate findings.

What was learned from the Kotrupi landslide investigation was that DInSAR could help capture the abrupt co-event deformation of the order of +/-60mm as the direct zone of impact. The MTInSAR processing revealed that slope movements continued after the failure at an average rate of -45 mm per year. The DEM of difference also reported significant terrain alteration, with elevation changes representing the mass movement and deposition of materials. The addition of SAR-based methods and optical data enabled a more detailed knowledge of landslide dynamics in time and space. In general, the multi-technique InSAR technique has given a stable and multi-temporal analysis of the Kotrupi landslides, which could be regarded as a valuable source of information about the pre-failure precondition and post-event slope evolution [16].

4.4.3 Mining-Induced Deformation: Jharia Coalfield

Jharia Coalfield in Jharkhand, India, is among the country's most economically significant yet environmentally fragile mining zones. The large-scale underground coal mining has also led to surface subsidence in the region, and this has posed a threat to the infrastructure and the settlements around. To monitor this, Thapa and Chatterjee [17] from IIRS–ISRO applied persistent scatterer interferometric SAR (PSInSAR) methods using L-band ALOS PALSAR data. They determined the presence of persistent scatterers or stable radar targets in the mining area using their methodology. In various acquisitions of SAR, the histories of the target periods were examined. Millimeter precision could be applied to detect ground movement by utilizing this PSInSAR technology. Its outcomes showed that it had local subsidence of over 15 millimeters in a month. These deformation areas of high deformation corresponded to known underground coal beds and active mining zones. It has been demonstrated that PSInSAR can be a valuable tool for surveying subsidence in vegetated or complex-structured mining areas where DInSAR is not feasible due to time decorrelation or atmospheric noise.

The Jharia Coalfield study revealed several main insights into ground deformation monitoring by using L-band PSInSAR. Rates of deformation have been determined to be over 15 millimeters per month in regions most correlated to active coal mining. The fact that PSInSAR has high spatial resolution makes it an effective tool in detecting hot spots of critical subsidence early enough, which is a useful proactive risk assessment tool. Results are significant in improving safety in mines, urban planning, or disaster mitigation planning in mining-related areas. Furthermore, research established that L-band PSInSAR is particularly well-suited for long-term deformation monitoring in Indian coalfields, where C-band SAR systems often face challenges due to vegetation-induced decorrelation [17].

4.4.4 Polarimetric Synthetic Aperture Radar-Based Forest Height Estimation and Digital Elevation Model Generation

Dr. Shashi Kumar is a senior-most scientist at IIRS-ISRO and a specialist in PolSAR, PolInSAR, and SAR tomography, with considerable contributions in forest structure estimation and the generation of DEM using SAR data. He performed detailed research employing dual-polarized PolInSAR techniques by utilizing ALOS PALSAR L-band data to estimate forest canopy height and generate DEMs in forested hills of Uttarakhand. The SAR backscatter signal was decomposed into a polarimetric form based on polarimetric decompositions, notably, the Yamaguchi four-component method and the Cloude–Pottier method, to distinguish the surface, the double-bounce, and volume scattering mechanisms. These decompositions enabled accurate retrieval of forest vertical structure. Canopy heights derived using PolInSAR were cross-validated with a ground inventory data and LiDAR airborne data, and proved the applicability of the method in forest monitoring, as well as in topographic modeling.

Key Findings and Outcome: PolInSAR retrieved canopy heights, with an accuracy of +/-5m in mixed forests, and the DEM that was generated had improved continuity within dense forest cover compared to optical techniques. It was determined that volume scattering was common in mixed vegetation. Overall, research demonstrated the capability of PolInSAR techniques to estimate forest structure and terrain elevation simultaneously. This is a practical approach that can significantly enhance SAR-based surveillance in forested and other topographically complicated sites [18].

4.4.5 Terrain Hazard Zonation: Rudraprayag–Srinagar Corridor

In the Uttarakhand Rudraprayag-Srinagar corridor, there is a high likelihood of landslides and slope instability. This is the case because it possesses an abrasive topography, tectonic activities, and intense monsoon rain. To address the need for spatial hazard assessment, Devara *et al.* [19] from IIRS–ISRO employed an MTInSAR approach, integrated with a GIS-based AHP model. The study has relied on the interferometric velocity maps from numerous SAR missions, including ERS-1/2, Envisat, and Sentinel-1, to assess long-term slope movements in the region. These displacement measurements were integrated with various thematic layers like the slope angle, lithology, land use, rainfall intensity, and drainage density. To evaluate the influence of each factor on slope instability, the AHP has been applied. This was accomplished by assigning a weight and ranking of each parameter based on the expert judgment and the impact of each parameter on the slope failure. The final output was a classified landslide susceptibility map, which subdivided the region into four hazard zones: Very Low, Low, Moderate, and High.

High-risk regions corresponded to known landslide scars or recent slope failures, confirming the procedure. The study exemplified the success of combining satellite-based deformation and multi-criteria geospatial analysis in mapping regional hazards. The study identified critical zones of persistent slope motion, corroborated by historical landslide records. The AHP-GIS fusion with InSAR displacement data improved the objectivity of hazard classification. The resulting susceptibility map serves as a vital decision-support tool for regional planning, infrastructure development, and disaster preparedness in the central Himalayas [19].

4.4.6 Joshimath Town Multi-Temporal Interferometric Synthetic Aperture Radar Analysis (2019–2023)

Using over 400 Sentinel-1 interferograms, Dr. Gangwar conducted a multi-year SBAS-InSAR analysis of Joshimath, Uttarakhand. Dr. Hari Shankar Gangwar is a scientist at the IIRS- ISRO, with research interests in SAR interferometry, deformation monitoring, and urban geohazard assessment. He has been a part of a few research studies that have applied Sentinel-1-derived MTInSAR approaches to long-term slope and infrastructure stability monitoring in the Himalayas. His research found LOS displacement rates between -20 and -90 mm/year, which are very similar to structural damage patterns experienced in the 2022 23 subsidence crisis.

The MTInSAR methodology enabled the detection of persistent deformation areas at high resolution, which helped authorities define risk zones, evaluate infrastructure, and develop evacuation strategies. The study identified the effectiveness of satellite monitoring in early warning and helping in hazard mitigation in mountainous regions [20].

4.5 Advanced Interferometric Synthetic Aperture Radar Techniques

The InSAR has turned out to be a set of more sophisticated techniques that perform far better as the necessity to keep an eye on the Earth and achieve high precision and constant time monitoring rises. These advancements have improved InSAR's sensitivity, spatial resolution, and applicability across diverse geographic or environmental settings. Some of the most important advancements in the field of interferometric remote sensing involve the development of techniques, including PSI and DInSAR [3, 12], PolInSAR [14], and MTInSAR [3, 8]. All these methods have been established to overcome some of the shortcomings of the traditional InSAR methods. Whereas DInSAR enhances the ability to detect surface deformation by eliminating the topographic effects, PSI will focus on identifying stable scatterers in urban environments to determine long-term displacements. While PolInSAR has enabled the extension of capability to vegetated and volume-scattering terrain using multiple polarizations, MT-InSAR has offered high temporal resolution and coherence over large areas using large SAR data stacks. Integrating these techniques has significantly increased the usability and credibility of InSAR to a wide range of geophysical and environmental settings. In one example, PSI is applicable in urban areas where stable scatterers may be reliably observed over time [3, 12].

Polarimetric InSAR enhances analysis in vegetated regions by utilizing polarimetric diversity [13].

The MTInSAR methods—such as SBAS—further improve temporal resolution and measurement reliability, especially in complex terrains [3]. In combination, the methods assist in minimizing the impacts of decorrelation in time and atmospheric errors, and maximize coherence in areas where changes are rapid or subtle. Their adoption in contemporary Earth observation systems has enabled more precise, certain, and scalable analysis of surface deformations, both locally and globally.

Differential InSAR (DInSAR)

Differential InSAR is one of the most vital methods for studying surface deformation.

It entails the production of two interferograms using SAR images acquired at diverse times. The topographic phase component is extracted by a reference DEM or a modelled surface. The outcome is an interferogram that reflects the surface displacement of the two acquisitions. Differential InSAR is mainly effective in detecting ground deformation associated with glacier movement, co-seismic events, volcanic inflation or deflation, mining operations, and land subsidence. Nevertheless, DInSAR's performance may be limited in vegetated regions or in cases where SAR acquisitions are divided by long temporal intervals, leading to coherence loss.

To address these challenges, PSI was introduced [3, 12]; it is concerned with finding stable radar targets, which are called persistent scatterers, because they can reflect radar signals at a certain time. Such scatterers, which may be located in urban environments, such as buildings, towers, and exposed rock surfaces, enable precise measurements of displacement by analyzing long time series of SAR acquisitions. Persistent scatterer interferometry measures the phase history of every target, allowing for the estimation of millimeter-order deformation. It also comes in handy when it comes to tracking infrastructure stability, urban subsidence in the long-term, landslide precursors, and impacts of tunnelling works. It has now been considered an important tool in assessing and managing geohazards in the urban environment due to its high reliability and accuracy.

Further improving the capabilities of interferometry, PolInSAR integrates radar polarimetry with interferometric techniques [13]. It is also distinct from conventional InSAR, since it measures multiple polarization channels (HH, VV, etc.) to investigate scattering properties of natural features. This is particularly useful in the analysis of volume scatterers like vegetated ground, snow cover, and forest canopies. By applying models like Random Volume over Ground, PolInSAR can distinguish among ground and volume scattering components, allowing for accurate estimation of parameters like snow depth and forest height. Therefore, PolInSAR is highly beneficial in environmental monitoring like biomass estimation, forest canopy structure, and mapping of semi-transparent layers of terrain.

Multi-temporal InSAR represents a class of algorithms that process long sequences of SAR images to improve temporal coherence or minimize noise, involving atmospheric distortions [3]. The MT-InSAR methods involve both point-target approaches, such as PSI, and distributed-target methods like the SBAS technique, which produces interferograms using SAR image pairs with small spatial and temporal baselines. The SBAS technique is particularly effective in non-urban and vegetated regions. The high resolution allows the use of MT-InSAR to investigate surface deformation in large areas, high-resolution time series, and is often applied in tectonic fault analysis, glacier flow, landslide processes, and subsidence of permafrost.

All these new InSAR techniques have significantly expanded the uses and accuracy of interferometric remote sensing. They enable high-accuracy, long-term monitoring of Earth's surface across a broad variety of environments or infrastructures. The applications will be in disaster risk, infrastructure health monitoring, and environmental assessment. The integration of InSAR with modern technologies involving cloud computing, AI, and big data platforms is pushing boundaries of Earth observation. This enhances our abilities to understand, predict, and respond to the dynamic processes being experienced on the planet; this is a new era of global geospatial intelligence.

4.6 Challenges and Future Trends in Interferometric Synthetic Aperture Radar

Despite its remarkable capabilities, InSAR faces several limitations that impact the accuracy and reliability of its measurements. Among the most prominent ones are atmospheric and ionospheric disturbances, complex data processing procedures, and coherence loss on some types of land covers. One of the primary issues is atmospheric delay, and it is explained by the fact that the radar signal must pass through the troposphere. A phase delay between SAR acquisitions may be caused by changes in moisture, air pressure, and temperature, and can simulate or mask real surface deformation. Similarly, phase artifacts due to ionospheric perturbations, which are particularly a nuisance to low-frequency SARs, including the L-band SARs, may also be caused by changes in electron content. The impacts are particularly strong in tropical and equatorial parts where atmospheric variability is high. These distortions may need correction of external data, such as tropospheric models based on GPS observations, weather measurements, and multi-frequency SAR data. Although these corrections increase the accuracy of measurements, they contribute to the computational cost of the InSAR processing chain. In addition to atmospheric effects, InSAR is also limited by spatial and temporal decorrelation, which may cause signal decoherence over time or in non-homogeneous terrain. Thus, it is not as effective in some environmental conditions. Temporal decorrelation occurs when the scattering behavior of a surface varies substantially between SAR acquisitions, including vegetated or snow-covered areas, leading to the loss of phase coherence. Spatial decorrelation, on the other hand, depends on the baseline separation between satellite passes; if the baseline is too large, the interferometric phase cannot be reliably estimated. These decorrelation effects limit the applicability of InSAR in dynamic or heterogeneous landscapes. Furthermore, phase unwrapping, which is the process of converting wrapped phase values into continuous deformation signals, remains a critical step that is computationally intensive and prone to errors in low-coherence areas.

Looking toward the future, several developments are poised to enhance the robustness and accessibility of InSAR. The upcoming generation of SAR satellite missions—such as NISAR (NASA-ISRO SAR), ROSE-L (ESA), and SAOCOM-CS (CONAE/ESA collaboration)—will offer higher spatial resolution, more frequent revisit times, and dual-frequency, full-polarimetric capabilities. These improvements will support more consistent time-series analysis, increase coherence in challenging terrains, and facilitate vegetation and biomass mapping through better penetration and polarimetric discrimination. Additionally, constellations of small SAR satellites being developed by private companies, such as Capella Space and ICEYE, are contributing to near-real-time SAR data availability, which will greatly benefit rapid disaster response and continuous urban monitoring [12].

A transformative shift is also emerging through the integration of AI and ML in InSAR data processing. The AI algorithms [11] are being trained to detect deformation patterns automatically, distinguish signal from noise, classify scattering mechanisms, and even correct for atmospheric and unwrapping errors without human intervention. Such modern methods are particularly useful in dealing with the large amounts of data that the recent missions of SAR satellites produce [12].

They assist in providing geospatial insights into action faster by shortening the time between data collection and its interpretation. Nowadays, ML algorithms are being introduced into cloud-based InSAR platforms. This facilitates scalable data processing so that the information on deformation can be accessed and analyzed by the researchers and policymakers in near real-time.

Although InSAR still suffers from the problems of atmospheric delays, coherence loss, and complexity of processing, the technology is advancing at a rapid pace. The InSAR is being turned into an effective, practical, and more accurate monitoring tool in the Earth's surface dynamics due to advancements being made in satellite systems, computing infrastructure, and intelligent analytics.

The combination of classical methods in remote sensing and contemporary data science is paving the way to innovation in Earth observation. This synergy allows massive progress in global geohazard evaluation, environmental surveillance, and sustainable infrastructure designs. The integration of the physical theory of remote sensing and the computational power of artificial intelligence, machine learning, and big data platforms enables researchers to derive more information from the complex geospatial data. The transdisciplinary approach, in addition to making monitoring systems more accurate and scalable, enables us to respond to the Earth in a nearly real-time manner to changing processes. Therefore, the integration of the classical Earth observation methods with modern developments in computations is being employed for a bright future of resilience-based geospatial applications in scientific, environmental, and policy-making realms.

4.7 Conclusion

Several research directions will characterize the future of the InSAR technology. One of the most promising avenues is the integration of InSAR with AI and ML algorithms. The tools provide the possibility of automating the interferogram interpretation process, filtering coherence, and classifying deformation patterns more efficiently. This kind of integration is particularly helpful in managing increasing amounts of SAR data acquired through regular acquisitions such as Sentinel-1 and the future NISAR. The other important area of emphasis is the generation of geohazard monitoring near-real-time InSAR applications. This is an early warning of landslides, earthquakes, volcanic activity, and instability within infrastructure, especially in vulnerable and poorly monitored areas. Moreover, as InSAR continues to expand into vegetated and cryospheric environments, there is a growing need for improved modeling of volume scattering effects and temporal decorrelation. Advanced PolInSAR techniques, along with dual- or quad-polarized acquisitions, can offer better insights into forest biomass estimation, snow cover analysis, and glacier mass balance studies. In urban contexts, the synergy of InSAR with smart infrastructure systems, 3D city models, and Internet of Things sensor networks is expected to enable dynamic monitoring of urban growth, subsidence, and structural health [12]. Finally, cloud-based processing platforms and open-data initiatives are democratizing access to InSAR technology, empowering a wider user community to engage in large-scale environmental and infrastructure monitoring. These advancements collectively represent a transformative phase in InSAR research, paving the way for highly accurate, scalable, and real-time Earth observation capabilities across scientific, environmental, and policy-making domains. One critical area is the improved mitigation of atmospheric and ionospheric artifacts through multi-frequency data fusion, data assimilation techniques, and external weather model integration. Additionally, there is a growing need to refine phase unwrapping algorithms and coherence estimation techniques to enhance the robustness of InSAR outputs in low-coherence and non-urban environments. Perhaps most transformative is the integration of artificial intelligence and machine learning into the InSAR workflow. The AI-driven approaches are expected to revolutionize deformation pattern recognition, noise filtering, and automatic hazard detection, thereby enabling real-time monitoring systems with minimal human intervention.

As both science and technology, InSAR continues to evolve with the confluence of radar physics, satellite engineering, and computational intelligence. With the proliferation of next-generation sensors and analytical platforms, it holds great promise not only for advancing scientific understanding of Earth processes but also for supporting sustainable development, disaster preparedness, and environmental resilience on a global scale.

AI disclosure statement: The author acknowledges the use of AI-based tools for minor language editing or grammatical correction. However, all intellectual contributions, analyses, and conclusions presented in this chapter are entirely the author's own.

References

1. Zebker, H.A. and Goldstein, R.M., Topographic mapping from interferometric synthetic aperture radar observations. *J. Geophys. Res.*, 91, B5, 4993–4999, 1986.
2. Torres, R., *et al.*, GMES Sentinel-1 mission. *Remote Sens. Environ.*, 120, 9–24, 2012.
3. Berardino, P., *et al.*, A new algorithm for surface deformation monitoring based on small baseline differential SAR interferograms. *IEEE Trans. Geosci. Remote Sens.*, 40, 11, 2375–2383, 2002.

4. Cumming, I.G. and Wong, F.H., *Digital Processing of Synthetic Aperture Radar Data: Algorithms and Implementation*, Artech House, Norwood, MA, USA, 2005.
5. Soumekh, M., *Synthetic Aperture Radar Signal Processing*, Wiley-Interscience, New York, USA, 1999.
6. Richards, J.A., *Remote Sensing with Imaging Radar*, Springer, Berlin, Germany, 2009.
7. Krieger, G., *et al.*, TanDEM-X: A satellite formation for high-resolution SAR interferometry. *IEEE Trans. Geosci. Remote Sens.*, 45, 11, 3317–3341, 2007.
8. Mogi, K., Relations between the eruptions of various volcanoes and the deformations of the ground surfaces around them. *Bull. Earthq. Res. Inst.*, 36, 99–134, 1958.
9. Franceschetti, G. and Lanari, R., *Synthetic Aperture Radar Processing*, CRC Press, Boca Raton, FL, USA, 1999.
10. Zebker, H.A., Rosen, P.A., Goldstein, R.M., On the derivation of coseismic displacement fields using differential radar interferometry. *J. Geophys. Res.*, 99, B10, 19617–19634, 1994.
11. Rosen, P.A., *et al.*, The NASA-ISRO SAR (NISAR) mission: Science and applications, IGARSS, 2017.
12. Ferretti, A., Prati, C., Rocca, F., Permanent scatterers in SAR interferometry. *IEEE Trans. Geosci. Remote Sens.*, 39, 1, 8–20, 2001.
13. Cloude, S.R. and Papathanassiou, K.P., Polarimetric SAR interferometry. *IEEE Trans. Geosci. Remote Sens.*, 36, 5, 1551–1565, 1998.
14. Massonnet, D., *et al.*, The displacement field of the Landers earthquake mapped by radar interferometry. *Nature*, 364, 138–142, 1993.
15. Sharma, R., Jaiswal, R.K., Rawat, V.K., Chatterjee, R.S., Relation of Damage Intensity of Nepal Earthquake in Indian Region with DInSAR Displacement Pattern and Site Factors. *IIRS Symposium*, Dehradun, 2016.
16. Niraj, K.C., Shukla, K., Chatterjee, R.S., Post-Landslide Deformation Monitoring Using DInSAR and MT-InSAR. *Geocarto Int.*, 2023.
17. Thapa, S. and Chatterjee, R.S., Advanced Subsidence Monitoring Using Persistent Scatterer Interferometry. *SPIE Proceedings*, 2017.
18. Kumar, S., *et al.*, Estimation of Forest Canopy Height and DEM Generation Using PolInSAR Techniques over Indian Forests. *IEEE J. Sel. Top. Appl. Earth Obs. Remote Sens.*, 2021.
19. Devara, N., Chatterjee, R.S., Katiyar, V.S., Landslide Susceptibility Mapping Using MT-InSAR and AHP-GIS. *Nat. Hazards*, 2021.
20. Gangwar, H.S., *et al.*, Multi-Temporal InSAR Analysis for Land Subsidence in Joshimath Region Using Sentinel-1 Data. *Geocarto Int.*, 2023.

5

Calibration and Validation Techniques for Disaster Response

Surmadhur Pant* and Yashender Sharma

Department of Electronics & Communication Engineering, Shivalik College of Engineering, Dehradun, India

Abstract

Disaster response game changer, synthetic aperture radar (SAR), can offer high-resolution imagery under varied environmental conditions. Yet, strong calibration and verification processes are important to its performance in accurately sensing, assessing, and responding to disasters. Given dynamic and fluctuating disaster processes, conventional strategies often fail to perform, making it essential that advanced methods be incorporated. This chapter analyzes how disaster response SAR calibration and validation can be revolutionized using artificial intelligence (AI) and machine learning (ML) algorithms. The chapter begins with a discussion on the inherent challenges of calibrating SAR data, including complex ground conditions, atmospheric effects, and sensor noise. The discussion then turns to AI-based approaches for enhancing the accuracy of SAR data and extracting meaningful information. Using labelled datasets to recognize patterns indicative of floods, landslides, and other disasters, supervised learning algorithms are introduced for classifying disaster-stricken areas. The assessment of disaster impact is facilitated at a quick pace through the use of unsupervised learning methods, including clustering and anomaly detection, to identify unexpected changes in SAR imagery. To dynamically optimize SAR system calibration and ensure continuous adaptability in evolving situations, reinforcement learning methods are proposed. The chapter also highlights how SAR validation can be enhanced by combining data from various sources such as optical imagery, ground truth, and real-time sensor feeds. There is an in-depth explanation of validation frameworks that involve AI-based simulations and cross-validation with auxiliary datasets. To illustrate the actual application of these measures and their contribution to improving response times and decision-making accuracy, case studies of disaster

**Corresponding author*: surmadhur.pant@sce.org.in

Pooja Joshi, Ashutosh Bhatt, Neelam Sharma and Kapil Joshi (eds.) Synthetic Aperture Radar for Disaster Management: Bridging Science and Solutions, (117–140) © 2026 Scrivener Publishing LLC

scenarios are presented. By the end of this chapter, readers will better comprehend how AI and ML algorithms can help overcome the limitations of traditional SAR calibration and validation toward a paradigm shift in disaster response. The methods discussed here enable swift and effective crisis intervention, aside from enhancing the reliability of SAR systems.

***Keywords*:** Model calibration, model validation, hydrological modeling, performance metrics, Nash–Sutcliffe efficiency (NSE), root mean square error (RMSE), synthetic aperture radar (SAR)

5.1 Introduction

Synthetic aperture radar (SAR) remote-sensing technology is an independent, all-season, day-and-night imaging tool. It is useful in disaster response and management. Whereas in optical sensing, remission from aerosols, cloud cover, dust, smoke, various contaminations, and smoke may restrict visibility, the SAR remote sensor can pierce clouds, smoke, or even small patches of vegetation in view of its smaller wavelengths. Such a property is advantageous during flood, landslide, and post-earthquake assessments when timely data is inelastic and the atmosphere is rarely conducive.

With the increasing SAR missions, such as Sentinel-1, RISAT, and TerraSAR-X, the global stockpile of satellite data is steadily increasing. Having said that, the interpretability issues of SAR data often hinder the operational use of SAR in disaster scenarios, notwithstanding the high level of technology involved. The operational extraction of useful information depends on a strong calibration and validation system to ensure that SAR measurements honestly reflect the conditions on the ground.

Calibration/Validation (Cal/Val) is indeed an opportunity to reduce biases in any one measurement, enhance the geometric and radiometric consistency across data, and ensure interoperability across sensors and platforms. However, these activities are quite difficult in dynamic, high-stakes arenas such as disaster management, where decisions must be made quickly and there is no time for actual field validation in some cases.

Recent advances in artificial intelligence (AI) and machine learning (ML) present transformative opportunities in this space. Algorithms trained on large and diverse datasets can help automate the interpretation of SAR signals, compensate for calibration gaps, and even predict errors in real-time. These technologies promise not only to enhance the reliability of SAR-derived information but also to make the process more adaptive and scalable.

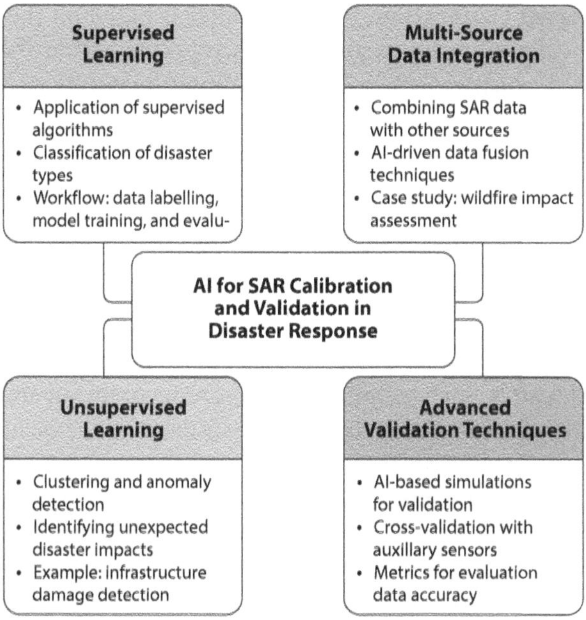

Figure 5.1 Overview of AI's role for SAR calibration and validation.

Insights are drawn from contemporary research – such as the detailed evaluation of land use/land cover (LULC) changes and ecosystem services in the Indian Himalayan Region [1] and the urban heat island effects linked to LULC patterns using SAR and optical data [2]. This chapter aims to explore the nuances of SAR calibration and validation, highlighting their limitations and the emerging role of AI and ML in overcoming them. Figure 5.1 illustrates how artificial intelligence (AI) techniques, including supervised, unsupervised, and reinforcement learning, contribute to enhancing SAR calibration and validation processes in disaster response scenarios.

5.2 Fundamentals of Synthetic Aperture Radar Calibration and Validation

5.2.1 Calibration Techniques: Internal, External, and Cross-Calibration

Synthetic aperture radar calibration adjusts the raw backscatter data to reflect the actual surface properties better. It can broadly be classified into three types:

- **Internal Calibration:** Its dependence is based on the internal calibration loops and reference points within the SAR system, such as noise sources or calibration pulses, to correct for system-specific distortions. Internal calibration, however, preserves internal consistency with a given mission, but it cannot compensate for external atmospheric variations or platform-induced ones. All these disturbances are converted into absolute backscatter values.
- **External Calibration:** It involves the use of ground targets whose radar cross sections are known; since in practice, such targets undergo strong reflection of radar signals and backscatter. Such techniques are more accurate but are intolerably complicated to implement in inaccessible or hazardous disaster-hit areas.
- **Cross-Calibration:** The harmony of data from various SAR satellites in a multi-platform environment is essential. Through cross-calibration methods, inter-sensor consistency can be established, thereby facilitating data fusion from multiple platforms. This proves to be an advantage in long-term change detection and multi-temporal studies, as reflected in LULC monitoring of Indian cities [2].

5.2.2 Validation Processes and Their Role

Validation complements calibration by assessing the accuracy and reliability of derived SAR products. This often involves comparing SAR outputs with *in-situ* measurements, other sensor data, such as optical or light detection and ranging (LiDAR), or modeled results. For example, in evaluating LULC changes in the Indian Himalayan Region, ground truth data were critical in confirming SAR-based classifications [1, 3].

In disaster response, validation helps in ensuring that flood extents, landslide zones, or damaged infrastructure detected *via* SAR genuinely represent on-ground conditions [3–6]. It supports decision-makers in trusting the remote data and responding with precision.

5.2.3 Limitations of Traditional Methods in Disaster Scenarios

While traditional Cal/Val methods are well-established, they face several constraints in emergency contexts.

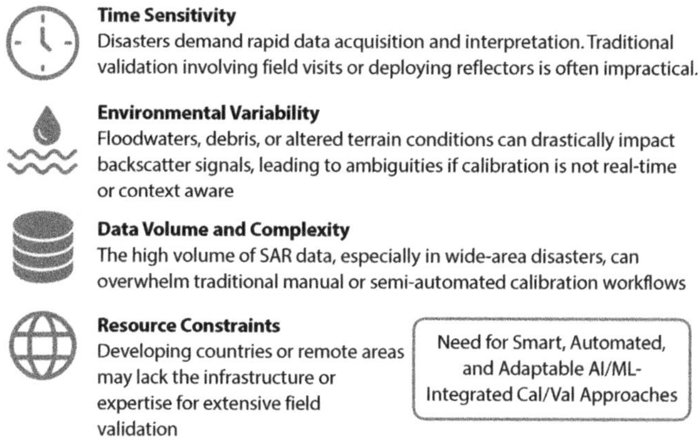

Figure 5.2 Limitations of existing methods in disaster scenarios.

Considering these limitations, there is a pressing need to develop smart, automated, and adaptable Cal/Val approaches, ideally integrated with AI/ML tools that can learn from multi-modal datasets and continually refine the SAR data interpretation process [7, 8]. Figure 5.2 depicts these constraints, including time sensitivity, environmental variability, data volume, and resource limitations, emphasizing why AI- and ML-integrated calibration and validation approaches are essential for effective disaster response.

5.3 Supervised Learning in Synthetic Aperture Radar for Disaster Response

Supervised learning techniques have rapidly become instrumental in SAR-based disaster response. These algorithms rely on annotated datasets – pairs of input SAR data and corresponding ground truth labels – to train models capable of automatic feature recognition [9]. In disaster contexts, this translates to the accurate classification of affected areas, identification of specific hazards, and damage assessment [10–14].

5.3.1 Application of Supervised Algorithms

Among the most effective supervised learning algorithms used in SAR image classification are:

Table 5.1 Comparative analysis of supervised ML in SAR.

Algorithm	Application in SAR image classification	Advantages	Limitations
Random Forest (RF)	Flood detection, land cover classification, and burned area mapping	○ Handles high-dimensional data well ○ Robust to noise ○ Less overfitting	○ Requires parameter tuning (e.g., tree number) ○ May be less interpretable
Support Vector Machine (SVM)	Terrain classification, damage assessment post-disaster	○ Effective in high-dimensional spaces ○ Works well with limited samples	○ High computational cost ○ Performance depends on kernel choice
Artificial Neural Networks (ANN)	Classification of urban, vegetation, and water surfaces	○ Can model complex relationships ○ Flexible architecture	○ Prone to overfitting ○ Requires large datasets and tuning
Gradient Boosting Machines (GBM)	Change detection in disaster zones	○ High prediction accuracy ○ Handles missing data well	○ Slow training process ○ Sensitive to noise and outliers
Deep Convolutional Neural Networks (CNNs)	Urban object classification, flood extent mapping, SAR-optical fusion	○ Extracts spatial and textural features ○ High accuracy with large datasets	○ Requires extensive training data ○ Computationally expensive
Extreme Gradient Boosting (XGBoost)	Damage estimation from SAR time-series	○ Efficient implementation of gradient boosting ○ Built-in regularization	○ Requires feature engineering ○ Not suited for highly imbalanced datasets
Ensemble Learning (e.g., Voting Classifiers)	Landslide detection, vegetation mapping	○ Combines strengths of multiple models ○ Increases generalization ability	○ Interpretation becomes difficult ○ Higher complexity

- **Random Forests:** These ensemble-based classifiers are known for their robustness and ability to handle high-dimensional SAR data. They excel in separating land cover classes and identifying flooded zones, urban debris, or landslides, especially when trained with textural and polarimetric SAR features [15].
- **Ensemble Learning Methods:** Combining multiple classifiers can significantly boost performance. Techniques like Gradient Boosting or Bagging enhance generalization across heterogeneous disaster-affected regions, accommodating varying backscatter patterns in SAR images [6, 15].

Table 5.1 summarizes key supervised ML algorithms applied to SAR image classification, including their advantages and limitations. These algorithms have been successfully integrated into operational workflows to detect flood extents, assess damage post-earthquake, or monitor urban expansion in hazard-prone areas.

5.3.2 Classification of Disaster Types

Supervised models are well-suited for:

- Flood Detection: Water has a distinct low backscatter signature in SAR, making it identifiable when combined with elevation and land cover data [6].
- Landslide Mapping: By detecting changes in slope geometry and surface roughness, landslide-affected zones can be classified.
- Collapsed Building Detection: In high-resolution SAR images, abrupt shifts in radar shadows and texture patterns often signal structural damage [4].

5.3.3 Workflow: Data Labeling, Model Training, and Evaluation

A typical supervised SAR classification pipeline involves:

- Data Labeling: Ground truth collection – either manually or *via* auxiliary data like unmanned aerial vehicles (UAVs), optical imagery, or records.

Table 5.2 Comparative analysis of unsupervised ML in SAR.

Algorithm	Application in SAR image classification	Advantages	Limitations
K-Means Clustering	Initial segmentation of SAR scenes, land cover partitioning, and flooded region delineation	o Simple and fast o Effective for spherical clusters	o Requires a predefined number of clusters (k) o Sensitive to noise and outliers
Density-Based Spatial Clustering of Applications with Noise (DBSCAN)	Detection of anomalies and infrastructure damage in disaster zones	o No need to predefine cluster number o Good at identifying outliers	o Struggles with varying cluster densities o Sensitive to parameter settings
ISODATA	Iterative self-organizing clustering for terrain and surface classification	o Flexible number of clusters o Handles SAR intensity variability	o Computationally intensive o May converge to local minima
Autoencoders	Feature extraction and dimensionality reduction for change detection	o Learns complex representations o Useful for denoising SAR images	o Requires careful architecture design o Needs tuning
Self-Organizing Maps (SOM)	Cluster analysis and visualization of SAR backscatter variations	o Preserves topological structure o Effective for high-dimensional data	o Performance depends on grid size o Less scalable to huge datasets
Principal Component Analysis	Change detection, noise reduction, and initial data compression	o Reduces data dimensionality o Highlights important variance patterns	o Assumes linearity o Difficult to interpret components in SAR context
Gaussian Mixture Models (GMM)	Classification of terrain types and seasonal changes in SAR time-series	o Probabilistic clustering o Models' data with overlapping distributions	o Sensitive to initialization o May require many components

- Model Training: Feeding the labeled SAR data into ML algorithms, often with feature engineering (e.g., coherence, backscatter intensity, textural metrics).
- Model Evaluation: Using metrics like Overall Accuracy, Kappa Coefficient, and F1-score, along with cross-validation to ensure reliability [1–3].

5.3.4 Example: Flood Extent Mapping Using Synthetic Aperture Radar Imagery

In the aftermath of monsoon-induced floods, Sentinel-1 SAR data were used in conjunction with labeled flood boundaries from historical records. A Random Forest classifier trained on VV polarization bands successfully mapped inundated zones, validated against optical MODIS data and field observations. Such use cases highlight how supervised learning can dramatically reduce the time and human effort required in disaster mapping [2]. Table 5.2 compares common unsupervised ML techniques for SAR data analysis.

5.4 Unsupervised Learning Techniques for Synthetic Aperture Radar Data Analysis

Unsupervised learning eliminates the dependency on ground truth, making it ideal for disaster scenarios where labelled data may be scarce or unavailable. These algorithms uncover hidden patterns, clusters, or anomalies in SAR imagery without prior knowledge, making them highly adaptable for rapid response.

5.4.1 Overview of Clustering and Anomaly Detection

- K-Means Clustering: This method groups pixels based on similarity in backscatter values or polarimetric features. It is often used for change detection - e.g., differentiating pre- and post-disaster landscapes.
- Density-Based Spatial Clustering of Applications with Noise (DBSCAN): Especially suited for irregular and noisy SAR data, DBSCAN can identify clusters of anomalous pixels (e.g., new water bodies, landslides) while ignoring irrelevant areas.

- Autoencoders and Principal Component Analysis-Based Methods: These dimensionality reduction techniques help in detecting subtle anomalies by learning compressed feature representations and identifying deviations.

5.4.2 Use Cases: Identifying Unexpected Disaster Impacts

- Rapid Flood Mapping Without Labels: By comparing SAR images from before and after a disaster using unsupervised methods, newly flooded regions can be inferred from changes in radar backscatter.
- Landslide or Debris Flow Detection: Sudden topographical changes manifest as clusters of anomalous pixels, especially in mountainous terrain, as evidenced in parts of the Indian Himalayan Region [1, 4].

5.4.3 Example: Detection of Infrastructure Damage

An unsupervised workflow using Sentinel-1 imagery and DBSCAN was employed to detect infrastructural anomalies in an earthquake-hit urban area. Post-event SAR images showed changes in shadow lengths and texture, which, when clustered, pointed to possible building collapses. The algorithm required no training labels, making it fast and scalable across large regions.

5.5 Reinforcement Learning for Adaptive Calibration

Reinforcement learning (RL) introduces an exciting frontier in SAR data calibration, offering the potential for dynamic adaptation in unpredictable disaster environments [2, 13]. Unlike supervised or unsupervised learning, RL focuses on learning through interaction with the environment, making it well-suited for real-time optimization tasks.

5.5.1 Fundamentals of Reinforcement Learning and Its Relevance to Synthetic Aperture Radar

In an RL framework, an agent (e.g., a calibration model) observes the environment (incoming SAR data and system performance), takes actions (e.g., adjusting calibration parameters), and receives feedback (reward signals based on accuracy or reliability). Over time, the agent learns an optimal calibration policy that maximizes the long-term reward - essentially minimizing SAR data errors [5, 8, 11].

This approach is particularly relevant for:

- Adaptive Noise Filtering: Adjusting filter parameters in real time to suppress speckle or system noise based on scene complexity.
- Sensor Drift Compensation: Identifying and correcting for performance degradation in sensors due to aging or environmental exposure.

5.5.2 Dynamic Optimization of Synthetic Aperture Radar Calibration

In disaster scenarios, environmental conditions are volatile - floodwaters rise or retreat, land cover changes rapidly, and atmospheric factors shift. An RL-based system can dynamically recalibrate the SAR signal interpretation in response to these changes, outperforming static calibration methods. Table 5.3 presents a comparative overview of reinforcement learning frameworks for SAR calibration.

For example, during a flood, a SAR sensor may experience variable backscatter due to shifting water depths and sediment concentrations. An RL agent, continuously evaluating prediction accuracy using external sources (e.g., hydrological models or IoT sensors), can adjust its calibration in near real-time to maintain data reliability [16–20].

5.5.3 Example: Continuous Adaptation to Changing Weather Conditions During Floods

An experimental RL model integrated into a flood monitoring system used Sentinel-1 SAR and real-time weather feeds. As storm intensity and cloud cover varied, the agent adapted the calibration parameters to maintain consistent flood mapping accuracy. This not only enhanced data fidelity but also improved early warning system responsiveness.

Table 5.3 Comparative analysis of reinforcement ML in SAR.

Reinforcement learning algorithm/ framework	**Application in SAR**	**Advantages**	**Limitations**
Q-Learning	Adaptive SAR system tuning (e.g., calibration parameters during floods)	o Simple to implement o Learns optimal policy over time	o Not suitable for large state/action spaces o Requires discrete environments
Deep Q-Networks (DQN)	Dynamic sensor control for optimizing SAR acquisition in disaster zones	o Handles high-dimensional input o Learns directly from SAR data frames	o Needs large data and computing o Can be unstable or slow to converge
Policy Gradient Methods	Continuous decision-making for SAR platform calibration	o Works well in continuous action spaces o Direct optimization of performance	o High variance in learning o Slower convergence than Q-learning
Actor-Critic Methods	Real-time terrain tracking or calibration under variable disaster conditions	o Combines the benefits of policy-based and value-based methods o Stable training	o Complex architecture o Requires tuning and parallel training strategies
Multi-Agent RL	Coordinated SAR and UAV data collection and calibration	o Supports teamwork among multiple sensors o Adaptive to spatially distributed data	o High complexity o Requires coordination and communication policies
Model-Based RL	Predictive calibration of SAR response under simulated weather scenarios	o Sample-efficient o Can simulate unseen disaster dynamics	o Needs an accurate environment model o Risk of compounding errors

5.6 Multi-Source Data Integration for Enhanced Calibration and Validation

To fully exploit the capabilities of SAR in disaster response, integration with complementary data sources is crucial. While SAR excels in all-weather, day-night imaging, its interpretability and calibration accuracy can be significantly improved through synergistic use of other data types [18].

5.6.1 Combining Synthetic Aperture Radar Data with Optical Imagery, Ground-Based Observations, and Unmanned Aerial Vehicle Inputs

- Optical imagery provides contextual and spectral information that is often missing in SAR. Although limited by weather and lighting conditions, optical data can enhance classification accuracy and offer intuitive visual confirmation of radar-inferred changes.
- Ground-based observations - from sensors, local reports, or *in-situ* measurements - serve as ground truth for model validation and fine-tuning. These inputs are invaluable in refining ML predictions, especially in early disaster phases. Table 5.4 provides a comparative summary of data integration techniques across multiple sources for SAR calibration and validation.
- The UAV inputs bridge the resolution gap between satellite SAR data and ground truth. The UAVs offer high-resolution, targeted imaging and can be rapidly deployed in localized disaster zones for real-time calibration checks.

5.6.2 Artificial Intelligence-Driven Techniques for Multi-Source Data Fusion

Advanced AI models, such as convolutional neural networks, graph neural networks, and attention-based transformers, are now employed for fusing SAR with other modalities [20, 23]. These techniques enable spatial–temporal alignment (matching datasets of varying resolutions and timestamps), feature–level fusion (integrating radiometric, textural, and spectral features across data types), and decision-level fusion (combining model predictions from different sources to form a consensus output).

Table 5.4 Comparative analysis of multi-source data integration for SAR.

Data source	Role in calibration/ validation	Integration techniques (AI/ML)	Advantages	Challenges
Optical Imagery (e.g., Sentinel-2)	Cross-verification of land cover, flood extent, and wildfire burn areas	o Feature-level fusion (CNNs, Transformers) o Late fusion *via* ensemble models	o Complements SAR with spectral info o Easier visual interpretation	o Cloud cover interference o Temporal misalignment
Ground Observations (In-situ)	Ground truth data for calibrating SAR backscatter and verifying ML classifications	o Regression models o Supervised ML (e.g., RF, SVM) using *in-situ* labels	o High reliability o Direct validation source	o Sparse coverage o Time-consuming collection
Unmanned Aerial Vehicle Imagery and LiDAR	High-resolution 3D terrain data to validate SAR elevation and change detection	o Sensor fusion *via* point cloud matching o AI-based co-registration	o High spatial accuracy o Ideal for urban/ post-disaster validation	o Costly operations o Weather and flight limitations
Multitemporal SAR Data	Temporal consistency checks for calibration drift and change detection	o Recurrent Neural Networks (RNNs, LSTMs) o Temporal stacking with ML models	o Captures progression of events o Useful for time-series analysis	o Complex to process o May require normalizing backscatter
Auxiliary Data (e.g., GPS, DEMs)	Geo-referencing support, terrain normalization, and calibration baselines	o Geospatial AI (GeoAI), rule-based models	o Improves geometric accuracy o Provides contextual information	o Quality and resolution differences

Fusion models not only improve calibration accuracy but also reduce uncertainty in SAR interpretation under chaotic disaster environments.

5.6.3 Case Study: Integrating Synthetic Aperture Radar and Optical Data for Wildfire Impact Assessment

In a recent wildfire event in the western United States, Sentinel-1 SAR data were integrated with Sentinel-2 optical imagery using a deep learning-based fusion model. While SAR helped detect structural changes through backscatter analysis, optical data revealed burn severity and vegetation loss. The fused dataset allowed for more accurate and timely damage assessment, enabling quicker resource allocation and mitigation efforts [1].

5.7 Advanced Validation Techniques

As SAR data becomes increasingly central to disaster response, its validation must keep pace with the growing complexity of scenarios and algorithms.

5.7.1 Artificial Intelligence-Based Simulations for Synthetic Aperture Radar Validation in Disaster Environments

Simulation frameworks using generative models (e.g., GANs) or synthetic datasets allow researchers to:

- Replicate Disaster Scenarios: Generate diverse and extreme disaster conditions for SAR model testing.
- Benchmark Algorithm Performance: Test SAR calibration under controlled variations of atmospheric, terrain, or structural changes.
- Train Models Where Data is Scarce: Especially useful in preparing AI for rare disasters like tsunamis or volcanic eruptions.

Such simulations not only help with model development but also in validating operational readiness before real-world deployment. Table 5.5 lists advanced AI-driven validation techniques used in SAR applications.

Table 5.5 Comparative analysis of advanced validation techniques being used in SAR.

Technique	Description	AI/ML involvement	Advantages	Limitations
Artificial Intelligence-Based Simulation Models	Simulates disaster scenarios (e.g., flooding, wildfires) to generate synthetic SAR responses	o Generative Models (e.g., GANs, VAEs) o Physics-informed neural networks	o Enables validation in hard-to-reach areas o Data augmentation tool	o Requires domain expertise o Limited realism in extreme scenarios
Cross-Validation with Auxiliary Sensors	Validates SAR-derived outputs against LiDAR, GPS, optical, or UAV datasets	o Feature alignment using ML o Supervised cross-validation pipelines	o High precision validation o Geospatial alignment improves reliability	o Temporal mismatch o Dependency on sensor availability
Ensemble Model Agreement Validation	Uses agreement/disagreement across multiple ML classifiers to estimate confidence in SAR classifications	o Random Forests, SVMs, Neural Nets in ensemble form	o No external ground truth needed o Fast and scalable	o Sensitive to bias in models o Cannot replace real validation
Bootstrap and Monte Carlo Methods	Statistical resampling and uncertainty quantification of SAR model outputs	o Used with ML for probabilistic interpretation	o Captures error margins o Useful in high-stakes disaster zones	o Computationally expensive
Pixel-Wise Confusion Matrix Analysis	Validates classification outputs at the pixel level with reference labels or manual annotations	o Classification error metrics (accuracy, f1-score, IoU, Kappa)	o Widely adopted o Clear performance indicators	o Depends on high-quality ground truth
Temporal Consistency Validation	Evaluates how SAR classifications hold over time (before, during, and after disaster events)	o Time series ML (e.g., LSTM, GRU) o Change detection AI pipelines	o Tracks disaster evolution o Detects calibration drift	o Needs well-aligned multi-temporal datasets

5.7.2 Cross-Validation with Auxiliary Sensors (LiDAR, GPS)

Validation is further strengthened by using:

- LiDAR: Provides accurate 3D surface data, ideal for cross-validating elevation changes post-landslide or earthquake.
- Global Positioning System (GPS): Provides geo-displacement and geo-movement data, supporting SAR interpretation in geodynamic disasters.

By combining these sensor signals, calibration models are not only better validated but also real-time corrected for increased accuracy.

5.7.3 Metrics for Evaluating Synthetic Aperture Radar Data Accuracy and Reliability

Key performance indicators for SAR validation include:

- Root mean square error and mean absolute error for regression-based evaluations.
- Overall accuracy, precision, recall, and f1-score for classification outcomes.
- Structural similarity index and Kappa Statistics for image comparison and spatial consistency.

These metrics guide researchers in refining AI models and ensuring reliable SAR interpretations in critical applications.

5.8 Case Studies and Practical Applications

Integrating SAR data with AI has greatly progressed disaster management capability. Through advanced calibration and validation procedures, SAR applications powered by AI have successfully been implemented for different disaster cases globally. The following section analyzes real-world implementations and case studies that demonstrate the effect of such technologies.

5.8.1 Real-World Applications of Artificial Intelligence-Enhanced Synthetic Aperture Radar Calibration and Validation

- **Urban Flood Monitoring in Southeast Asia**

In Southeast Asia's flood-prone areas, AI-aided SAR data has played a significant role in observing and regulating urban flooding. For example, on

Vietnam's central coast, a multi-scale change detection model based on SAR image time series was implemented to study flood dynamics, supplying valuable information to emergency response groups. Likewise, the Mekong Delta has been aided by quick surface water and flooding mapping capabilities that aid water management and risk reduction strategies.

- **Earthquake Damage Mapping in Turkey**

In the wake of the destructive earthquakes in Syria and Turkey in February 2023, SAR analysis based on AI was pivotal in estimating building damage. A study suggested a method of integrating SAR amplitude and phase coherence change detection to assess seismic building damage in urban environments to enable effective resource allocation for relief and rescue missions.

- **Hurricane Impact Analysis in the Caribbean**

The Caribbean has utilized AI to integrate SAR data to evaluate the impacts of hurricanes efficiently. NASA generated damage proxy maps from SAR data to identify areas of probable damage in Puerto Rico during Hurricane Maria, helping relief organizations prioritize their responses. Remote sensing methods were also employed to quantify coastal erosion due to the hurricane, offering insights into the environmental impact of the storm.

5.8.2 Case Studies of Specific Disasters

- **Floods – Brahmaputra Basin, India**

Artificial intelligence-supported SAR analysis has played a crucial role in monitoring floods in the Brahmaputra River Basin. One study applied Sentinel-1 SAR data on the Google Earth Engine platform to map inundation from floods, allowing for rapid evaluation of flood damage and supporting mitigation measures. Similar approaches have been proposed for real-time flood mapping using SAR and deep neural networks [21] and for adaptive SAR calibration under variable disaster scenarios [22].

- **Earthquakes – Santiago, Chile**

Seismic activity in Chile requires efficient damage assessment techniques. Studies utilizing differential interferometric SAR methods tracked ground deformations in the Santiago basin and delivered high-resolution information vital to urban design and infrastructure robustness.

- **Hurricanes – Puerto Rico**

After Hurricane Maria, AI-aided SAR systems were applied to surface inundation mapping in Puerto Rico. Assessment of the extent of flooding *via* SAR imagery and ML algorithms enabled officials to comprehend the effects of the hurricane better and organize the response accordingly.

5.9 Impact Assessment: Speed, Accuracy, and Decision-Making Improvements

The application of AI in SAR processes has resulted in several improvements in disaster response:

- Faster Turnaround Times: Robust analysis pipelines enabled by ML have made damage assessments more rapid, making aid and resources available faster.
- Accuracy Increased: Deep learning models tuned using multi-source data have minimized damage assessment inaccuracy, making them less susceptible to false negatives and positives under disaster conditions.
- Data-Informed Decision Making: Real-time, accurate damage maps provided by AI-enabled SAR data have enabled officials to make well-informed, fact-based decisions that maximize the effective use of resources and responses.

5.10 Conclusion and Future Directions

The convergence of AI, SAR technology, and multi-source data analytics is a paradigm shift in how we monitor, respond to, and build resilience against disasters. By integrating sophisticated computation with heterogeneous data sources, we are moving towards an age of intelligent, automated, and highly responsive remote sensing systems. This change is especially revolutionary for time-critical and resource-limited disaster situations.

5.10.1 Summary of Artificial Intelligence and Machine Learning Contributions to Synthetic Aperture Radar Calibration and Validation

The application of AI and ML methods has significantly improved the accuracy, scalability, and efficiency of SAR calibration and validation processes:

- **Supervised Learning Techniques** (e.g., support vector machines, random forests, convolutional neural networks): These methods have excelled in tasks like land cover classification, flood mapping, and change detection, provided sufficient labeled data is available. Their deterministic learning frameworks allow for high precision and reduced misclassification rates in well-annotated environments.

- **Unsupervised and Reinforcement Learning Models**: These approaches are particularly valuable where labeled data is sparse or unavailable. Unsupervised clustering and dimensionality reduction techniques have supported anomaly detection and landscape segmentation. Reinforcement learning, though still emerging in SAR contexts, shows promise in dynamically adjusting calibration strategies based on real-time environmental feedback.
- **Artificial Intelligence-Driven Data Fusion and Physical Modeling**: Hybrid models that integrate AI with physical simulations (e.g., radiative transfer models) are reshaping both calibration and validation stages. By fusing SAR data with optical imagery, meteorological data, and historical event profiles, these methods generate more holistic and trustworthy geospatial insights.

5.10.2 Challenges and Opportunities for Future Research

While substantial progress has been achieved, several technical and logistical hurdles persist:

- **Limited Availability of High-Quality Ground Truth**: Many regions, especially those most vulnerable to natural hazards, lack sufficient *in-situ* measurements for model training and validation. This scarcity constrains the reliability of supervised learning approaches.
- **Model Transferability and Generalization**: AI models often demonstrate location-specific performance. A network trained on European flood patterns may underperform when applied to South Asian monsoon floods due to differences in topography, infrastructure, and hydrology. Domain adaptation and transfer learning remain key research frontiers.
- **Scalability and Real-Time Processing Constraints**: The computational cost of processing high-resolution SAR data in near real-time is significant. There is a growing demand for edge-computing solutions and cloud-based infrastructures that can handle the data deluge without compromising responsiveness.

5.11 Outlook: Vision for Artificial Intelligence-Driven Synthetic Aperture Radar Systems in Disaster Management

The future of disaster resilience lies in fully autonomous SAR systems empowered by AI:

- **Onboard Intelligent Processing:** Future satellite constellations will host lightweight AI models capable of on-orbit SAR data analysis. With this capability, tremendous latency will be eliminated from data acquisition to actionable insight, resulting in rapid alerts during earthquakes, floods, and storms.
- **Crowdsourced and Participatory Validation:** User-generated data may consist of photos, social media posts, or GPS traces, and can be incorporated to enrich validation workflows, especially for inaccessible or data-scarce areas. Democratizing data integration will enhance AI-based insight and contextual relevance.
- **Self-Adaptive and Continually Learning Models:** Resilient AI-SAR platforms will be based on systems that grow older, learning from false positives and missed detections. The models will recalibrate their performance by themselves, depending on patterns in new data, changes in their environment, and user feedback.
- **Global Accessibility and Equity:** This feature aims to bring benefits accrued from AI-SAR systems to local governments, non-governmental organizations, and citizen responders, besides national agencies. These tools can support grassroots-level disaster preparedness and response through intuitive dashboards and open-data initiatives.

AI disclosure statement: The author acknowledges the use of AI-based tools for minor language editing or grammatical correction. However, all intellectual contributions, analyses, and conclusions presented in this chapter are entirely the author's own.

References

1. Khan, S.M., Shafi, I., Butt, W.H., De la Torre Díez, I., López Flores, M.A., Galán, J.C., Ashraf, I., A systematic review of disaster management systems: Approaches, challenges, and future directions. *Land*, 12, 8, 1514, 2023. doi: 10.3390/land12081514.
2. Amitrano, D., Di Martino, G., Di Simone, A., Imperatore, P., Flood detection with SAR: A review of techniques and datasets. *Remote Sens.*, 16, 4, 656, 2024. doi: 10.3390/rs16040656.
3. Othman, S.H. and Beydoun, G., A metamodel-based knowledge sharing system for disaster management. *Expert Syst. Appl.*, 63, 49–65, 2016. doi: 10.1016/j.eswa.2016.07.005.
4. Ban, Y. and Jacob, A., Object-based fusion of multitemporal multiangle ENVISAT ASAR and HJ-1B multispectral data for urban land-cover mapping. *IEEE Trans. Geosci. Remote Sens.*, 51, 4, 1998–2006, 2013. doi: 10.1109/TGRS.2012.2219539.
5. Bian, L. and Yu, L., Deep learning for synthetic aperture radar (SAR) image classification: A review. *Remote Sens.*, 14, 11, 2666, 2022. doi: 10.3390/rs14112666.
6. Chini, M., Hostache, R., Matgen, P., Schumann, G., Giustarini, L., A hierarchical split-based approach for parametric thresholding of SAR images: Flood inundation as a test case. *IEEE Trans. Geosci. Remote Sens.*, 55, 12, 6975–6988, 2017. doi: 10.1109/TGRS.2017.2737664.
7. Cui, Y., Wu, X., Huang, L., He, Y., Yang, W., Flood detection using Sentinel-1 SAR data *via* image analysis and deep learning techniques. *Remote Sens.*, 12, 9, 1456, 2020. doi: 10.3390/rs12091456.
8. Di Martino, G. and Brunner, D., SAR data for damage assessment in the context of disaster response: A review. *Remote Sens.*, 13, 12, 2347, 2021. doi: 10.3390/rs13122347.
9. Ghamisi, P., *et al.*, Advances in hyperspectral image and signal processing: A comprehensive overview of the state of the art. *IEEE Geosci. Remote Sens. Mag.*, 5, 4, 37–78, 2018. doi: 10.1109/MGRS.2018.2860060.
10. Guo, H., Huang, J., Zhang, L., SAR image analysis using attention-based deep learning models. *ISPRS J. Photogramm. Remote Sens.*, 172, 62–79, 2021. doi: 10.1016/j.isprsjprs.2020.12.010.
11. Kang, J., Li, Y., Zhao, Z., Damage detection from SAR imagery using deep convolutional neural networks. *Remote Sens.*, 12, 5, 832, 2020. doi: 10.3390/rs12050832.
12. Li, W., Fu, H., Yu, L., Gong, P., Enhancing pixel-level flood mapping from SAR images through supervised classification with multi-feature fusion. *Remote Sens. Environ.*, 173, 402–412, 2016. doi: 10.1016/j.rse.2015.11.009.
13. Liu, Y., Shi, Z., Zou, Z., Unsupervised deep clustering for SAR image change detection. *IEEE Geosci. Remote Sens. Lett.*, 15, 9, 1386–1390, 2018. doi: 10.1109/LGRS.2018.2845676.

14. Lu, P., Liu, Y., Ma, J., Deep learning for SAR image classification: Recent advances and future trends. *IEEE Trans. Geosci. Remote Sens.*, 59, 8, 6586–6605, 2021. doi: 10.1109/TGRS.2020.3047894.
15. Luo, L., Li, Q., Xie, Y., Zhao, H., Multi-source data fusion for wildfire assessment using SAR and optical imagery. *Remote Sens.*, 14, 3, 588, 2022. doi: 10.3390/rs14030588.
16. Nogueira, K., Penatti, O.A.B., Santos, J.A., Towards better exploiting convolutional neural networks for remote sensing scene classification. *Pattern Recognit.*, 61, 539–556, 2017. doi: 10.1016/j.patcog.2016.07.001.
17. Papathanassiou, K.P. and Cloude, S.R., Single-baseline polarimetric SAR interferometry. *IEEE Trans. Geosci. Remote Sens.*, 39, 11, 2352–2363, 2001. doi: 10.1109/36.964973.
18. Persello, C. and Stein, A., Deep fully convolutional networks for the detection of informal settlements in VHR SAR images. *IEEE Geosci. Remote Sens. Lett.*, 14, 12, 2325–2329, 2017. doi: 10.1109/LGRS.2017.2760399.
19. Schumann, G.J.-P., Di Baldassarre, G., Bates, P.D., The utility of spaceborne radar to render flood inundation maps based on multialgorithm ensembles. *IEEE Trans. Geosci. Remote Sens.*, 47, 8, 2801–2807, 2009. doi: 10.1109/TGRS.2009.2017937.
20. Singh, P. and Bhatia, S., A review on applications of deep learning in synthetic aperture radar imagery. *Arch. Comput. Methods Eng.*, 27, 287–312, 2020. doi: 10.1007/s11831-019-09341-5.
21. Ullah, S. and Ahmad, M., Real-time flood mapping using SAR and deep neural networks. *Remote Sens. Appl. Soc. Environ.*, 21, 100439, 2021. doi: 10.1016/j.rsase.2020.100439.
22. Zhang, H. and Wang, C., Reinforcement learning for adaptive SAR calibration under variable disaster scenarios. *Remote Sens. Lett.*, 12, 5, 425–434, 2021. doi: 10.1080/2150704X.2021.1905989.
23. Zhu, X.X., *et al.*, Deep learning in remote sensing: A comprehensive review and list of resources. *IEEE Geosci. Remote Sens. Mag.*, 5, 4, 8–36, 2017. doi: 10.1109/MGRS.2017.2762307.

6

Artificial Intelligence-Driven Insights from Synthetic Aperture Radar Data: Transforming Disaster Preparedness and Response

Nidhi Sharma[1] and Gopal Krishna[2*]

[1]Uttaranchal College of Health Sciences, Uttaranchal University, Dehradun, Uttarakhand, India
[2]Uttaranchal Institute of Technology, Uttaranchal University, Dehradun, Uttarakhand, India

Abstract

Artificial intelligence (AI) and synthetic aperture radar (SAR) have transformed disaster management by providing real-time monitoring, quick decision-making, and effective resource allocation. Natural and human-made disasters threaten lives and can harm infrastructure and the environment, thus requiring the deployment of new technologies for better preparedness and response. Processing SAR data using AI improves its predictive capability, assisting in risk assessment, early warning systems, and infrastructure preparedness. Artificial intelligence also enhances community awareness and prepositioning of resources to aid in successful disaster preparedness. From a methodological perspective, AI utilizes machine learning (ML) and deep learning (DL) algorithms to analyze SAR data, enabling automation of feature extraction, noise elimination, change detection, and real-time prediction of disaster scenarios. Artificial intelligence can be used alongside SAR to automate disaster monitoring, optimize resource allocation, and enhance emergency response with real-time situational awareness and damage assessment. As evidenced through the outcomes, integrating AI with SAR data leads to a considerable enhancement in predicting disasters, shortening the time taken for response and optimizing the allocation of resources, which in turn mitigates the loss of life and damage to the economy. This chapter discusses SAR data

**Corresponding author*: cse.gopalkrishna@gmail.com

Pooja Joshi, Ashutosh Bhatt, Neelam Sharma and Kapil Joshi (eds.) Synthetic Aperture Radar for Disaster Management: Bridging Science and Solutions, (141–180) © 2026 Scrivener Publishing LLC

analysis methods utilizing AI, particularly DL, with its convolutional neural networks for feature extraction, generative adversarial networks for image processing, and recurrent neural networks for time series disaster monitoring. Artificial intelligence is also being used in emergency coordination, including rescue missions with unmanned aerial vehicles, robotics, and improved logistics planning. While these benefits are tremendous, data privacy, AI bias, the expensive infrastructure in underdeveloped areas, and ethical issues pose challenges for AI-SAR integration and, therefore, need more evidence for implementation. However, through AI-SAR synergy, disaster resilience is becoming increasingly feasible with proactive risk assessment and mitigation, real-time response to disasters, and improved recovery after such events. In the latter case, future systems will integrate seamlessly using AI-powered SAR satellite constellations, Internet of Things (IoT) technologies, and digital twin technology to foil disasters when they occur and contribute to humanitarian efforts. The key to leveraging AI-SAR technology in effective disaster management will be to tackle ethical concerns and promote global collaboration.

Keywords: Artificial intelligence, systemic aperture, disaster management, real-time monitoring, predictive analytics

6.1 Introduction

6.1.1 Overview of Disaster Preparedness and Response

Natural and man-made disasters can have devastating effects on life, infrastructure, and the environment. Effective disaster preparedness and response strategies are critical to minimizing losses and allowing for rapid recovery. Artificial intelligence (AI) and synthetic aperture radar (SAR) have revolutionized disaster management processes through instantaneous monitoring, swift decision-making, and optimal resource management capabilities. Machine learning (ML) algorithms will be developed for pre-processing-related AI-generated analyses, tracking, and forecasting products. Proactive accident preparedness, based on risk analysis and prediction using historical data and predictive modeling, early warning systems (EWS) aided by AI analytics, and readiness of infrastructure, can contribute to resilience [1]. Furthermore, community awareness and training educate the population on safety, and resource prepositioning prepares key supplies for immediate availability. Through quantitative-based data extraction from satellite SAR images, AI-driven real-time monitoring and damage identification of affected areas are being achieved, speeding up the evaluation process and enabling authorities to assess the damage quickly. Artificial intelligence-assisted rescue, relief, emergency communication,

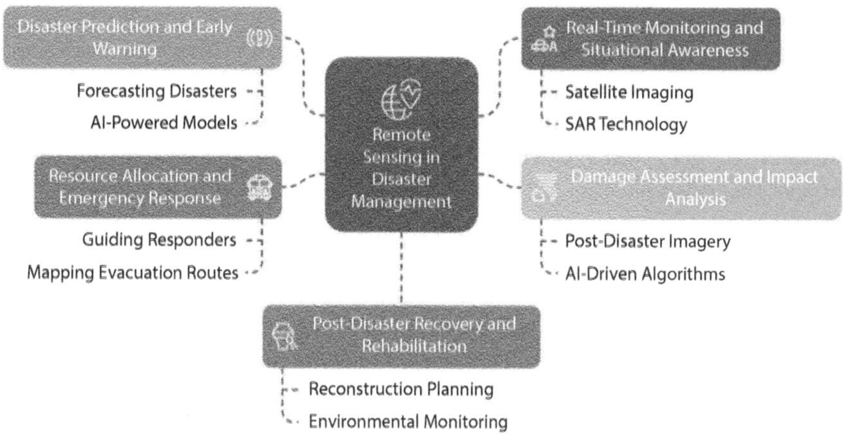

Figure 6.1 Remote sensing in disaster management.

and coordination enable automated drone deployment and robotic assistance for quick response efforts. It further underpins medical and humanitarian assistance, allowing for prioritization of care and logistics in support of displaced populations. Then, after a disaster, AI can aid in recovery through rehabilitation planning for remodeling infrastructure and returning to normalcy. The combination of AI and SAR data revolutionizes disaster management by enabling accurate predictions, effective coordination, and optimized resource deployment. By leveraging technology, this framework fortifies disaster resilience, reduces disaster perils, and introduces a transformative shift toward a data-centric and proactive disaster management system [2]. Figure 6.1 depicts the role of remote sensing in disaster management.

6.1.2 Importance of Artificial Intelligence in Enhancing Synthetic Aperture Radar Data Analysis

Remote sensing is a crucial component of disaster management processes, generating timely, accurate, and large-scale data for disaster monitoring, assessment, and response. Using satellite or aerial sensors, remote sensing obtains the needed information to monitor, assess, and respond to both natural and man-made disasters. The SAR can overcome the limitations of these traditional observation methods and is therefore indispensable in disaster preparedness and response. Remote sensing plays a key role in disaster prediction and early warning, as meteorological and environmental conditions can be analyzed for hurricanes, floods, and wildfires. Continuous imaging of the environment provides real-time monitoring and support for situational awareness, with SAR providing uninterrupted day-and-night monitoring in all types of weather. Post-disaster imagery and AI-driven algorithms evaluate the damage and prioritize response efforts. Artificial intelligence-powered mapping helps in the allocation of resources and emergency response, aiding responders in deploying help in identifying evacuation routes. Post-disaster recovery or rehabilitation also uses satellite images for rebuilding infrastructure and long-term environmental monitoring. With features as wide coverage, rapid data acquisition, multispectral and radar imaging, and non-intrusive monitoring, remote sensing plays an important role in disaster management. With AI, enhancements in predictive modeling and situational awareness can lead to improved response efforts, especially if SAR is harnessed to deliver better human and economic loss mitigation results [3].

6.1.3 Importance of Artificial Intelligence in Enhancing Synthetic Aperture Radar Data Analysis

Synthetic aperture radar data analysis has transformed remote sensing with its high-resolution, all-weather, and day-and-night imaging abilities, which are essential components for disaster management. The immense spatio-temporal volume and heterogeneity of the SAR data pose challenges to its interpretation, demanding sophisticated computational methods to analyze it effectively. Deep learning (DL) and ML techniques contribute significantly to SAR data through increased automation of key tasks, such as data processing, and improved accuracy, allowing real-time decision-making [4]. Table 6.1 gives an insight into the existing literature on AI and SAR in disaster preparedness and management [6–18].

One major advantage of AI lies in automated feature extraction and classification, in which deep learning models such as convolutional neural networks (CNNs) detect terrain changes, infrastructure damage, and flood extents with broader accuracy than passive methods. Noise reduction and image enhancement algorithms like GANs and deep denoising networks also contribute to improved image clarity by filtering speckle noise and increasing contrast. Artificial intelligence also tracks changes for disaster monitoring, implementing time-series analysis with recurrent neural networks to help monitor the progression of a disaster and assess damage as it occurs. Moreover, ML-based real-time disaster prediction and EWS allow for the incorporation of SAR data with meteorological and seismic information to predict floods, landslides, and earthquakes.

Table 6.1 Evidence from the literature indicating existing evidence, as well as the roles of AI and SAR in disaster preparedness and management.

Reference	Objective	Methodology	Key findings	Implications
Creative Learning for Disaster Preparedness	To explore how educational resources can enhance students' responsibility in disaster preparedness.	Film project, 'Go-Bag' project, online questionnaire (n = 176).	School-based learning positively impacts families' disaster preparedness; students gain deeper cognitive understanding and increased self-efficacy.	Highlights the role of active learning in building community resilience and suggests further research on educational interventions.
Emergency Response Challenges in Hurricane Katrina	To analyze bureaucratic inefficiencies and organizational challenges in disaster response.	Case study of Hurricane Katrina response, review of agency effectiveness.	Bureaucracies hinder emergency response; flexible, resilient organizations (e.g., Coast Guard) perform better.	Recommends restructuring emergency agencies for flexibility, political support, and proactive disaster planning.
Seismic Risk and Peak Ground Acceleration Discretization in Indonesia	To enhance earthquake damage assessment through Peak Ground Acceleration discretization.	Modified K-means and K-medoids clustering algorithms; average silhouette width for assessment.	K-means clustering effectively discretizes peak ground acceleration for improved damage assessment.	Supports data-driven disaster planning and earthquake-resistant infrastructure development.

(*Continued*)

Table 6.1 Evidence from the literature indicating existing evidence, as well as the roles of AI and SAR in disaster preparedness and management. (*Continued*)

Reference	Objective	Methodology	Key findings	Implications
Artificial Intelligence and SAR for Disaster Management	To explore AI-SAR integration for real-time disaster monitoring and response.	Literature review, analysis of AI applications in SAR.	Artificial intelligence enhances SAR capabilities in disaster prediction, response, and post-disaster recovery.	Advocates for ethical AI governance, global collaboration, and further research in AI-SAR integration.
Library Emergency Funding During COVID-19	To examine the impact of federal emergency funding on an academic library.	Case study on library funding utilization.	Emergency funding supported library operations during closure.	Suggests integrating financial preparedness into disaster planning for libraries.
Emergency Mobility Facilities (EMFs) for Disaster Preparedness	To optimize EMF allocation and road network design under disaster scenarios.	Two-stage stochastic programming, non-convex mixed integer non-linear programming, was transformed into mixed integer linear programming.	Efficient EMF allocation improves disaster response capabilities.	Supports data-driven emergency planning and infrastructure resilience.

(*Continued*)

Table 6.1 Evidence from the literature indicating existing evidence, as well as the roles of AI and SAR in disaster preparedness and management. (*Continued*)

Reference	Objective	Methodology	Key findings	Implications
Community-Based Disaster Risk Management (CBDRM)	To explore the role of CBDRM in reducing vulnerabilities and promoting resilience.	Case studies from the Philippines, critical reviews of resilience definitions.	Having CBDRM enhances disaster prevention, preparedness, response, and recovery by involving local communities. People-led initiatives contribute to resilience and social justice.	Strengthening community capacity and participatory governance is essential for effective disaster management and sustainable development.
Flood Early Warning Systems (FEWS) & Civil Society Organizations (CSOs)	To analyze challenges in the implementation of FEWS and the role of CSOs in addressing them.	Literature review, global online survey, and experiential knowledge.	Challenges include poor communication, low community participation, a lack of disaster risk reduction knowledge, and inadequate gender inclusion. The CSOs play a key role in improving awareness and response.	Strengthening community involvement, gender inclusivity, and local governance is crucial for improving the effectiveness of FEWS.

(*Continued*)

Table 6.1 Evidence from the literature indicating existing evidence, as well as the roles of AI and SAR in disaster preparedness and management. *(Continued)*

Reference	Objective	Methodology	Key findings	Implications
Future-Ready Education and Preparedness	To explore the transformation of learning landscapes to prepare students for an uncertain future using concepts of preparedness from disaster management.	Conceptual analysis integrating trends in education and disaster preparedness.	Conventional learning landscapes lack future readiness. Uncertainty preparedness can be developed through adaptive educational strategies.	Educators must integrate preparedness concepts into curricula to enhance resilience and adaptability among students.
Humanitarian Logistics and Emergency Response	To develop a multi-objective stochastic model for balancing efficiency, effectiveness, and fairness in humanitarian logistics.	Mathematical modeling, statistical transformation of constraints, augmented s-constraint method.	The proposed model optimizes resource allocation, improving demand forecasting and flexibility in emergency logistics.	Decision-makers can use the model for better planning and resource distribution in disaster scenarios.

(Continued)

Table 6.1 Evidence from the literature indicating existing evidence, as well as the roles of AI and SAR in disaster preparedness and management. (*Continued*)

Reference	Objective	Methodology	Key findings	Implications
Healthcare Emergency Preparedness Coalitions	To examine the role of healthcare emergency preparedness coalitions in community health resilience.	Case studies of a large healthcare system and a public safety agency in Pennsylvania.	Traditional coalitions focus on hospitals rather than community-wide health. Inclusive approaches enhance overall community preparedness.	Expanding coalition focus beyond hospitals to include community health systems strengthens disaster response.
Geographic Information Technologies in Disaster Management	To develop an interoperable geographic data model and analysis tools for disaster management.	Scenario-based analysis of 39 activities, development of an open-source geographic data model.	The development of an open-source geographic data model facilitates data exchange and enhances disaster preparedness, though data transformation remains labor-intensive.	Standardized geographic data models improve coordination and efficiency in disaster response efforts.

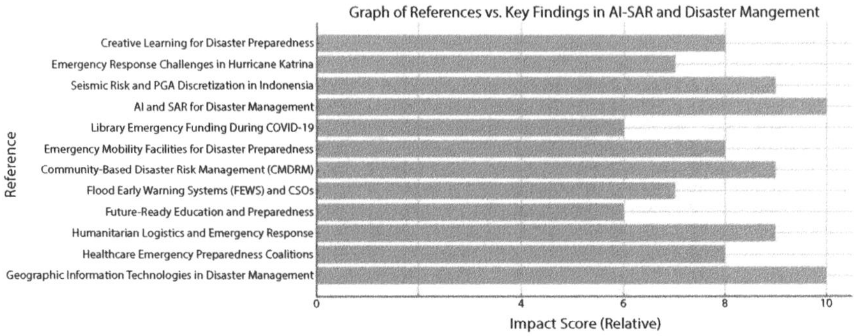

Figure 6.2 Graph indicating key findings in AI-SAR and disaster management.

Artificial intelligence-based resource allocation and response plans opti mize the measures taken in emergencies by spotting severely affected areas and deciding the prioritization of relief distribution. Artificial intelligence-driven SAR analysis offers speed and efficiency, processing data from hours-long to minutes; accuracy and precision in detecting potential disasters; scalability to monitor multiple disaster-prone regions; and automation to limit human error. The combination of this AI with advanced SAR satellites provides disaster management authorities an unprecedented efficiency in prediction, monitoring, and responding to a disaster. It is this synergy that is paving the future for disaster resilience and preparedness [5]. Figure 6.2 represents a graph indicating key findings in AI-SAR and disaster management.

6.2 Fundamentals of Synthetic Aperture Radar in Disaster Management

6.2.1 Basics of Synthetic Aperture Radar Technology

Advanced remote sensing technology, such as SAR, provides transformative capabilities in disaster management. Apart from conventional optical sensors that depend on visible light, SAR works on the microwave bands, allowing the creation of high-resolution images of the Earth's surface regardless of weather conditions or time. This makes it an essential tool for monitoring disasters, as SAR can see through clouds, smoke, and even vegetation, providing valuable information to assist in emergency response and mitigation. The SAR works by using sensors onboard satellites, aircraft, and drones to emit microwave pulses. These pulses make their way down to the Earth's surface, where they reflect to the sensor, and the time delay and intensity of the reflected signals provide information to generate detailed images of the terrain. The "synthetic aperture" method draws on multiple radar signals over time, mimicking a much larger antenna and yielding high spatial resolution. Some of the key features of SAR are all-weather imaging and day-night imaging, which allow it to work in adverse conditions and produce high-resolution imaging, which can detect land deformation, infrastructure damage, and water levels. Its penetration capability depends on its wavelength, but it serves to detect structures beneath our land and the extent of landslides and floods. More advanced techniques such as interferometric SAR (InSAR) are used to monitor ground displacement due to quakes and subsidence, while polarimetric SAR (PolSAR) can classify land cover and assess disaster damage. Speckle SAR can be deployed for disaster management systems as it is used for flood detection, earthquake damage, landslide monitoring, wildfire tracking, and cyclone impact assessment. Combined with the power of AI, SAR can also significantly improve response to disasters by automating the process of analyzing images, providing better estimation of disasters, and faster decision making. As SAR capabilities allow for high-definition, real-time, all-weather imaging, it is an essential tool for disaster preparedness and resilience planning [19].

6.2.2 Advantages of Synthetic Aperture Radar over Optical Imaging

As SAR technologies do not rely on light, they do not face the same problems, and therefore, SAR has several advantages over the traditional optical imaging system. The SAR works by emitting microwave signals rather than light, as optical sensors do, so it can operate in conditions where optical systems fail. It possesses unique characteristics that make it ideal for providing reliable, high-resolution observations of disaster-prone and affected regions, constituting a valuable asset for emergency response and mitigation. Unlike optical imaging, which is dependent on sunlight and requires clear skies, SAR is operational day or night and in inclement weather. Moreover, SAR has radar wavelength-dependent penetration ability (e.g., L-band, P-band), enabling it to penetrate through vegetation, soil, and some specific construction materials to detect hidden structures and subsurface deformations. Optical imaging, by comparison, is restricted to surface detail. The key advantage of SAR is its high-resolution imaging capabilities with excellent clarity, as the synthetic aperture techniques used can maintain clarity of vision over large areas, even from high-orbit satellites.

Optical imaging requires low-altitude imaging for high-resolution images but tends to be disturbed by atmospheric elements. Another one is surface deformation and change detection, in which the radar technology known as InSAR can detect minute shifts in the ground caused by events like earthquakes, landslides, and subsidence—something optical imaging cannot do. Moreover, structural and materials classification is improved with PolSAR, with the possibility of material discrimination by their dielectric nature that contributes to the damage assessment. Lastly, SAR enables superior disaster monitoring and response, allowing real-time high-accuracy disaster mapping capability (even in extreme conditions). The AI-optimized SAR processing allows the automated detection of floods, assessment of damage, and coordination of response, resulting in timely and effective disaster management. Optical imaging has been strongly associated with color-coded imaging, but it is generally unreliable in emergencies due to the limitations of the environment [20].

6.2.3 Key Applications of a Synthetic Aperture Radar in Disaster Monitoring

The SAR is an important technology for disaster monitoring because of its high-resolution, all-weather, and day-night imaging capability. It can penetrate clouds, vegetation, and some buildings unlike many other sensors, making it an invaluable tool for disaster risk assessment, early warning detection, and post-disaster assessment. Artificial intelligence shows how SAR can be used to respond to disasters, especially when these two fields are combined. One of its main uses is flood detection and monitoring, where SAR can monitor the extent of flooding even in cloudy areas, distinguish between land and water, and detect submerged infrastructure. Real-time flood mapping has also been performed using Sentinel-1 SAR imagery, to give an example. The ability of InSAR to detect ground displacement and damage to infrastructure provides important information for disaster managers to use in assessing areas that have been affected to offer the most immediate assistance to them, and provides crucial information in cases where disaster surges such as earthquakes occur. SAR Satellite data is used in landslide detection and monitoring soil movement and slope instability, and helps in EWS in regions like the Himalayas, which are prone to landslides. A key application is in wildfire detection and assessment of burn severity, where SAR can penetrate smoke and cloud cover to detect burned and unburned vegetation, aiding in recovery efforts. Lastly, SAR has significant importance in analyzing impacts from hurricanes and cyclones to monitor losses from these storms, such as in Hurricane Harvey (2017), where SAR detected storm damage in SAR-derived digital elevation models. It is also useful for tsunami and coastal disaster monitoring, which can be used to evaluate alterations to coastlines and water intrusion to facilitate long-term rebuilding efforts. Similar characteristics enable SAR to help monitor volcanic activity, allowing for the detection of land deformation and predictions of eruptions. Finally, SAR supports the assessment of infrastructure and urban damage, including the identification of collapsed buildings and the support of resource allocation [21].

6.3 Artificial Intelligence Techniques for Synthetic Aperture Radar Data Processing

6.3.1 Machine Learning and Deep Learning in Synthetic Aperture Radar Image Analysis

Integrating AI with SAR disaster monitoring and management. Conventional SAR image analysis depends on manual interpretation; it is time-consuming and vulnerable to error. Yet, ML and DL techniques have brought a ground-breaking technology for SAR data autonomous processing, faster, more accurate, and better real-time decision capability. Automated classification, pattern recognition, and change detection in SAR images are also made possible through ML. However, supervised learning techniques like support vector machines (SVM), random forest (RF), and k-nearest neighbors have primarily been utilized for land cover classification, flood extent mapping, and damage assessment. On the other hand, unsupervised learning methods (e.g. k-means clustering and autoencoders) are used to discover anomalies in SAR images, making them viable for analyzing terrain after the occurrence of a disaster. Advanced processing of the raw SAR data can be achieved with reinforcement learning, which can optimize the feature extraction strategies according to historical disaster events. Convolutional neural networks learn spatial patterns from SAR imagery to support applications like building damage assessment and flood mapping. Time-series SAR data are analyzed using recurrent neural networks and long short-term memory (LSTM) networks for the tracking of disasters, e.g., landslide movement and flood evolution. These GANs offer advanced capabilities to improve the quality of SAR images by greatly downscaling noise for sharper analyses making them valuable for clearer disaster assessment. Another key application involves using AI for change detection, in which solutions like CNNs, autoencoders, and transformer-based models compare pre-disaster and post-disaster SAR images, allowing users to identify changes in terrain, vegetation, and infrastructure. The efficient processing of SAR images and algorithms based on vision transformers are suitable for tasks involving segmentation and classification, which can aid in the continuous monitoring of disasters (landslides) and advance early warning mechanisms. This includes mechanisms for automation and scalability of SAR image processing, high precision in disaster detection, real-time processing for emergency response operations, and improved image quality through AI-based denoising approaches. The algorithms developed enable automated SAR analysis and efficient damage mapping in disaster- and conflict-affected countries, making AI-integrated SAR analysis a powerful tool to inform better preparedness and response strategies globally [22].

6.3.2 Data Preprocessing and Feature Extraction

The effectiveness of AI in SAR data analysis heavily relies on the quality of preprocessing and feature extraction. There is usually noise, distortions, and inconsistencies in SAR data, which are caused by environmental conditions and sensor limitations. While data preprocessing cleanses raw SAR images and optimizes them for the consumption of AI models, the feature extraction lays more focus on key characteristics of the data for improved classification, segmentation, and change detection. Data preprocessing of SAR data, including noise removal, geometric correction, radiometric calibration, and normalization, is an important method for preparing SAR data for this AI-based analysis. Lee and Frost filters, wavelet transformations, and AI-based DL denoiser methods (DNNs and GANs) are applied to reduce the speckle noise in medical imaging data; these techniques significantly improve the quality, contrasted with the original image, while maintaining important structures. Radiometric calibration consequently provides a standard format for backscatter intensity values, enabling correction of sensor inconsistencies and terrain variability, and geometric correction and terrain flattening correct distortions caused by SAR's side-looking geometry leveraging methods, such as the range-Doppler algorithm. To explain further, surface flattening is important in finding the critical collapse of buildings in an earthquake damage evaluation. After preprocessing, emphasis is laid on feature extraction, where the key textures, structures, and patterns are identified for classification and segmentation using AI. Texture analysis through techniques such as gray level co-occurrence matrix, Gabor filters, and deep CNNs is applied to differentiate between land types such as water, vegetation, and built-up areas.

The PolSAR retrieves information about the interaction of the electromagnetic wave from several different polarization channels, increasing the value of information used to assess disasters (e.g., collapsed buildings in earthquake areas or waterlogged areas during floods) [23]. Multi-view change detection (all interferometric SAR (InSAR) and differential InSAR (DInSAR) purchased, with the higher frequency, by shortening the unit time to model surface displacement over time with the help of AI methods; for example, PCA and autoencoders, LSTM networks to predict the occurrence of events. An AI-based segmentation approach clusters pixels into meaningful objects within the data, offering higher accuracy and improved assessments for post-disaster damage. Self-supervised learning minimizes the need for labeled datasets, and means of processing SAR data have not remained static; several AI-based engineering techniques continue to drastically improve SAR data processing, including graph neural networks, which allow the capturing of spatial relationships between terrain features, and vision transformers, which extract long-range dependencies more efficiently than CNNs. Such developments related to artificial intelligence-backed SAR analysis can considerably improve disaster monitoring, prediction, and response [24].

6.3.3 Supervised vs. Unsupervised Learning for Disaster Prediction

The use of AI in SAR data processing presents exceptional opportunities for disaster prediction and response. Also, SAR imagery can utilize ML, which allows supervised and unsupervised learning from the obtained images to extract meaningful insights that help in disaster preparedness, allowing SAR imagery to contribute to automatic and dynamic classification, anomaly detection, and predictive modeling of disasters. In this technique, there exist some labeled SAR datasets, which means that the input images are related to a known outcome (disaster-related in this case). This makes it applicable in several areas such as Classification, segmentation, and regression. Major supervised learning algorithms involve SVM for differentiating between affected and unaffected areas, RF for classifying terrain features and areas affected by disaster, and CNNs for extracting intricate spatial patterns in SAR images. Such techniques have been applied to multiple disaster situations: CNN-based classifiers have been employed to delineate flood boundaries, deep learning models have been developed to assess earthquake damage (e.g., detecting collapsed buildings during the 2015 Nepal Earthquake), and random forest and gradient boosting machines have been used to create landslide susceptibility maps.

Moreover, deforestation and ecological alteration monitoring also utilize supervised learning techniques to make all processes, which are based on the analysis of Sentinel-1 SAR data performed by CNNs [25]. Supervised learning approaches in SAR disaster analysis have the following benefits:

(i) Higher accuracy can be obtained with datasets that contain labelled information.
(ii) It yields better interpretation, as the relationships are well-established for properly labelled datasets.
(iii) Good effectiveness in structural problems, especially when historical records of disasters are present.

On the other hand, unsupervised learning works on SAR data without pre-defined labels; therefore, it is suited for anomaly detection, clustering, and trend analysis, all of which aid disaster forecasting. Essential algorithms like k-means clustering categorize pixels based on similarities to perform terrain classification, autoencoders create an output that compares normal patterns to detect anomalies, and self-organizing maps identify clustering in disaster-prone areas. Techniques like autoencoders have been tried for identifying ground deformation followed by pre-quakes, and k-means has been used for monitoring multi-temporal changes in SAR images, and clustering techniques for segmenting flood and water body areas. Unsupervised models are also used for processing InSAR data to label slope areas that are vulnerable to landslides, where self-organizing maps show good detection of territory instability in mountainous regions. The key strengths of unsupervised learning techniques in SAR disaster analysis are the independence

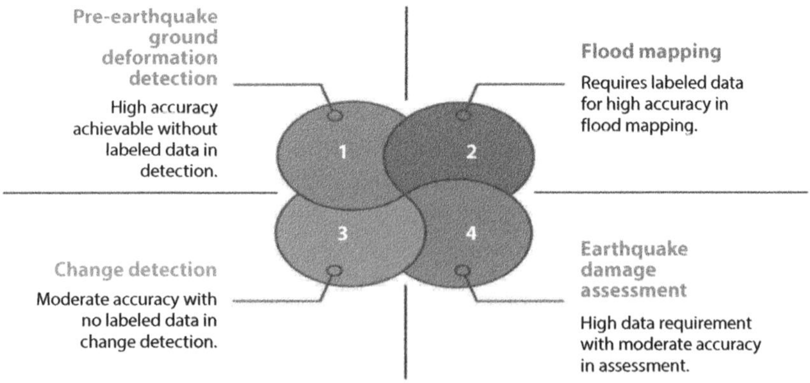

Figure 6.3 Artificial intelligence techniques for SAR data processing.

of labeled datasets, the capability of revealing hidden patterns, which may be used as early warnings for disasters, and the scalability to handle the processing of large SAR datasets efficiently. Artificial intelligence-based SAR techniques use supervised and unsupervised learning methods together to improve disaster prediction, response, and mitigation [26]. Figure 6.3 represents AI techniques for SAR data processing.

6.4 Artificial Intelligence-Driven Disaster Preparedness Using Synthetic Aperture Radar Data

6.4.1 Early Warning Systems with Artificial Intelligence and Synthetic Aperture Radar Integration

Consequently, EWS can play a key role in disaster preparedness and risk reduction by delivering alerts that enable people to prevent loss of life and property. Artificial intelligence and SAR data-driven disaster detection systems provide reliable solutions to disaster monitoring for real-time monitoring, pattern recognition, and predictive analytics for disaster forecasting increasing their accuracy and efficiency in both implementations. Conventional EWS are based on ground-level sensors, meteorological information, and historical records, which are usually limited by coverage, latency, and trustworthiness. Yet, with advantages of wide-area, all-weather, and high-resolution disaster monitoring, AI-SAR-based systems can efficiently address such challenges. This is a structured approach that begins with data acquisition and real-time monitoring. It is enabled by SAR satellites acquiring continuous high-resolution images, Internet of Things (IoT) sensors and weather stations supplying complementary environmental data, and drone-based SAR extending cover to localized disaster-prone areas. This is followed by AI-based data processing involving pre-processing (removing noise and distortions), feature extraction (detecting critical changes in terrains and water levels), and employing ML and DL models for detecting anomalies and predicting disaster risks. Using this analysis, disaster prediction and early warning generation occur, with AI models determining disaster occurrence probability and issuing automated real-time alerts to authorities and the general public while assisting with decision-making surrounding evacuation and resource allocation. In disaster preparedness, AI-SAR-based EWS have various practical applications. The effectiveness of SAR comes in handy in predicting and monitoring floods because it detects water, including

Table 6.2 Areas of application of AI-SAR techniques in disaster management.

Disaster type	AI-SAR techniques	Data processing methods	Key applications	Major benefits
Flood Management	- Real-Time Flood Mapping: AI-enhanced SAR detects water spread in urban and rural areas. - Predictive Flood Modeling: ML analyzes SAR images, rainfall data, and topographical models. - Damage Assessment and Recovery Planning: AI processes SAR images to estimate infrastructure and agricultural damage.	- Artificial intelligence-based SAR image segmentation distinguishes flooded and non-flooded zones. - Time-series SAR analysis tracks flood progression. - Supervised learning models predict future flood risks.	- Rapid flood detection and response. - Improved flood modeling and early warnings. - Post-flood recovery planning.	- Accurate flood detection in all weather conditions. - Faster response for evacuation and relief. - Efficient recovery resource planning.
Earthquake Impact Analysis	-Interferometric SAR Analysis: AI models detect ground displacement and fault movement. - Automated Damage Assessment: AI-SAR identifies collapsed buildings and infrastructure failures. - Seismic Risk Prediction: ML analyzes historical SAR data to forecast seismic activity.	- Artificial intelligence processes differential SAR images for deformation mapping. - Convolutional neural networks classify damage severity. - Deep learning models analyze pre- and post-earthquake structural shifts.	- Post-earthquake damage assessment. - Seismic risk analysis and forecasting. - Prioritization of emergency response.	- Faster and more accurate post-earthquake damage assessment. - Improved rescue and relief prioritization. -Enhanced seismic hazard preparedness.

(Continued)

Table 6.2 Areas of application of AI-SAR techniques in disaster management. (*Continued*)

Disaster type	AI-SAR techniques	Data processing methods	Key applications	Major benefits
Wildfire Prediction and Management	- Early Fire Risk Detection: AI processes SAR data to assess vegetation dryness and temperature anomalies. - Real-Time Fire Spread Analysis: ML models track wildfire movement. - Post-Fire Damage Assessment: AI maps burned areas to support recovery efforts.	- Artificial intelligence-driven thermal anomaly detection. - Fire spread prediction using SAR image patterns. - Random Forest and DL models classify risk zones.	- Pre-fire risk mapping. - Fire containment planning. - Post-fire recovery support.	- Early detection of fire-prone areas. - Improved wildfire containment strategies. - Effective post-fire rehabilitation planning.
Landslide Prediction and Early Warning	- Ground Movement Analysis: AI detects slope shifts using InSAR data. - Landslide Susceptibility Mapping: ML integrates SAR images with seismic and environmental data. - Real-Time Monitoring: AI evaluates risk zones and triggers automated alerts.	- Artificial intelligence-based SAR interferometry detects ground deformation. - Topographical modeling combined with ML assesses landslide risk. - Temporal SAR analysis tracks terrain stability.	- Real-time landslide detection. - Risk assessment for infrastructure planning. - Early warning systems for communities.	- Accurate landslide prediction and early warnings. - Reduced loss of life and property damage. - Better infrastructure resilience planning.

under cloud cover, together with AI analyzing river flow patterns to predict the flooding before flooding happens for earthquake hazard assessment, AI models track potential earthquake-prone areas based on past and real-time seismic patterns while Interferometric SAR (InSAR), which detects ground deformation, offers early reflection of seismic activity. The cyclone and hurricane impact forecasting combine well with SAR's wind speed and sea surface roughness tracking ability, enabling AI to quantify storm intensity and enhance evacuation planning. Still, AI-SAR-based EWS encounter challenges, such as high computational demand for real-time data processing, SAR satellite revisit times that do not allow continuous monitoring, and high-quality training datasets that must be used to reduce false alarms [27]. Future developments focus on overcoming these challenges through Edge AI and cloud computing for faster disaster prediction and integration with 5G networks for real-time SAR data transmission and hybrid AI models combining supervised and unsupervised learning for more accurate results. Thus, integrating these technological advancements, AI-SAR-based EWSs will transform disaster preparedness, response, and mitigation efforts globally. Table 6.2 describes areas of application of AI-SAR techniques in disaster management.

6.4.2 Predictive Analytics for Disaster Risk Assessment

Data passes through a pipeline until predictions are made, so data processing is one of many systems that are put in place to calculate, for example, disaster risk and use that information to better prepare governments, emergency responders, and communities to provide services, prevent interventional spread, and enable effective management of hazards. Risk assessment has been substantially altered using AI techniques coupled with SAR data, as predictive analytics allows for early warning and risk reduction of disaster hotspots, and maintains the prediction of the areas that are vulnerable to hazards or disasters. Predictive analytics employs AI and ML models to analyze SAR data and predict disaster risks by using historical disaster data for pattern recognition, real-time SAR imagery for monitoring environmental changes, and AI models for estimating and predicting the occurrence of potential disasters. Traditional risk assessment methods can also become slow, incomplete datasets and subject to human error. Therefore, it is in this way that AI-driven predictive analytics can break through these limitations and provide automated, scalable, and far more accurate disaster risk assessments. The AI-SAR predictive disaster risk assessment process consists of several steps. The SAR satellites, such as Sentinel-1 and RADARSAT, are employed for data collection and preprocessing to obtain images of terrain, water bodies, and urban areas at high resolution, followed by data preprocessing techniques to reduce noise, calibrate, or geo-reference the image to improve its quality. Artificial intelligence identifies key risk indicators from SAR data in the feature extraction stage, including ground deformation for earthquake risk, water accumulation patterns for flood risk, vegetation dryness for wildfire risk, and slope instability for landslide risk. Supervised learning methods, such as RF and Gradient Boosting, are used by AI-enabled risk prediction models to classify disaster-sensitive areas; CNNs are used by models to identify structural vulnerabilities before earthquakes. Unsupervised learning models, including K-Means clustering, categorize high-risk regions according to similar patterns observed in SAR data; in contrast, autoencoders are capable of recognizing anomalies, such as unexpected ground shifts [28]. Deep learning techniques augment seismic risk assessment, for example, by integrating InSAR for hybrid AI-backed approaches. The last step, risk mapping and visualization, allows for AI-generated risk maps — where SAR data is layered with the likelihood of hazards happening in a specific area — which helps governments apply geographic information systems (GIS) to plan effective disaster response measures. There are plenty of use cases for predictive analytics in disaster risk assessment. Also, AI assists in predicting earthquake risk by identifying patterns of ground displacement from paired SAR images, which can predict seismic events, while the technology called InSAR can register earth deformations before seismic quakes so that fewer casualties could be predicted and avoided. Then come flood susceptibility mapping, which can leverage SAR's detection of water accumulation regardless of all-weather conditions, while AI uses the river flow patterns for zones that experience flooding based on past river data. ML algorithms analyze SAR-based digital elevation alterations to predict landslide-prone regions. In the context of wildfire prediction, SAR data exploits dry vegetation, say, and land surface temperature anomalies to allow classification of potentially dangerous areas through AI models, enabling early intervention, as demonstrated with respect to AI-driven SAR wildfire prediction models [29].

6.4.3 Artificial Intelligence-Based Mapping and Real-Time Monitoring

Conventional disaster response relies on old or incomplete data, rendering the humanitarian response inefficient. On the other hand, AI-enabled mapping and monitoring systems enable prompt, high-resolution insights that greatly improve disaster prediction, assessment, and response. Artificial intelligence helps in disaster mapping and real-time monitoring *via* the better processing of SAR images, enabling automatic classification of landforms, water bodies, and infrastructure. Hazard maps produced from ML models indicate regions at risk, GIS-based AI integration creates interactive maps for disaster risk analysis. Using object detection algorithms to identify flooded regions, landslides, and damaged infrastructure, AI also allows for real-time monitoring. The SAR data is used to track geographical changes, and continuous stream of radar images provides disaster monitoring capability even in bad weather. The workflow of AI-SAR for use in disaster mapping and monitoring includes several critical stages. Preprocessing of these images is done, and SAR satellites, such as Sentinel-1 and the RADARSAT, can be used to perform these operations. The AI-based pre-processing methods for noise cancellation, adjustment, and feature enhancement also enhance image quality. During the feature extraction stage, DL models are implemented using CNN and GAN to determine affected features caused by the disaster (e.g., flooded area, ground deformation area, burnt area, and landslide area) [30]. Based on SAR data, a process is employed that automatically calculates how bad the affected region is, allowing for accurate damage assessment, and dynamic risk maps during the response planning stage. The AI-assisted change detection techniques also assist in assessing the extent of damage over a given time scale. Lastly, real-time monitoring and alerting systems leverage ML algorithms to continuously analyze SAR images and detect ongoing disasters. Alerting systems powered by AI are activated as soon as potentially hazardous forecasts are detected, allowing the development (cloud-based) of AI products to be executed in real time by disaster management teams. The utility of AI-based SAR mapping and monitoring has numerous important applications in disaster management. Flood detection and mapping applications leverage AI to identify rising water in SAR images and to forecast risk likelihood (i.e., where flooding is likely to occur). AI-SAR analysis improves the ability to assess damage from earthquakes by measuring the displacement of buildings due to shaking and estimating the total effects of the earthquake on our disrupted lives. Although AI-SAR mapping and monitoring are advantageous, there are challenges associated with it. Real-time AI processing has high computational requirements that can strain existing resources, and the limited revisit frequency of SAR satellites makes continuous Earth monitoring challenging. The reason behind this is that AI models still need to be optimized to reduce false negatives in capturing disaster detection. Giving way to situation awareness, with multi-sensor fusion SAR (Synthetic aperture radar), Li (Light detection and ranging), thermal imagery will ensure that the situation of such disasters can be monitored effectively [31].

6.4.4 Artificial Intelligence-Driven Synthetic Aperture Radar Strategies for Sudden-Onset vs. Slow-Onset Disasters

The AI-SAR applications are highly varied for sudden-onset disasters such as earthquakes, and slow-onset disasters like droughts. Sudden disasters come with the need for real-time assessment of damage, whereas slow-onset disasters require long-term monitoring and analysis of trends. Table 6.3 depicts distinctive approaches to AI-SAR.

Table 6.3 The AI-SAR strategies for sudden-onset disasters and slow-onset disasters.

Aspect	Sudden-onset disasters (earthquakes)	Slow-onset disasters (droughts)
Objective	Rapid damage assessment, emergency response	Early detection, long-term monitoring, and mitigation planning
Synthetic Aperture Radar Data Processing	Change detection using before-and-after images	Time-series analysis of soil moisture, vegetation health
Artificial Intelligence Techniques Used	Deep learning for building collapse detection, CNNs for debris mapping	Machine learning models (RF, LSTMs) for drought trend prediction
Response Time	Immediate (within hours)	Gradual (weeks to months)
Satellite Data Used	High-resolution SAR (RADARSAT-2, TerraSAR-X) for urban damage analysis	Multi-temporal SAR (Sentinel-1, SMAP) for hydrological analysis
Challenges	Cloud interference, rapid data processing	Seasonal variations, data sparsity
Decision-Making Impact	Resource allocation for search & rescue	Drought mitigation policies, irrigation planning

6.5 Artificial Intelligence-Enabled Response and Recovery Using Synthetic Aperture Radar Data

6.5.1 Damage Assessment and Post-Disaster Analysis

Accurate damage assessment for fast response, resource allocation, and planning for recovery is vital in the aftermath of disasters and not only at their occurrence. Conventional assessment methods based on ground surveys and manual inspections are generally time-consuming, labor-intensive, and hazardous in high-risk areas. Artificial intelligence has been effectively used in conjunction with SAR data for damage assessment in post-disaster analysis, facilitating large-scale, automated, and high-resolution damage assessment. The AI-SAR technology considerably improves disaster recovery with faster damage assessments, less reliance on human resources for monitoring, and the ability to monitor vast areas, even in remote or otherwise inaccessible areas. By making sure destruction classifications are correct, however, resources can be allotted such that the response to disasters can be more effective. The AI-SAR pipeline for disaster damage assessment is made up of several stages. This stage includes fetching pre- and post-disaster SAR images from selected satellites (Sentinel-1, RADARSAT, TerraSAR-X, etc.) and data preparation (extraction, conversion, etc.). Three additional SAR preprocessing methods that use AI, like denoising, georeferencing, and image calibration, improve SAR data quality. In the damage detection and classification phase, AI-based models compare the differences between pre- and post-disaster SAR images for the detection of building collapses, flooded areas, landslide-affected areas, and burned areas caused by wildfires. A novel application is the use of InSAR combined with AI for monitoring land subsidence and deformations caused by earthquakes. Artificial intelligence in post-disaster damage mapping and visualization automates damage map generation that shows affected areas. Geographic information systems that can provide real-time visuals to disaster responders, and cloud-based AI groups that allow remote international collaboration between scientists, are just some examples of how advancements in technology are informing and aiding in relief efforts. There are many applications of AI-SAR technology in post-disaster analysis. In the identification of surface displacement caused by earthquakes and the identification of building collapse, AI-InSAR can advance this capability of measuring surface and building collapse in some cases by applying AI algorithms to characterize SAR imagery [32]. Although AI-SAR damage assessment holds significant transformative potential, it is limited by several issues. The CNNs used to analyze SAR images require complex AI models, and training these models is limited by the small number of labeled disaster datasets available. Moreover, improving model performance will need to address the integration challenges between SAR, light detection and ranging (LiDAR), and optical data. It is important to note that AI-enabled SAR-based damage assessment is transforming the landscape of disaster recovery, where top-notch accuracy and efficiency are crucial. Real-time SAR imaging, ML, and DL are enabling rapid, automated, and accurate damage evaluations to be carried out like never before. Figures 6.4 and 6.5 depict SAR-based AI assessment.

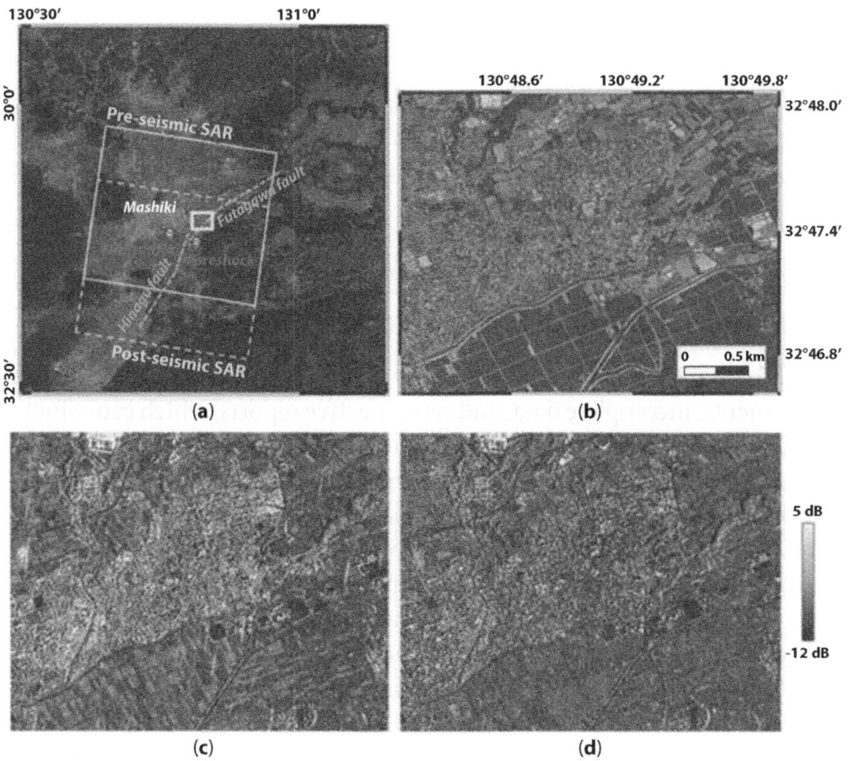

Figure 6.4 The SAR-based assessment of damaged buildings. Source: [34] (Licensed under CC BY 4.0).

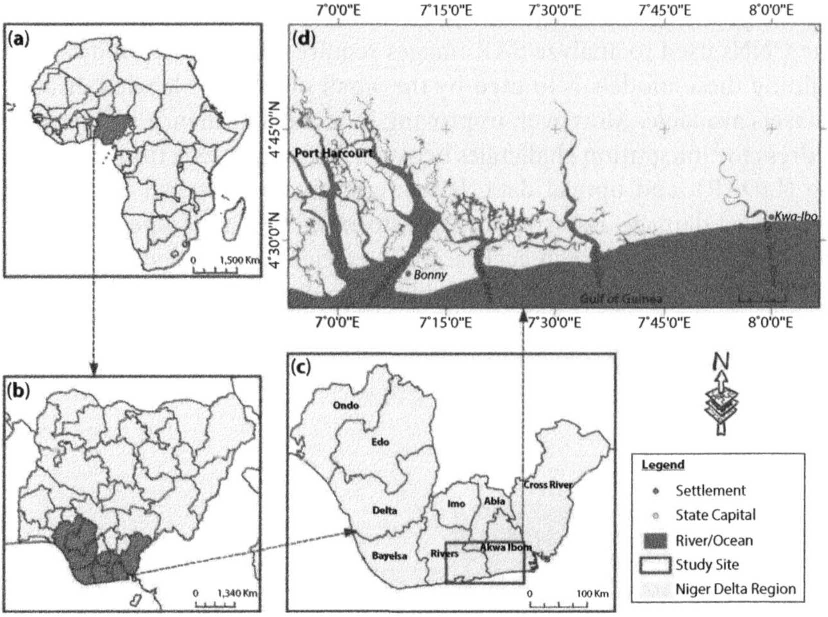

Figure 6.5 Monitoring shorelines at high and low tide using SAR. Source: [35] (Licensed under CC BY 4.0).

6.5.2 Resource Allocation and Decision Support Systems

Traditional methods for allocating resources are usually based on manual assessments, incomplete data, and retrospective reports, which cause inefficiencies in relief distribution and logistical setups. Integrating AI with SAR data revolutionizes disaster management through automated data-centric decision-making tools. The AI-SAR models expedite damage assessment for affected areas, optimize the distribution of relief materials based on predicted severity of damage, and automate prioritization of high-risk areas for response. Also, AI-powered analytics allow for real-time disaster management, which shortens response time over traditional analytics, leading to more real-time coordination between emergency agencies, NGOs, and government agencies, and limits human error. Optimization algorithms such as genetic algorithms help ensure that resources for food, medical aid, and emergency shelter are distributed efficiently; in contrast, reinforcement learning enables dynamic adaptation of resource allocation in response to unfolding disaster scenarios. Predictive modeling improves logistics and relief operations by predicting transportation bottlenecks based on SAR-detected damaged roads, ensuring the matching of demand and supply for crucial resource delivery. Cloud-based AI dashboards with GIS platforms merge maps of structural damage with key parameters for real-time tracking of disasters in AI-SAR decision support systems. The AI-based virtual assistants and chatbots allow emergency response teams to make quick decisions. The AI-SAR technology has been applied to many disaster scenarios. In earthquake response and recovery, AI-SAR identifies at-risk buildings and infrastructure ahead of a pending earthquake for impacted areas, aiding recovery efforts by guiding emergency personnel to affected points. While AI-SAR has many benefits, it also poses several challenges, such as the necessity of processing large amounts of SAR data in real-time, limited revisit rates of SAR satellites resulting in lower quality for continuous monitoring, and the lack of standardized AI frameworks that enable multi-agency coordination. When integrated with technologies like Edge AI and 5G connectivity, AI-SAR technology solutions for SAR data collection allow for in-real-time disaster intelligence, autonomous drones and UAVs with SAR payloads provide SAR dataset localized assessment through motion, and portable SAR solutions, and multi-source AI fusion enables smarter decision-making in disaster scenarios combining SAR, optical imagery, IoT, and social media data. These advancements will enable AI systems to improve disaster preparedness, response efficiency, and resilience planning globally. Figure 6.6 depicts AI-enabled response and recovery using SAR data.

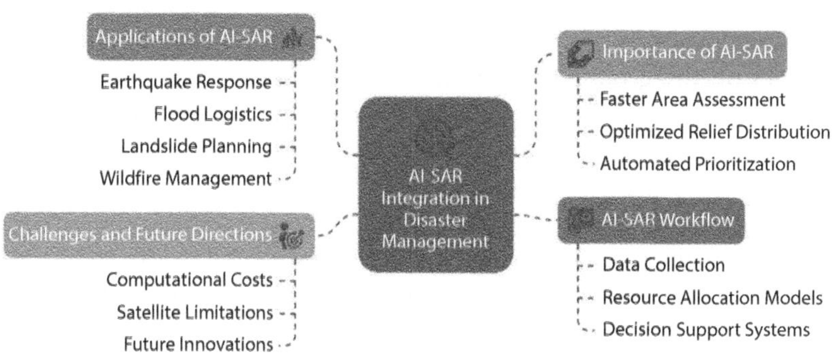

Figure 6.6 The AI-enabled response and recovery using SAR data.

6.5.3 Artificial Intelligence-Powered Rapid Response Mechanisms

In such disaster scenarios, time-sensitive decision-making is critical for saving lives and minimizing damage. Standard disaster relief approaches tend to struggle due to problems such as late data capture, poor coordination, and flawed assessments, which constrain the success of emergency measures. The SAR data combined with AI is transforming rapid response approaches by automating disaster assessment, centralizing information, expediting the allocation of aid, and improving situational awareness in real-time. Artificial intelligence-assisted SAR monitoring allows for quicker detection of disasters, automatic scoring of rescue efforts based on the severity of the destruction, and coordination between emergency response agencies. Moreover, AI enables swift identification of affected populations for tailored medical assistance and evacuation, as well as enhancing situational awareness through AI insights. Sequentially, the AI-SAR workflow for quick disaster mitigation is carried out. Using predictive analytics in search and rescue operations, AI-assisted rescue coordination is where ML models are used to understand high-risk zones and the likely location of human survivors during a structure collapse. The SAR sensors mounted on drones become a real-time situational analysis tool. Automated resource mobilization uses reinforcement learning algorithms to improve the efficiency with which relief is distributed, and logistics AI models map the safest routes of transport for emergency supplies. The AI-driven allocation models help ensure that these resources are distributed to areas where they are most needed. Integrating AI and GIS enables real-time disaster mapping with AI for situational awareness and communication. The AI chatbots, employed in emergency management of disasters, use natural language processing models on social media, news, and IoT data points to verify reports of disasters, assisting responders in virtually every on-the-ground decision made in response to the event [33]. The AI-SAR technology has done wonders in the context of different disasters. Following earthquake response and immediate rescue, it identifies collapsed buildings, infrastructure damage, and potential survivor locations. Emerging as a promising solution, AI-enabled rapid response systems come with their own set of challenges. Processing SAR data in real-time requires high computational power, which restricts its immediate applications. Moreover, SAR satellites have limited revisit times, impacting continuous monitoring, and the integration of SAR with optical images and IoT sensor networks is a crucial technical barrier. But future innovations are designed to circumvent these restrictions. Also, multi-modal AI fusion for SAR, LiDAR, and real-time social media data will enable enhanced situational awareness to make disaster response more efficient and effective.

6.6 Case Studies

6.6.1 Case Study 1: Machine Learning-Based Flash Flood Susceptibility Assessment in Northeastern Bangladesh

The northeastern part of Bangladesh is facing devastating impact due to recurring flash floods, which heavily affect the lives and livelihoods of people. An effective flood management and mitigation strategy is needed because of the growing frequency and severity of these floods. Defining flood hazard areas is the essential first step for that type of planning. Although ML methodologies have been employed in flood susceptibility models, there is a lack of research on the effectiveness of the model at a local scale in Bangladesh. Additionally, most evaluations are conducted using only statistical measures without accounting for the spatial consistency of the resulting susceptibility maps. This case study compares the performance of two machine learning algorithms to locally evaluate the flash flood susceptibility for RF and SVM using statistical and spatial metrics.

Study Area: This study is focused on a flood-prone area of northeastern Bangladesh with low topography, high seasonal rainfall, and riverine systems that frequently overflow, causing what are commonly referred to as flash floods. The region is largely agricultural, and frequent flooding threatens food security, infrastructure, and community welfare in the region.

Methodology

Data Collection and Preprocessing

Factors influencing flood susceptibility were considered, including:

- Topographical features: Elevation, slope, curvature
- Hydrological factors: Distance to rivers, drainage density
- Land use/land cover (LULC): Vegetation cover, built-up areas
- Soil and geological factors: Soil type, permeability

The dataset was compiled from remote sensing sources, GIS databases, and ground surveys.

Model Implementation

- Random forest and SVM algorithms were employed for classification.

- The dataset was split into training (70%) and testing (30%) subsets.
- Feature importance analysis was conducted to identify key contributing factors to flood susceptibility.

Model Evaluation

Both statistical and spatial metrics were used to assess model performance:

- Statistical Metrics: Area under the curve, accuracy, kappa coefficient
- Spatial Metrics: Spatial agreement between susceptibility maps

Results

The RF model also outperformed SVM, with an area under the ROC curve value, accuracy, and kappa coefficient of 0.964, 92.1%, and 0.894, respectively. Whereas the SVM model yielded poorer performance with lower agreement and accuracy. In susceptibility mapping, a natural breaks classification method was used to classify the study area into five susceptibility zones (i.e., very high susceptibility, high susceptibility, moderate susceptibility, low susceptibility, and very low susceptibility). Overall concordance between RF and SVM maps was 73.3%, with maximum concordance being reported for the class of very low susceptibility (97.15% RF-SVM, 81.54% SVM-RF).

Implications and Recommendations

The results demonstrate that ML techniques are effective in determining local-scale flash flood susceptibility. The model showed comparatively higher predictive power and outperformed others, and is a favorable model option for flood risk mapping at similar geographic settings. Tracking susceptibility maps for land-use planning could be used to mitigate damage from floods to people, property, and the ecosystem, while the identification of high-risk areas in the study could be translated to community stakeholder ideation and contributions to the adoption of policies to mitigate future flooding risk.

6.6.2 Case Study 2: Earthquake Vulnerability Assessment Using Long Short-Term Memory and Geographic Information System Techniques in India

Earthquakes are among the most destructive natural hazards, inflicting enormous physical, psychological, and economic damage globally. Over the last century, 1100 such earthquakes have been damaging, and they have caused around 15 lakh deaths. The Himalayan region is highly susceptible to seismic activity, and studies estimate that a magnitude 7.5–8.0 earthquake in the future may cause more than 200,000 deaths and damage worth over $150 billion. The DL techniques for predictive modeling, including LSTM networks, have outperformed traditional methods in predicting and evaluating earthquake vulnerability. Here, we propose an open-source LSTM model fused with GIS to determine seismic vulnerability on a Pan-India level.

Methodology

Data Collection and Preprocessing

- Geospatial Data: Seismic activity records, fault lines, soil type, and geological formations were collected from national and international databases.
- Socioeconomic Data: Population density, urban infrastructure, and land use data were integrated to assess social and structural vulnerability.
- Geographic Information Systems-Based Mapping: GIS techniques were employed to develop layers of geotechnical, structural, and social vulnerability.
- Deep Learning Implementation: The LSTM model was trained using historical earthquake data and real-time seismic activity to predict future vulnerabilities.

Model Development and Training

- The LSTM model was trained using past earthquake occurrences, seismic intensity, and geospatial parameters.
- The model was validated using historical earthquake events, achieving an accuracy of 87.8%, sensitivity of 90%, and specificity of 84.9%.

Vulnerability Mapping

Using the trained model, seismic vulnerability maps were generated for the entire country. The vulnerability index was calculated based on:

- Social Vulnerability: Population exposure, housing density, and infrastructure resilience.
- Structural Vulnerability: Building codes, construction quality, and urban expansion.
- Geotechnical Vulnerability: Soil composition, fault proximity, and topographical features.

Results

The seismic vulnerability was assessed as high to very high in many places. The Delhi NCR region is a highly populated, urbanized region that tends to have high structural vulnerability. Northeastern India is a seismically active zone, with weak infrastructure that makes it especially vulnerable to earthquakes. Gujarat is a highly seismically active area with a history of major earthquakes, including the 2001 Bhuj earthquake. Soils are soft in the West Bengal Plains, leading to seismic amplification and increasing the risk. We need immediate seismic risk mitigation and planning for preparedness in these regions to build resilience in their socio-ecological environments and minimize ruinous damage.

Implications and recommendations

The proposed LSTM-GIS framework offers a strong methodology for earthquake vulnerability assessment with several advantages. Assisted with DL-based spatio-temporal analysis, which outperforms traditional techniques, thereby improving tracer predictive accuracy. In contrast, the framework provides comprehensive risk assessment by incorporating social, structural, and geotechnical variables, supporting integrated vulnerability mapping. Moreover, since the high-risk areas can be revealed through the study, it can also deliver useful insights to financiers of disasters by supporting hazard preparedness initiatives, urban planning regulations, and strengthening of infrastructures, thereby enhancing resilience to seismic hazards.

6.7 Challenges and Future Prospects

6.7.1 Technical Limits to Artificial Intelligence-Driven Synthetic Aperture Radar Data Analysis

While the use of AI-driven SAR analysis has transformative potential, several technical challenges pose scalability and effectiveness issues, especially in disaster preparedness and response.

<u>Noise and Interference in SAR Data</u>: SAR images are susceptible to noise, such as speckle, with atmosphere distortion, and signal interference, which can lead to inaccurate results of the AI models. Due to coherent scattering, SAR data is inherently afflicted by the presence of speckle noise, which deteriorates image quality and feature extraction performance for disaster assessment. The high temporal sampling frequency of SAR imagery

necessitates advanced denoising methods, including those based on wavelet filters and deep-learning-driven despeckling algorithms, to improve SAR imagery before AI-driven interpretation.

Model Generalizability Across Geographies: AI models taught using SAR datasets from certain target-prone localities generally do not generalize well in different terrains. Differences in soil composition, vegetation density, and infrastructure in urban areas cause inconsistencies in disaster prediction. Domain adaptation methods and transfer learning can enhance model robustness in various environments.

Administrative Fees and Compute Fees: Processing data requires HPC resources, sufficient storage space, and specialized AI frameworks for processing large-scale datasets. Cloud-based approach offloads AI workload to an external data processing hub, circumventing local infrastructure limitations could be challenging for real-time disaster response due to latency and targeted network limitations. Optimizing AI models for edge computing can help address cost and scalability issues.

6.7.2 Limitations of Synthetic Aperture Radar and Artificial Intelligence Integration

Challenges of SAR Data Processing: SAR data will, to some extent, be complex, inaccessible, and challenging to integrate into existing disaster management systems. The complexity of the data and the challenges associated with processing SAR images are that such images are inherently impacted by speckle noise, which hinders the clarity of images and affects the precision of AI-based classification. Moreover, the processing of SAR needs a lot of computational resources, making it difficult to process in real-time, particularly in areas with inadequate infrastructure. Artificial intelligence model implementation is also complicated by the extensive data preprocessing (filtration, calibration, feature extraction, etc.) that is required for AI models.

Limited Data Availability and Accessibility: Many high-resolution SAR datasets are proprietary, making them unsuitable for disaster response. Moreover, the limitations on the consistent imaging modes and resolutions of SAR satellites make it challenging to train portable AI models. The absence of large historical records for AI training also restricts the precision of disaster prediction models. Limited data for training is a reason for

AI models to show bias and overfitting, which will limit their effectiveness during new disaster scenarios. Also, though SAR can catch large swathes of damage, AI models can miss smaller changes to infrastructure or terrain that could be early warning signs.

Integration with Existing Disaster Management Frameworks: Governments and disaster management authorities have largely resisted integrating new AI tools into their disaster management frameworks due to minimal awareness and knowledge of AI, and instead continue to rely on more traditional methods. The AI-SAR systems cannot be seamlessly implemented alongside current systems due to inadequacies in interoperability and standardization. It hinders the adoption of the AI-SAR technology due to large policy and regulatory barriers.

Cost and Infrastructure Constraints: These limit the widespread use of AI-SAR, making high-resolution SAR imagery from commercial satellites costly to use and beyond the research capacity of disaster-prone developing areas. The financial sustainability problem is exacerbated by the demand for high-performance computing resources, and limited real-time data transmission capabilities hinder the effective deployment of SAR imagery in remote-disaster settings. These challenges can be overcome with the development of AI-SAR processing, data-sharing policies, computational infrastructure, and supportive regulatory frameworks for effective disaster management.

6.7.3 Ethical and Privacy Concerns in Artificial Intelligence-Based Disaster Management

The utilization of AI algorithms on SAR in this domain has evolved alongside, leading to improved early alert systems, timely damage evaluation, and immediate action. Yet, the growing dependence on AI-powered technologies brings up ethical and privacy issues that need to be resolved for responsible and equitable application. Major challenges comprise data privacy, AI determination, ethical choice creation, and socio-economic disparities in AI adjustment to disaster response.

Data Privacy and Security Risks: AI-driven disaster management systems require large amounts of SAR imagery, geospatial data, and personal information from impacted populations. The safety challenges this poses, as this data is stored and processed, include breach, unauthorized access, and misuse of data. Moreover, the ability of AI-enhanced SAR imaging to give real-time information on disasters makes it possible to continuously monitor regions prone to disasters, and this has the potential for mass surveillance if governments or other organizations misuse the technology for initiatives other than disaster response. Weak data anonymization could expose sensitive details such as where households live and how they move. In practical terms, SAR data usage for disaster response must be governed with anti-hacking protocols, which is the focus of this time series data exploration. Encryption techniques can guard sensitive information from unauthorized access, while privacy-preserving models—such as federated learning—enable AI systems to use data without ever seeing the raw details.

AI Bias and Ethical Decision-Making: Disparities in biases in datasets may lead to certain geographic regions being favored more than others by AI models trained on biased datasets, with consequences for future disaster preparedness and response. If SAR-based AI models are trained more on urban environments, they will not predict disaster impact in rural or underdeveloped areas. Moreover, AI-powered resource allocation could favor urban centers at the expense of marginalized communities. Another cause of algorithmic discrimination could be historical bias present in the data on which the algorithms were trained, thereby potentially leaving some demographic groups out of an adequate response for disaster relief. To tackle these challenges, we must train AI models on more diverse and well-rounded datasets, incorporate human oversight in AI decision-making processes, and regularly audit AI for fairness using explainable AI techniques.

6.7.4 Future Directions in Artificial Intelligence and Synthetic Aperture Radar Research for Disaster Resilience

Artificial intelligence has emerged as a groundbreaking advancement in recent years that revolutionized disaster resilience through its integration with SAR, equipping organizations with unique capabilities like real-time monitoring, predictive analysis, and significant response mechanisms. With the evolution of technology, upcoming research should focus on tackling the mentioned challenges, improving AI algorithms, and increasing the applicability of SAR data for better management of disasters. Significant improvements have recently been made in SAR data analysis due to gains in AI algorithms. Moreover, future AI-SAR will utilize self-learning AI models (e.g., subject directly use real disaster data to adapt continuously) with RL and federated learning. As AI models take more roles in disaster response AI systems, it will become crucial to increase their transparency and interpretability for Explainable AI techniques to be able to guarantee trust from policymakers, emergency responders. The incorporation of AI at the edge will allow for on-device SAR data analysis performed on drones, satellites, and IoT sensors, which will not only lessen the need for cloud processing but also quicken the response time in remote disaster-affected regions. The next generation of SAR technologies will enhance even more upon real-time disaster monitoring capabilities. The study will highlight multi-frequency and multi-polarization SAR systems using various frequency bands (X-band, L-band, C-band) for precise detection. The AI-enabled SAR constellations—networks of small satellites, each outfitted with an AI-driven sensor—will allow for constant global coverage and autonomous target selection in a world of disasters. This will also integrate SAR with hyperspectral imaging, allowing precise exploration with material composition, which will further enhance damage evaluation for infrastructures in events such as earthquakes or floods. Future studies will also focus on the integration of AI-SAR with IoT and smart infrastructure for disaster preparedness and response enhancement. Intelligent good disaster management systems based on AI will process the SAR data from ground-based sensors, UAVs, and satellites to provide real-time alerts for floods, seismic events, and vulnerability in the structures. By creating a virtual simulation of real-world disaster scenarios, predictive models of infrastructure failures, population movements, and optimal evacuation routes can be executed with a digital twin technology to allow authorities to proactively make strategic plans to respond to disasters. As climate change leads to more frequent and severe natural disasters, AI-SAR will play an essential role in climate resilience and humanitarian assistance. AI-controlled SAR systems will discover transitions caused by phenomena including sea level rise, permafrost thaw (melting of glacial ice), and drought patterns, and help in long-term assessments of climate change impacts. However, humanitarian agencies will leverage AI-SAR to identify the most affected areas, as well as deploy SAR drones to the most affected terrain for damage assessment and delivery of medicine to remote places. And finally, it will be critical for future AI-SAR research to address ethical, privacy, and policy challenges. The AI governance frameworks that prioritize guidance on ethical use will guarantee equitable and unbiased predictions of disasters and ultimately ensure vulnerable populations are prioritized. To counter the negative impact of sensitive geospatial data, privacy-preserving AI techniques like differential privacy, secure multi-party computation will come in hand. Furthermore, global consensus and collaboration are essential to develop and implement worldwide regulations and standardized protocols for AI-SAR data sharing around the world to enable integrated and efficient disaster response initiatives worldwide. Promoting AI algorithms, enhancing SAR capabilities, integrating IoT-based smart infrastructure, and addressing ethical nuances will contribute toward improving the disaster resilience, effective response, and equitable disaster response practices globally. The AI-powered algorithms can analyze SAR data to identify changes in land use, damage patterns, and environmental challenges caused by natural disasters, leading to more efficient disaster preparedness, response, and recovery efforts. Together, the combination of AI-powered insights with SAR imagery is becoming not only a powerful motion but a critical desegregated way of helping to overcome the calling of natural disasters. Generating high-volume recorded data, SAR can penetrate clouds, be used in various weather conditions, and be exploited as an important tool for monitoring and assessing damage from various disasters. The analysis of SAR data by the AI model greatly improves the ability to identify and analyze image data for potential disaster manifestation and prediction. Algorithms based on AI enhance data preprocessing, extraction of features, and automated change detection, assisting real-time decision-making. This showcases the potential of AI-SAR integration for EWS, flood management, earthquake impact assessment, and even predicting areas prone to wildfire. With the advancement of AI and SAR technologies, there would be increased automation in the disaster management process, and real-time analytics would allow for proactive measures to combat future disasters. Advanced AI models capable of self-learning can be implemented for adaptive and timely disaster prediction, real-time event mapping, and AI-powered SAR satellite constellations can be deployed for continuous monitoring. To conclude, AI-SAR will significantly contribute to climate change adaptation through environmental monitoring, predicting extreme weather patterns, and efficient allocation of humanitarian assistance to affected areas. However, to utilize the potential of AI-SAR, it is essential to embrace ethical AI governance, enhance global collaborations, and increase investment in AI-SAR research. As technology advances, AI-SAR will remain at the forefront of reimagining disaster resilience, more effectively safeguarding lives and communities from future catastrophes.

AI disclosure statement: The author acknowledges the use of AI-based tools for minor language editing or grammatical correction. However, all intellectual contributions, analyses, and conclusions presented in this chapter are entirely the author's own.

References

1. Al Shafian, S. and Hu, D., Integrating Machine Learning and Remote Sensing in Disaster Management: A Decadal Review of Post-Disaster Building Damage Assessment. *Buildings*, 14, 8, 2344, 2024, doi: 10.3390/buildings14082344.
2. Linardos, V., Drakaki, M., Tzionas, P., Karnavas, Y.L., Machine Learning in Disaster Management: Recent Developments in Methods and Applications. *Mach. Learn. Knowl. Extr.*, 4, 2, 446–473, 2022, doi: 10.3390/make4020020.
3. Chinmayi, H.K., Flynn, K.C., Ashworth, A.J., Advancements in Remote Sensing Techniques for Earthquake Engineering: A Review. *Earthquake Res. Adv.*, 2024, 100352, 2024.
4. Yang, S., Huang, Q., Yu, M., Advancements in remote sensing for active fire detection: A review of datasets and methods. *Sci. Total Environ.*, 31, 173273, 2024.
5. Durlik, I., Miller, T., Kostecka, E., Tuński, T., Artificial Intelligence in Maritime Transportation: A Comprehensive Review of Safety and Risk Management Applications. *Appl. Sci.*, 14, 18, 8420, 2024, doi: 10.3390/app14188420.
6. Tan, M.C.J., Building Resilience and Community-Based Disaster Risk Management (CBDRM): Experiences and Lessons from Communities in the Philippines, in: *International Handbook of Disaster Research* [Online], pp. 845–861, 2023, doi: 10.1007/978-981-19-8388-7_52.
7. Perera, D., Agnihotri, J., Djalante, R., Siedou, O., Identifying societal challenges in flood early warning systems. *Int. J. Disaster Risk Reduct.*, 51, 101794, 2020.
8. LAM Man-Ho, A., Re-Envisioning the Learning Landscape: Equipping Our Students with Uncertainty Preparedness for the Future World of Crises. *Asia Pac. J. Educ.*, 43, 3, 692–705, 2023, doi: 10.1080/02188791.2023.2231643.
9. Tian, X., Liao, H., Pcng, X., A Bio-Objective Location-Inventory Model for Pre-Disaster Humanitarian Logistics, in: *Proc. 37th Chinese Control Conf. (CCC)*, Wuhan, China, pp. 2156–2160, 2018, doi: 10.23919/ChiCC.2018.8483870.
10. Cormier, S., Wargo, M., Winslow, W., Transforming Health Care Coalitions From Hospitals to Whole of Community: Lessons Learned From Two Large Health Care Organizations. *Disaster Med. Public Health Prep.*, 9, 6, 712–716, 2015, doi: 10.1017/dmp.2015.146.
11. Aydinoglu, A.C. and Bilgin, M.S., Developing an open geographic data model and analysis tools for disaster management: landslide case. *Nat. Hazards Earth Syst. Sci.*, 15, 335–347, 2015, doi: 10.5194/nhess-15-335-2015.
12. Sharpe, J., Learning to Be Practical: A Guided Learning Approach to Transform Student Community Resilience When Faced with Natural Hazard Threats, in: *Observing the Volcano World*, C.J. Fearnley, D.K. Bird, K. Haynes, W.J. McGuire, G. Jolly (Eds.), Springer, Cham, Advances in Volcanology, 2017, doi: 10.1007/11157_2017_1.

13. Haynes, K., McGuire, W.J., Jolly, G. (Eds.), *Observing the Volcano World*, Springer, Cham, Advances in Volcanology, 2017, doi: 10.1007/11157_2017_1.
14. Bier, V., Hurricane Katrina as a Bureaucratic Nightmare, in: *On Risk and Disaster: Lessons from Hurricane Katrina*, R. Daniels, D. Kettl, H. Kunreuther (Eds.), pp. 243–254, University of Pennsylvania Press, Philadelphia, PA, 2006, doi: 10.9783/9780812205473.243.
15. Sari, D.P. and Rosadi, D., Exploring the Discretization of Peak Ground Acceleration Variables: a Comparative Analysis of K-means and K-medoids. *Philipp. J. Sci.*, 153, 5, 2024.
16. Nishijima, S., *et al.*, Superconductivity and the environment: A roadmap. *Supercond. Sci. Technol.*, 26, 11, 113001, 2013.
17. Gilmore, T.J., Duong, K., Travis, T., Transforming Collection Development through Three Rounds of Emergency Funding. *Collect. Manage.*, 48, 3, 178–191, 2023.
18. Gong, H. and Yang, X., A two-stage stochastic programming for the integrated emergency mobility facility allocation and road network design under uncertainty. *Networks Spatial Econ.*, 1–42, 2024.
19. Richards, J.A., *Remote Sensing with Imaging Radar*, vol. 1, pp. 172–173, Springer, Berlin/Heidelberg, Germany, 2009.
20. Sommervold, O., Gazzea, M., Arghandeh, R., A survey on SAR and optical satellite image registration. *Remote Sens.*, 15, 3, 850, 2023, doi: 10.3390/rs15030850.
21. Plank, S., Rapid damage assessment by means of multi-temporal SAR — A comprehensive review and outlook to Sentinel-1. *Remote Sens.*, 6, 6, 4870–4906, 2014, doi: 10.3390/rs6064870.
22. Gu, Z. and Zeng, M., The use of artificial intelligence and satellite remote sensing in land cover change detection: Review and perspectives. *Sustainability*, 16, 1, 274, 2024, doi: 10.3390/su16010274.
23. Choi, H. and Jeong, J., Speckle noise reduction technique for SAR images using statistical characteristics of speckle noise and discrete wavelet transform. *Remote Sens.*, 11, 10, 1184, 2019, doi: 10.3390/rs11101184.
24. Zhu, L., Suomalainen, J., Liu, J., Hyyppä, J., Kaartinen, H., Haggren, H., A review: Remote sensing sensors, in: *Multi-purposeful Application of Geospatial Data*, vol. 19, pp. 19–42, 2018.
25. Mirmazloumi, S.M., *et al.*, Supervised machine learning algorithms for ground motion time series classification from InSAR data. *Remote Sens.*, 14, 15, 3821, 2022, doi: 10.3390/rs14153821.
26. Grewal, R., Singh Kasana, S., Kasana, G., Machine Learning and Deep Learning Techniques for Spectral Spatial Classification of Hyperspectral Images: A Comprehensive Survey. *Electronics*, 12, 3, 488, 2023, https://doi.org/10.3390/electronics12030488.
27. Chintalapati, B., Precht, A., Hanra, S., Laufer, R., Liwicki, M., Eickhoff, J., Opportunities and challenges of on-board AI-based image recognition for small satellite Earth observation missions. *Adv. Space Res.*, 2024.

28. Shen, G., Fu, W., Guo, H., Liao, J., Water Body Mapping Using Long Time Series Sentinel-1 SAR Data in Poyang Lake. *Water*, 14, 12, 1902, 2022, https://doi.org/10.3390/w14121902.
29. Sorboni, N.G., Flood Hazard and Vulnerability Mapping using Deep Learning and Earth Observation Data, Ph.D. dissertation, Univ. Western Ontario (Canada), 2024.
30. Uddin, K., Matin, M.A., Meyer, F.J., Operational Flood Mapping Using Multi-Temporal Sentinel-1 SAR Images: A Case Study from Bangladesh. *Remote Sens.*, 11, 13, 1581, 2019, https://doi.org/10.3390/rs11131581.
31. Park, S.-E. and Jung, Y.T., Detection of Earthquake-Induced Building Damages Using Polarimetric SAR Data. *Remote Sens.*, 12, 1, 137, 2020, https://doi.org/10.3390/rs12010137.
32. Tomaszewski, B., Judex, M., Szarzynski, J., Radestock, C., Wirkus, L., Geographic information systems for disaster response: A review. *J. Homel. Secur. Emerg. Manag.*, 12, 3, 571–602, 2015.
33. Calle Müller, C., Lagos, L., Elzomor, M., Leveraging Disruptive Technologies for Faster and More Efficient Disaster Response Management. *Sustainability*, 16, 23, 10730, 2024, https://doi.org/10.3390/su162310730.
34. Kim, M., Park, S.E., Lee, S.J., Detection of Damaged Buildings Using Temporal SAR Data with Different Observation Modes. *Remote Sens.*, 15, 2, 308, 2023. https://doi.org/10.3390/rs15020308.
35. Dike, E.C., Oyetunji, A.K., Amaechi, C.V., Shoreline Delineation from Synthetic Aperture Radar (SAR) Imagery for High and Low Tidal States in Data-Deficient Niger Delta Region. *J. Mar. Sci. Eng.*, 11, 8, 1528, 2023. https://doi.org/10.3390/jmse11081528.

7

Interferometric Synthetic Aperture Radar-Based Assessment of Snow Water Equivalent Trends in the Dry Snow Zone of Dhundi, Himachal Pradesh

Dilsa Nasar and Geetha Priya M.*

Centre for Incubation, Innovation, Research and Consultancy (CIIRC), Jyothy Institute of Technology, Bengaluru, India

Abstract

This study investigates the spatial and temporal variability of snow water equivalent (SWE) in the Dhundi region, located within the Chandra basin of the Western Himalayas, using Sentinel-1 data and the interferometric synthetic aperture radar (InSAR) technique validated with field measurements. The SWE values obtained from snow pits during fieldwork in January and February of 2024 were compared with estimates obtained from InSAR. Dry snow patches were identified by mapping snow cover using Sentinel-2 data. Results showed a strong correlation (R^2 = 0.85) between InSAR-derived SWE changes and ground truth data, with a mean bias error (MBE) of 0.93 cm, a root mean square error of 0.95 cm, and a relative uncertainty of 13.44%. The study emphasizes the importance of microtopography, weather, and snowpack properties in SWE variability. This study highlights the integration of remote sensing and field techniques for furthering snow hydrology research and validates the efficacy of InSAR in estimating SWE in snow-fed basins such as the Chandra Basin.

***Keywords*:** Sentinel-1, dry snow, Himalayas, snow water equivalent, InSAR

**Corresponding author*: geetha.sri82@gmail.com

Pooja Joshi, Ashutosh Bhatt, Neelam Sharma and Kapil Joshi (eds.) Synthetic Aperture Radar for Disaster Management: Bridging Science and Solutions, (181–204) © 2026 Scrivener Publishing LLC

7.1 Introduction

Snowfall influences freshwater availability year-round, making it vital to the Earth's hydrological and climatic processes [1]. Mountains and high-altitude areas get lots of snow in winter. It will start melting in spring and summer, keeping rivers flowing. Rivers provide food for farming, people, and animals. One of the best ways to figure out how much seasonal snow adds to the water cycle is to look at the snow water equivalent (SWE). It estimates how much liquid water is stored in the snow pack [2]. The accurate estimation of SWE is essential for water resource management, climate studies, flood forecasting, and hydroelectric power generation, particularly in regions where snow melt is a dominant source of freshwater. Cependant, getting accurate and spatially continuous SWE data is still difficult, especially in remote and rough areas like the Himalayas [3].

The Western Himalayas are one of the most important snowy areas in the world. They are a key source of water for major river systems that support millions of people [4]. Because of changes in the climate, the snowpack in this area changes temporally. This makes it important to monitor the changes in SWE to estimate how much water is available and what effects it might have on those living downstream [5]. The melting snow from the Himalayas has a big impact on the hydrology of rivers like the Indus, Ganges, and Brahmaputra, changing the timing and size of streamflows [6]. The catchments in high-elevation areas have varying temperatures, precipitation, and seasonal snow. This makes the processes of snow accumulation and ablation quite complicated, and changes over time and space [7]. To predict long-term water resource availability and mitigate rapid snow melt-induced risks, SWE dynamics in high-altitude zones must be understood [8].

Ground-based observations and empirical models are the standard approaches for estimating SWE. The utilization of snow pits, snow pillows, and coring devices for measuring SWE in the field provides reliable estimates in certain areas; however, these techniques are demanding of labor and impractical for extensive monitoring [9]. Empirical and physically based models attempt to ascertain SWE from meteorological data such as temperature, precipitation, and wind speed; however, their accuracy is significantly influenced by the precision of the input data [10]. Optical, passive microwave, and active radar sensors are frequently utilized to remotely assess snow characteristics. Remote sensing methods have emerged as an effective alternative for monitoring extensive snow-covered regions [11]. People often use optical sensors, such as MODIS and Sentinel-2, to map snow-covered areas. However, these sensors have issues in estimating SWE because they only work on light that reflects off the surface and cannot penetrate through snow layers [12]. Both AMSR E and SSM/I are examples of passive microwave sensors that use microwave emissions from the snow pack to estimate SWE. Their low spatial resolution makes them less useful in complicated terrains [13].

The synthetic aperture radar (SAR) is one of the most important remote sensing techniques for monitoring snow; it can penetrate clouds, work in all weathers, and give high-resolution images. The interferometric synthetic aperture radar (InSAR) uses the phase differences between two SAR images taken at different periods to find changes in snowpack characteristics and surface movement. People have utilized InSAR a lot to find minor changes in elevation, glacier movement, and ground deformation. In recent years, it has gotten a lot of attention for estimating SWE [14]. The basic idea behind InSAR-based SWE estimation is that radar waves slow down when they go through dry snow. The amount of time they take to do this depends on the depth, density, and moisture content of the snow. You can use the phase shift to measure how SWE changes over time by comparing two SAR acquisitions performed before and after a snow event [15].

When estimating SWE using InSAR, using Sentinel-1 SAR data has many benefits. Sentinel-1 can capture high-resolution images in all-day, all-weather conditions due to its operation in the C-band. Operating in the C-band, Sentinel-1 gathers very good resolution data regardless of the weather. Interferograms generated using single look complex (SLC) data show the phase changes that happen when snow falls and melts [16]. The process of producing SWE estimates involves correlating SAR images, producing interferograms, unwrapping phase data, and applying conversion models. For areas where more traditional methods fall short, the InSAR, having millimeter-level accuracy in detecting changes in surface elevation, makes it a useful tool for SWE assessment [17].

Significant progress has been made in the use of InSAR for SWE estimation since recent studies have shown that it can provide accurate and widely distributed monitoring. Changes in SWE have been effectively connected to phase delay in SAR signals caused by snow accumulation, and field validation has shown that InSAR-derived estimations are reliable [20]. Combining radar remote sensing with ground-based data gives us a better picture of the snowpack features. This fills the void between localized field observations and large-scale satellite-based monitoring. Calibration and validation are necessary to ensure accuracy because variables, including snow quantity, stratification, and surface roughness, affect the sensitivity of SAR signals to variations in SWE [21].

In the Western Himalayas and other high-altitude places where field data are difficult to obtain due to difficult terrain and severe weather, InSAR-based SWE estimation is very useful. The Dhundi area in Himachal Pradesh is a great place to evaluate the potential of this method. An adequate water supply for Himachal Pradesh and adjacent states supports the Beas River. Accurate monitoring and estimation of SWE in Dhundi is essential for water resource management, avalanche forecasting, and assessing the effects of climate change on the area [19].

A significant problem in InSAR-based SWE estimation is temporal coherence decorrelation, which arises when surface properties change between SAR observations, therefore impacting the interferometric signal. The coherence decorrelation may occur due to snowfall, snow compaction, and surface metamorphism in snow-covered areas, necessitating the selection of SAR pairs with a small temporal baseline [22]. The penetration depth of the radar signal varies with snow cover conditions, which requires careful consideration of radar frequency and acquisition parameters [21, 23]. Advances in multi-temporal InSAR processing, machine learning-based SWE retrieval models, and data fusion techniques with optical and passive microwave sensors contribute to improving the accuracy of SWE estimation [21].

Understanding SWE dynamics in the Himalayas is critical for assessing the potential impacts of climate change on regional water resources. Rising temperatures and changing precipitation. It is essential to comprehend SWE dynamics in the Himalayas to evaluate how climate change may affect water resources in the region. Snowfall patterns have been impacted by changing precipitation patterns and rising temperatures. Some regions have seen a decrease in seasonal snow cover, while others have seen changes in when snow accumulates [24]. These changes have a direct effect on the amount of water available, the stability of glaciers, and the downstream flow of water. Sustainable water management techniques and climate change adaptation strategies can benefit from the capacity to track SWE using cutting-edge remote sensing technologies, which in turn shed light on the patterns in snowpack properties over the long term [25].

By utilizing high-resolution, all-weather, and all-day radar observations to evaluate snow accumulation and melting trends, this study estimates SWE in the Dhundi region using Sentinel-1 SAR data and InSAR algorithms. This study intends to understand the distribution of SWE and regional hydrology by combining field data with measurements taken from satellites. The results will give valuable outputs to make snow hydrology models better, flood forecasting systems better, and help people make choices about how to plan for water resources in the Western Himalayas.

7.2 Study Area

Positioned at 76°44'49"E and 31°59'32"N, the study location is in Dhundi, a scenic section of the Kullu district in Himachal Pradesh, India (Figure 7.1). The hydrology of the Western Himalayas is majorly impacted by diverse river systems and seasonal snow cover patterns in Dhundi, which make this region famous. As a result of this environment, it is better able to maintain its equilibrium, and the people who live downstream can make a living. The elevation of the study region ranges from 1,096 to 6,108 masl. Dhundi has a broad variety of climates with each of which affects the movement of snow. Snowfall at high altitudes adds to the snowpack, giving the area a natural freshwater reservoir. Dhundi is important for more than just its water supply. It is also home to the Snow and Avalanche Study Establishment, which is a major research center for the Defence Research and Development Organization. This shows how important it is to study snow and avalanches in the Himalayas. Dhundi is not only a place of scientific significance, but it is also a gateway to the natural and cultural wealth of the region. It is surrounded by natural icons like the peaceful Beas Kund, Hanuman Tibba,

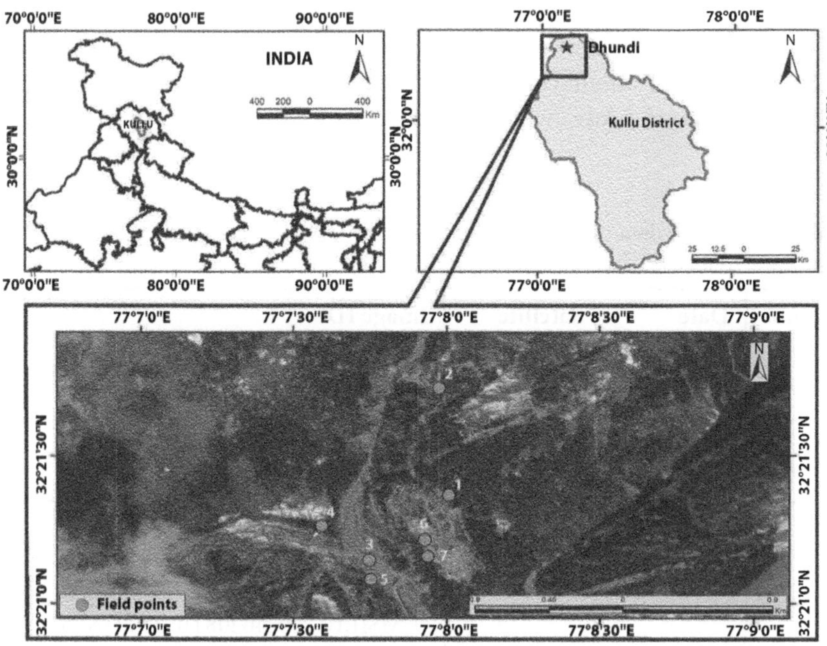

Figure 7.1 Study area map of Dhundi, Kullu District. The map highlights the selected field sample point sites (1–7) for the study.

and the Manali Leh Tunnel, which is an engineering marvel. This study takes place in a place with distinct landforms and weather patterns that make it perfect for studying SWE to learn more about how snow changes and what that means for managing water resources in the region.

7.3 Data Used

The present study estimated SWE changes in the Dhundi region of the Chandra Basin in the Western Himalayas. We employed satellite remote sensing datasets and field measurements taken from the ground. Sentinel-1 and Sentinel-2 satellites were mostly used for remote sensing analysis [26, 27], and snow pits were used to check the results. The Sentinel-1 data in IW mode has a spatial resolution of 10 m and a 250 km swath width. The data, which were collected between January 2024 and February 25, 2024, were processed using InSAR techniques for the estimation of SWE. The Sentinel-1 mission is part of the Copernicus program, and it delivers dual-polarized C-band SAR data with consistent temporal coverage, which is perfect for cryosphere applications. On February 8, the optical Sentinel-2 data, which are also part of the Copernicus program, were obtained as Level-1 C products. These multispectral data have a resolution of 10 m in visible and near infrared bands with a swath width of 290 km [28]. In addition, a digital elevation model (DEM) with a spatial resolution of 12.5 m from the Advanced Land Observing Satellite PALSAR was used

Table 7.1 List of data used for the study, including Sentinel-1 and Sentinel-2 datasets.

SI no.	Date	Satellite	Image ID
1	20-01-2024	Sentinel-1A	S1AA_20240120T005142_VVP036_INT80_G_weF_4D7
2	25-02-2024	Sentinel-1A	S1AA_20240225T005141_VVP036_INT80_G_weF_4D7
3	08-02-2025	Sentinel-2B	S2B_MSIL1C_20240208T053009_N0510_R105_T43SFR_20240208T071950.SAFE & S2B_MSIL1C_20240208T053009_N0510_R105_T43SGR_20240208T071950.SAFE

for terrain correction in InSAR processing and to analyze the influence of microtopography. The data used for the study are given in Table 7.1. On January 21, 2024, and February 25, 2024, snow pits were used to collect ground-based field data. We adopted SWE and snow depth measurements at seven-point sites in the study area. These observations taken in the field were very important for checking the accuracy of the satellite-derived SWE estimations and helped with uncertainty analysis.

7.4 Methodology

7.4.1 Sentinel-1 Data Processing Using Interferometry

Sentinel-1 SAR data were processed to estimate SWE changes in the Dhundi region using interferometric techniques [29]. Initially, Sentinel-1 SLC images were co-registered to align multiple acquisitions pixel-by-pixel, ensuring spatial consistency [30]. Following co-registration, the SLC images were stacked to reduce noise and enhance signal coherence, creating a robust dataset for interferometric analysis [31]. Interferograms were generated by calculating the phase differences between two temporally spaced radar acquisitions, capturing surface changes related to snowpack dynamics [32]. During this process, a coherence image was derived to evaluate the quality of the interferometric phase, with a threshold applied to retain high-quality data [33]. Topographic phase correction was performed using a high-resolution DEM to remove terrain-induced phase contributions [22]. The corrected interferograms were processed using Goldstein phase filtering to suppress noise while preserving signal integrity [34], followed by phase unwrapping to resolve ambiguities and retrieve '$\Delta\phi$' representing changes in snowpack properties such as accumulation or compaction [35]. Multi-looking was applied to improve the signal-to-noise ratio, and geocoding ensured accurate spatial alignment with geographical coordinates [36]. These steps enabled the precise acquisition of $\Delta\phi$, which served as the foundation for SWE estimation [29].

7.4.2 Sentinel-2 Data Processing and Snow Water Equivalent Estimation

In addition to the interferometric analysis, Sentinel-2 MSI data were utilized to classify snow cover and dry snow patches. We employed the normalized difference snow index (NDSI) with a threshold of > 0.4 to delineate snow-covered regions and eliminate non-snow pixels. We utilized near-infrared (NIR) band reflectance with a threshold of 0.5 to delineate dry snow patches and identify locations suitable for measuring SWE [27, 37]. These guidelines ensured accurate classification and minimized the presence of mixed or wet snow. We used the dry snow cover area from the combination of NDSI and NIR classifications and Sentinel-1 to figure out how SWE changed. The dry snow cover area showed how far the snowpack changed, while the physical changes in the snowpack showed how SWE varied. This method uses Sentinel-1 SAR's ability to observe through structures and Sentinel-2 MSI's

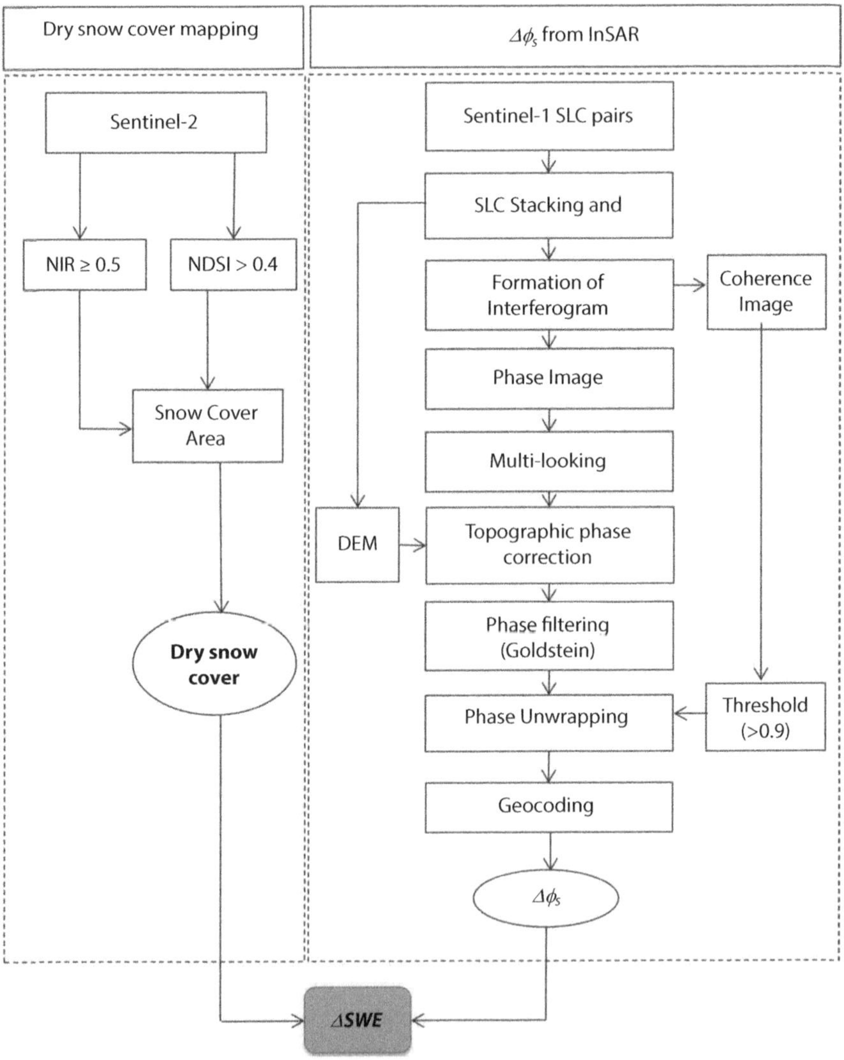

Figure 7.2 Process flow of the methodology used in the study. The figure illustrates the sequence of steps followed for data processing and analysis.

ability to observe through snow to sort it into different types [29]. By combining these datasets, the workflow offers a comprehensive workflow for SWE estimation, emphasizing the importance of early co-registration and stacking for ensuring high-quality interferometric analysis in the snow-fed Dhundi region [32]. The complete methodology is explained in Figure 7.2.

The NDSI is defined as:

$$NDSI = \frac{Green\ Band\ Reflectance - SWIR\ Band\ Reflectance}{Green\ Band\ Reflectance + SWIR\ Band\ Reflectance} \tag{7.1}$$

7.4.3 Estimation of Snow Water Equivalent from the Interferometric Synthetic Aperture Radar Phase

The Interferogram is generated by the complex conjugate product of the images, with the amplitude multiplied and the phase difference between the two images representing the interferometric phase. The computed interferogram exhibits phase variations '$\Delta\phi$' influenced by several factors, such as the flat-earth phase $\Delta\phi_{flat}$, noise $\Delta\phi_{noise}$, delay caused by the atmosphere $\Delta\phi_{atm}$, topographic phase $\Delta\phi_{topo}$ and phase shift due to dry snow $\Delta\phi_s$ [20] as given by Equation 7.2.

$$\Delta\phi = \Delta\phi_{flat} + \Delta\phi_{noise} + \Delta\phi_{atm} + \Delta\phi_{topo} + \Delta\phi_s \tag{7.2}$$

These contributions can be identified, removed, or mitigated by eliminating the flat-earth phase and applying noise-filtering techniques like the Goldstein filter. The parameter $\Delta\phi_s$ is the residual phase, which is associated with the bidirectional propagation of microwaves within the snow layer. By separating all other factors, the interferometric phase resulting from the microwave signals in the snow layer can be used to estimate temporal variations in SWE [20]. This approach uses the radar receiver's sensitivity to changes in the dielectric properties of snowpack, which are directly related to SWE. By analyzing the phase shift over time, the temporal variations in SWE were quantified by providing valuable insights into snowpack dynamics and hydrological processes [23]. The relationship between snow depth and the measured interferometric phase [15] is given by Equation 7.3:

$$\Delta\phi_s = \frac{4\pi}{\lambda}\left(cos\ \theta - \sqrt{\varepsilon - sin^2(\theta)}\right)\Delta d_s \tag{7.3}$$

Where, 'λ' is the wavelength of the radar signal at the angle of incidence θ, and ε is the permittivity of snow and Δd_s is the change in snow depth.

The permittivity of snow ε is a critical parameter that reflects the electromagnetic properties of snow influenced by its density, temperature, and water content. These factors determine the ability of snow to transmit, absorb, and reflect electromagnetic waves, making permittivity a key variable in remote sensing studies. SWE is defined by Equation 7.4:

$$SWE = \frac{\rho}{\rho_w} * d_s \tag{7.4}$$

where ρ is the density of snow, and ρ_w is the density of water. The snow permittivity ε as a function of snow density (ρ) is described using approximations proposed by Mätzler 1987. For $\rho < 0.4$ g/cm^3 and $\rho \geq 0.4$ g/cm^3, Mätzler's model assumes a linear relationship, expressed as Equations 7.5 and 7.6, respectively.

$$\varepsilon(\rho) = 1 + 1.5995\rho + 1.861\rho^3 \tag{7.5}$$

$$\varepsilon(\rho) = \left(1 - \frac{\rho}{0.917}\right) + \left(1.4759 * \frac{\rho}{0.917}\right) \tag{7.6}$$

Combining Equations 7.3, 7.4, 7.5, and 7.6, a relationship between the temporal variation of SWE and the interferometric phase can be derived as given by Equation 7.7.

$$\Delta SWE = SWE(t_2) - SWE(t_1) = \frac{\lambda}{2\pi} \frac{1}{\gamma(1.6 + \theta^{2.5})} * \Delta\phi_s \tag{7.7}$$

The acquisition times of the two SAR images, denoted by t_1 and t_2, respectively, are considered. Additionally, γ is a parameter that is approximately equal to 1 and is held by the angle of incidence θ [20]. The local incidence angle is derived from DEM. Sentinel-1 SAR, with a wavelength of $\lambda = 5.5cm$ and a mean incidence angle of $\theta = 43°$, we assume $\gamma \approx 1$. So, Equation 7.7 can be written as Equation 7.8:

$$\Delta SWE = SWE(t_2) - SWE(t_1) = 0.0041 * \Delta\phi_s \tag{7.8}$$

Equation 7.8 indicates that phase variation $\Delta\phi_s$ typically varies within the range of $[-\pi, \pi]$ for wrapped phase data. The use of the unwrapped phase in the methodology eliminates phase ambiguities. Unwrapping ensures that phase shifts beyond $[-\pi, \pi]$ range are captured accurately, which is essential for estimating temporal changes in SWE over larger spatial and temporal scales where significant variations in snowpack depth occur [38]. Also, the local incidence angle (θ) is very important for changing the phase response because it affects the dielectric characteristics of the snow and the link between phase change and snow depth [23].

7.5 Results and Discussion

The estimation of SWE is critical for hydrological and climate modeling, particularly in regions like the Dhundi area of Himachal Pradesh, which is known for its seasonal snow cover. This study integrates field-based SWE measurements with SAR interferometry techniques to evaluate the feasibility of remote sensing for SWE estimation. The results and discussion focus on comparing field-measured and InSAR-derived SWE values, assessing spatial and temporal variability, and evaluating the accuracy and reliability of the remote sensing techniques. Notably, these results align with prior studies emphasizing the utility of SAR in snow monitoring [38, 39]. The work also explores uncertainties associated with both field measurements and remote sensing methodologies, offering insights into their complementary roles in SWE monitoring.

7.5.1 Overview of Field Measurements

Field measurements were conducted on January 21, 2024, and February 25, 2024, during the peak winter snow accumulation and the onset of the melting season. Snow pits were dug at seven locations, carefully chosen to represent varying snowpack conditions in the region, as shown in Figure 7.3. At each site, the SWE was calculated using density and depth measurements (Figure 7.4) obtained from standardized snow sampling techniques [40]. During the study period, the region was covered by fresh snow at an approximate depth of 29 cm, and snow samples were collected to calculate snow water equivalent. As snow depth increases, the SWE was also observed to rise, indicating higher water content and greater density within the snowpack [24]. The change in SWE (ΔSWE) between the two field campaigns was used to assess temporal variability and validate InSAR-derived estimates.

Figure 7.3 Digging of the snow pit for field measurements. The figure shows the process of digging the snow pit using a shovel to estimate the SWE and snow depth.

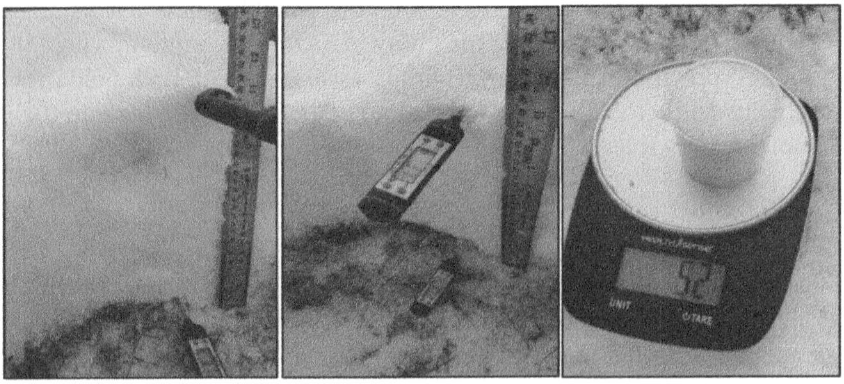

(a) (b) (c)

Figure 7.4 Fieldwork snapshots of snow measurements. (a) and (b) show snow depth measurement, and (c) illustrates snow mass calculation for density estimation.

The results showed that the ΔSWE values measured in the field ranged from 6.45 cm to 7.92 cm across the seven sites (Table 7.2). This variability reflects the heterogeneity of snowpack due to local factors such as topography, vegetation cover, and microclimatic influences [41]. For instance, sites with higher snow depth and density recorded greater ΔSWE, which aligns with expectations based on snow accumulation and ablation dynamics [42].

Table 7.2 Location information and calculated ΔSWE from field measurements. The table also includes InSAR-estimated ΔSWE for the corresponding locations.

Site. no.	Location	ΔSWE_{InSAR} (cm)	ΔSWE_{field} (cm)	Absolute difference (cm)	Relative difference (%)
1	Lat:32°21'22.08"N Lon:77°15'05.04"E	5.62	6.57	0.95	14.45
2	Lat:32°21'43.83"N Lon:77°07'58.47"E	6.93	7.92	0.99	12.5
3	Lat:32°21'08.90"N Lon:77°07'44.73"E	5.90	6.96	1.06	15.22
4	Lat:32°21'15.81"N Lon:77°07'35.49"E	5.41	6.45	1.04	16.12
5	Lat:32°21'04.93"N Lon:77°07'45.03"E	5.73	6.63	0.9	
6	Lat:32°21'12.97"N Lon:77°07'55.62"E	6.15	7.24	1.09	
7	Lat:32°21'09.60"N Lon:77°07'56.23"E	6.42	6.92	0.5	

7.5.2 Analysis of Interferometric Synthetic Aperture Radar-Derived Snow Water Equivalent Data

The application of Sentinel-1 SAR data involved processing two acquisitions, one from January and the other from February 2024, to derive InSAR coherence changes corresponding to snowpack variability. To enhance the accuracy of SWE estimation, snow-covered regions were identified using Sentinel-2 optical imagery by applying thresholds of the NDSI ≥ 0.4 and NIR ≥ 0.5 to isolate dry snow zones for analysis (Figure 7.5). This hybrid approach has demonstrated this improved spatial delineation of snow cover in previous studies [45]. The SWE change (ΔSWE) values obtained from InSAR ranged between 5.41cm and 6.93cm across the study sites with dry snow, as shown in Figure 7.5, which slightly underestimates the field measurements (Table 7.2). This aligns with findings by Leinss *et al.* [38], who reported a similar tendency for SAR to underrepresent SWE due to challenges in capturing sub-canopy snowpack variability and density effects.

The ΔSWE differences between field and InSAR estimates averaged approximately 0.93 cm, with site-specific variations potentially influenced by terrain slope, SAR penetration depth, and snowpack stratigraphy [15]. Correlation analysis yielded a strong relationship ($R^2 > 0.85$) between field

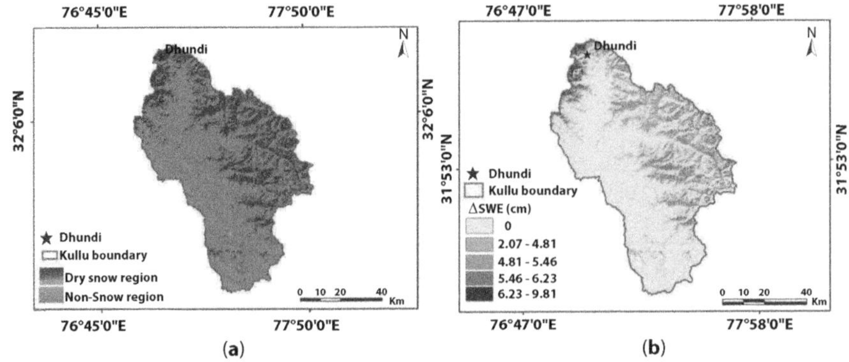

Figure 7.5 (a) Sentinel-2 derived dry snow region over Dhundi, Kullu District; (b) Sentinel-1 InSAR-derived ΔSWE for dry snow over Dhundi, Kullu District.

and InSAR data, demonstrating the reliability of InSAR for capturing snowpack changes under favorable conditions. However, uncertainties due to signal decorrelation, phase unwrapping errors, and atmospheric distortions were noted [43].

7.5.3 Percentage Change and Bias Analysis

The percentage change in SWE between the two methods was calculated to quantify discrepancies. The average relative difference was approximately 13.44%, with variations ranging from 12.5% (Site 2) to 16.12% (Site 4). This variability underscores the importance of incorporating site-specific factors, such as terrain slope, vegetation, and snowpack density, in SWE estimation models [40].

Bias analysis was performed using the Mean Bias Error (MBE) and Root Mean Square error (RMSE) Equations 7.9 and 7.10:

$$MBE = \frac{1}{n}\sum_{i=1}^{n}\left(\Delta SWE_{InSAR} - \Delta SWE_{field}\right) \tag{7.9}$$

$$RMSE = \sqrt{\frac{1}{n}\sum_{i=1}^{n}\left(\Delta SWE_{InSAR} - \Delta SWE_{field}\right)^2} \tag{7.10}$$

The MBE was 0.93 cm, reflecting an average underestimation by InSAR. The RMSE of 0.95 cm aligns with error margins reported by Hall and Riggs (2011) in similar alpine studies. These measurements highlight the systematic nature of deviations and the potential for refinement in SAR processing techniques.

7.6 Comparative Analysis of Field and Interferometric Synthetic Aperture Radar Results

Combining field-measured SWE values with those from InSAR gave us a lot of useful information about the pros and cons of using remote sensing to keep an eye on snow. Both datasets showed identical trends in SWE changes throughout the seven measurement sites, but there were disparities in the numbers and locations that showed how they may be used in different ways.

To assess the reliability and consistency of the remote sensing method, a correlation analysis was carried out between SWE values determined by InSAR and those measured in the field. There was a substantial positive correlation (0.85), which means that InSAR does a good job of showing the overall trend of SWE changes across sites, even though it tends to underestimate them (Figure 7.6). The relationship between the two datasets was modeled using a linear regression Equation 7.11:

$$y = a * x + b \tag{7.11}$$

Where y is ΔSWE_{InSAR}, a is the slope, x is ΔSWE_{field} and b is the intercept.

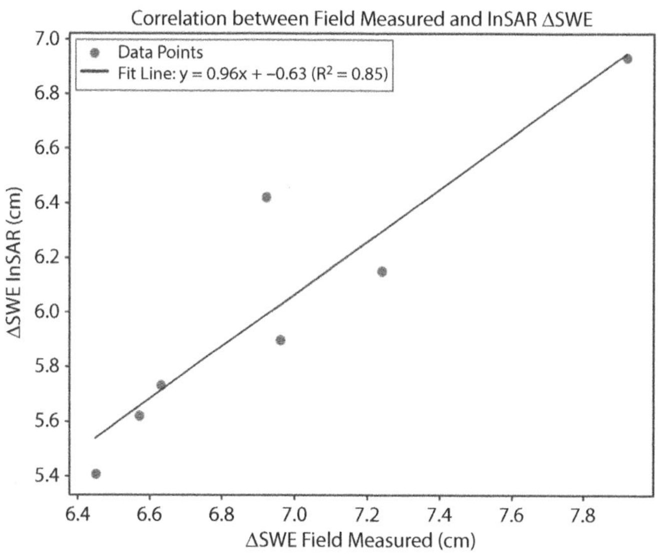

Figure 7.6 Scatter plot of $\Delta SWE_{\text{field}}$ vs. ΔSWE_{InSAR}. The plot includes the estimated regression equation for the relationship between field and InSAR-derived ΔSWE.

7.6.1 Quantitative Differences between Field and Interferometric Synthetic Aperture Radar Snow Water Equivalent

Field measurements of SWE changes (ΔSWE) were consistently higher than those derived from InSAR data across all sites. The ΔSWE values ranged from 6.45 cm to 7.92 cm for field data, while InSAR values ranged from 5.41 cm and 6.93 cm. The quantitative differences between the two datasets are summarized in Table 7.2. This discrepancy, averaging approximately 0.93 cm (or ~13.44% lower for InSAR), aligns with prior observations that SAR-based techniques can underestimate SWE due to factors such as limited penetration depth and signal attenuation in dense or wet snow conditions [38, 39].

These differences are primarily attributed to limitations in InSAR coherence analysis, particularly in densely vegetated or complex terrain, where phase unwrapping errors and atmospheric effects may reduce accuracy [43]. The correlation coefficient ($R^2 = 0.85$) between the two datasets indicates that InSAR can reliably capture SWE trends under suitable conditions. Equations 7.12 and 7.13 were used to calculate the absolute difference and relative difference between the field and InSAR-derived ΔSWE.

$$Absolute\ Difference = \left|\Delta SWE_{field} - \Delta SWE_{InSAR}\right| \tag{7.12}$$

$$Relative\ Difference = \left(\frac{Absolute\ difference}{\Delta SWE_{field}}\right) * 100 \tag{7.13}$$

7.6.2 Spatial Variability Across Measurement Sites

Spatial variability in SWE changes was evident across the seven measurement sites, reflecting the influence of microclimatic and topographic factors. Sites situated in protected locations or on roads, such as Sites 2 and 6, showed larger ΔSWE values because there was more snowfall and less wind redistribution [41]. On the other hand, sites with steeper slopes, like Sites 4 and 5, had lower ΔSWE values, probably because the snow eroded and sublimated more quickly. The SWE values from InSAR had a more even geographical distribution than the field observations. This could be because SAR backscatter signals average out over larger pixel regions [38]. This homogeneity can hide differences that are only present in some areas, especially when the snowpack is uneven. The patterns of snow distribution seen in this study are comparable to those seen in other studies of similar alpine areas, where topographical factors and localized meteorological events have a big effect on where snow falls [40, 44]. Also combining the Sentinel-2-derived NDSI and NIR thresholds helped to better define dry snow-covered areas, which lowered the chances of overestimating SWE in mixed land cover pixels [45].

7.7 Uncertainty Analysis

Field measurements of SWE inherently include errors due to instrumentation and procedural limitations. The uncertainty in snow depth (Δ_h) and snow density (Δ_ρ) measurements arises from the variability in probing or manual pit excavation techniques. For this study, Δ_h and Δ_ρ was assumed to be 0.1 cm and 0.05 g/cm^3 respectively to calculate the overall uncertainty E using Equation 7.14. The overall field SWE uncertainty (mean) was determined to be 0.36 cm. This value underscores the reliability of the field methods while acknowledging limitations such as operator bias and snowpack heterogeneity.

$$E = \sqrt{(\Delta_h)^2 + \left(\Delta_\rho * h\right)^2} \tag{7.14}$$

where 'h' represents the snow depth, directly proportional to the SWE in field measurements. Based on the field SWE values provided, the overall field SWE uncertainty (mean) was determined to be 0.36 cm. This value underscores the reliability of the field methods, while acknowledging limitations such as operator bias and snowpack heterogeneity.

7.7.1 Sources of Error in Interferometric Synthetic Aperture Radar Estimation

The InSAR-derived SWE estimations are affected by several factors:

1. Radar Penetration Depth: Assumptions about the depth to which the radar signal penetrates the snowpack depend on snow density and grain size, leading to potential errors [21].
2. Atmospheric Delays: Variations in atmospheric conditions can introduce phase errors, particularly in regions with dynamic weather patterns [46].

3. Snow Microstructure: Variability in snow grain size and wetness affects backscatter behavior, influencing SWE estimation accuracy.
4. Thresholding Assumptions: In this study, snow-covered regions were identified using thresholds of NDSI > 0.4 and NIR reflectance ≥ 0.5, assuming these parameters reliably isolate dry snow. These thresholds, while supported by literature, introduce some degree of classification uncertainty.

7.8 Comprehensive Analysis of Snow Water Equivalent Variability and Validation

This study comprehensively analyzed SWE variability and validated InSAR-derived estimates against field measurements in the Chandra Basin, highlighting the complex interplay of microtopography, meteorological conditions, and snowpack characteristics. Microtopography, including slope, aspect, and elevation, emerged as a critical factor in SWE distribution. Steeper slopes showed reduced SWE due to gravitational runoff and wind-driven redistribution, while north-facing slopes retained higher SWE, consistent with [47]. Meteorological factors, such as snowfall intensity, temperature fluctuations, and wind patterns, further shaped SWE variability. Temperature-driven freeze-thaw cycles contributed to heterogeneity, a phenomenon that was also observed [48].

Snow density and stratigraphy added another layer of complexity to SWE estimation. Dense snow layers, resulting from compaction and temperature gradients, influenced SWE variability consistently [49]. Stratigraphy, characterized by layers formed during snowfall events, contributed to localized variations in SWE, complicating both field and remote sensing measurements. Validation of InSAR-derived SWE against field data yielded an overall uncertainty (RMSE) of 0.95 cm and a relative uncertainty of 13.44%, aligning with findings by [38]. However, discrepancies were noted due to localized variations in snow properties and the use of thresholds for NDSI (≥0.4) and NIR reflectance (≥0.5), as highlighted by Nagler *et al.* [29].

The study shows how useful InSAR is for calculating SWE in high mountain regions, especially when combined with field observations. This combination makes SWE estimates more reliable and gives important information for managing water resources in basins that get their water from snow. But more research is needed on problems, including the effects of microtopography, categorization problems, and changes in the weather.

The results support what other research has found about SWE estimating and remote sensing methods, such as studies by Nagler *et al.* [29] and Matzler [23]. Also, look at the unique dynamics of the Chandra Basin, which is a less-studied area.

There are big effects on snow hydrology. Scalability and efficiency in data-scarce regions are offered by InSAR, making it a robust tool for monitoring SWE dynamics. Increasing the accuracy, however, still requires reducing categorization uncertainty and boosting spatial resolution. Future studies should look at combining long-term datasets with new technologies like UAV-based LiDAR and machine learning to get a better picture of how things change over time and at smaller scales.

7.9 Conclusion

The study demonstrates the ability of InSAR technology for calculating SWE across large remote regions, especially when combined with field observations. This combination makes SWE estimates more reliable and gives important information for managing water resources in basins from snowmelt and glaciermelt water. However, more research is needed to solve these problems, including the effects of microtopography, uncertainty in classification, and weather changes. The results address the unique dynamics of the understudied Chandra Basin and provide support to previous research on SWE estimation and remote sensing methods, such as those of Nagler *et al.* [29] and Matzler [23].

There are big effects on snow hydrology. With its scalability and efficiency, InSAR is a powerful instrument for tracking SWE dynamics even in areas with limited data. But enhancing accuracy still depends on improving spatial resolution and reducing uncertainty. Future studies must look at the ability to combine long-term datasets with new technologies like UAV-based LiDAR and machine learning to get a better picture of how things change over time and at smaller scales.

Acknowledgment

The authors express their sincere gratitude to the Director of CIIRC, Jyothy Institute of Technology, Bengaluru, and PTICL & TITCL, for their invaluable support. The authors sincerely acknowledge the support provided by the Border Roads Organisation (BRO) during the fieldwork conducted at Dhundi, Himachal Pradesh. Their assistance was instrumental in facilitating on-ground research activities, ensuring smooth logistical operations, and contributing to the successful execution of this study. The authors are grateful to the Copernicus program for providing them with the valuable dataset that was important to the success of this research.

AI disclosure statement: The author acknowledges the use of AI-based tools for minor language editing or grammatical correction. However, all intellectual contributions, analyses, and conclusions presented in this chapter are entirely the author's own.

Bibliography

1. Savoskul, O. and Smakhtin, V., Glacier Systems and Seasonal Snow Cover in Six Major Asian River Basins: Hydrological Role under Changing Climate Water, Land and Ecosystems, International Water Management Institute (IWMI), *(IWMI Research Report 150)*, p. 45, 2013.
2. Guyennon, N., Valt, M., Salerno, F., Petrangeli, A.B., Romano, E., Estimating the snow water equivalent from snow depth measurements in the Italian Alps. *Cold Reg. Sci. Technol.*, 167, 102859, 2019.
3. Singh, V.B. and Ramanathan, A., Hydrogeochemistry of the Chhota Shigri glacier meltwater, Chandra basin, Himachal Pradesh, India: solute acquisition processes, dissolved load and chemical weathering rates. *Environ. Earth Sci.*, 76, 5, 1–11, 2017.
4. Notarnicola, C., Observing Snow Cover and Water Resource Changes in the High Mountain Asia Region in Comparison with Global Mountain Trends over 2000–2018. *Remote Sens.*, 12, 23, 2020.
5. Tiwari, S., Kar, S., Bhatla, R., Interannual Variability of Snow Water Equivalent (SWE) over Western Himalayas. *Pure Appl. Geophys.*, 173, 2015.
6. Thapa, S., Li, H., Li, B., Fu, D.-L., Shi, X., Yabo, S., Lu, L., Qi, H., Zhang, W., Impact of climate change on snowmelt runoff in a Himalayan basin, Nepal. *Environ. Monit. Assess.*, 193, 2021.
7. Wang, R., Yao, Z., Liu, Z., Jiang, L., Wang, L., Snow cover variability and snowmelt in a high altitude ungauged catchment. *Hydrol. Process.*, 29, 2015.
8. Tewari, V., Verma, R., Gadow, K., Climate change effects in the Western Himalayan ecosystems of India: evidence and strategies. *For. Ecosyst.*, 4, 1–9, 2017.
9. Taheri, M. and Mohammadian, A., An Overview of Snow Water Equivalent: Methods, Challenges, and Future Outlook. *Sustainability*, 14, 18, 2022.
10. Aschauer, J., Michel, A., Jonas, T., Marty, C., An empirical model to calculate snow depth from daily snow water equivalent: SWE2HS 1.0. *Geosci. Model Dev.*, 16, 14, 4063–4081, 2023.

11. Schilling, S., Dietz, A., Kuenzer, C., Snow Water Equivalent Monitoring—A Review of Large-Scale Remote Sensing Applications. *Remote Sens.*, 16, 6, 2024.
12. Liu, Q. and Zhang, Y., Study on Snow Cover Change Based on The Fusion of Sentinel-2 and Modis Images. *ISPRS Ann. Photogramm. Remote Sens. Spat. Inf. Sci.*, V-3-2022, 333–338, 2022.
13. Murtaza, K.O. and Romshoo, S.A., Recent glacier changes in the Kashmir Alpine Himalayas, India. *Geocarto Int.*, 32, 2, 188–205, 2017.
14. Koskinen, J., Pulliainen, J., Hallikainen, M., The use of ERS SAR data for snow melt monitoring. *Geosci. Remote Sensing IEEE Trans.*, 35, 601–610, 1997.
15. Guneriussen, T., Hogda, K.A., Johnsen, H., Lauknes, I., InSAR for estimation of changes in snow water equivalent of dry snow. *IEEE Trans. Geosci. Remote Sens.*, 39, 10, 2101–2108, 2001.
16. Oveisgharan, S., Zinke, R., Hoppinen, Z., Marshall, H.P., Snow water equivalent retrieval over Idaho – Part 1: Using Sentinel-1 repeat-pass interferometry. *Cryosph.*, 18, 2, 559–574, 2024.
17. Irshad, F., Malik, J., Khalil, R.M.Z., Mapping wet snow using sar c-band through google earth engine, in: *6th Int. Conf. Aerosp. Sci. Eng. ICASE 2019*, pp. 1–5, 2019.
18. Negi, H., Kulkarni, A., Semwal, B., Estimation of snow cover distribution in Beas basin, Indian Himalaya using satellite data and ground measurements. *J. Earth Syst. Sci.*, 118, 525–538, 2009.
19. Varade, D., Maurya, A.K., Dikshit, O., Singh, G., Manickam, S., Snow depth in Dhundi: An estimate based on weighted bias corrected differential phase observations of dual polarimetric bi-temporal Sentinel-1 data. *Int. J. Remote Sens.*, 41, 3031–3053, 2019.
20. Conde, V., Mateus, P., Catalao, J., Gritsevich, M., On the Estimation of Temporal Changes of Snow Water Equivalent by Spaceborne SAR Interferometry: A New Application for the Sentinel-1 Mission. *J. Hydrol. Hydromech.*, 67, 93–100, 2019.
21. Tsai, Y.L.S., Dietz, A., Oppelt, N., Kuenzer, C., Remote sensing of snow cover using spaceborne SAR: A review. *Remote Sens.*, 11, 12, 2019.
22. Liu, H., Zhao, Z., Jezek, K.C., Correction of Positional Errors and Geometric Distortions in Topographic Maps and DEMs Using a Rigorous SAR Simulation Technique. *Photogramm. Eng. Remote Sens.*, 70, 9, 1031–1042, 2004.
23. Matzler, C., Microwave permittivity of dry snow. *IEEE Trans. Geosci. Remote Sens.*, 34, 2, 573–581, 1996.
24. Sturm, M., Taras, B., Liston, G.E., Derksen, C., Jonas, T., Lea, J., Estimating Snow Water Equivalent Using Snow Depth Data and Climate Classes. *J. Hydrometeorol.*, 11, 6, 1380–1394, 2010.
25. Mishra, S.K., Veselka, T.D., Prusevich, A.A., Grogan, D.S., Lammers, R.B., Rounce, D.R., Ali, S.H., Christian, M.H., Differential impact of climate

change on the hydropower economics of two river basins in high mountain Asia. *Front. Environ. Sci.*, 8, 26, 2020.

26. Neha, S., Geetha Priya, M., a R, D., Classification of Dry and Wet Snow of Glaciers in Western Himalayas Using SAR Polarimetric Decomposition—An Application of Satellite Image Processing, in: *Emerging Technologies in Electrical Engineering for Reliable Green Intelligence*, pp. 359–369, Springer, Singapore, 2024.
27. Nagajothi, V., Geetha Priya, M., Sharma, P., Krishnaveni, D., Classification of Dry/Wet Snow Using Sentinel-2 High Spatial Resolution Optical Data, in: *Intelligent Data Engineering and Analytics. Advances in Intelligent Systems and Computing*, pp. 1–9, Springer, Singapore, 2021.
28. Shashank, B. and Geetha Priya, M., Monitoring of Melting Glaciers of Ny-Ålesund, Svalbard, Arctic Using Space-Based Inputs. *Futur. Commun. Netw. Technol.*, 381–390, 2023.
29. Nagler, T., Rott, H., Ripper, E., Bippus, G., Hetzenecker, M., Advancements for snowmelt monitoring by means of Sentinel-1 SAR. *Remote Sens.*, 8, 4, 1–17, 2016.
30. Hanssen, R., Radar Interferometry Data Interpretation and Error Analysis, in: *PhD thesis, in Remote Sensing and Digital Image Processing*, Kluwer Academic Publisher, New York, 20012001.
31. Ferretti, A., Prati, C., Rocca, F., Permanent scatterers in SAR interferometry. *IEEE Trans. Geosci. Remote Sens.*, 39, 1, 8–20, 2001.
32. Gabriel, A.K., Goldstein, R.M., Zebker, H.A., Mapping small elevation changes over large areas: differential radar interferometry. *J. Geophys. Res.*, 94, B7, 9183–9191, 1989.
33. Zebker, H.A. and Villasenor, J., Decorrelation in interferometric radar echoes. *IEEE Trans. Geosci. Remote Sens.*, 30, 5, 950–959, 1992.
34. Goldstein, R.M. and Werner, C.L., Radar interferogram filtering for geophysical applications. *Geophys. Res. Lett.*, 25, 21, 4035–4038, 1998.
35. Chen, C.W. and Zebker, H.A., Network approaches to two-dimensional phase unwrapping: intractability and two new algorithms: erratum. *J. Opt. Soc. Am. A*, 18, 5, 1192, 2001.
36. Rosen, P.A., Hensley, S., Joughin, I.R., Li, F.K., Madsen, S.N., Rodriguez, E., Goldstein, R.M., Synthetic aperture radar interferometry. *Proc. IEEE*, 88, 3, 333–382, 2000.
37. Salomonson, V.V. and Appel, I., Estimating fractional snow cover from MODIS using the normalized difference snow index. *Remote Sens. Environ.*, 89, 3, 351–360, 2004.
38. Leinss, S., Wiesmann, A., Lemmetyinen, J., Hajnsek, I., Snow Water Equivalent of Dry Snow Measured by Differential Interferometry. *IEEE J. Sel. Top. Appl. Earth Obs. Remote Sens.*, 8, 8, 3773–3790, 2015.
39. Takala, M., Luojus, K., Pulliainen, J., Derksen, C., Lemmetyinen, J., Kärnä, J.-P., Koskinen, J., Bojkov, B., Estimating northern hemisphere snow water equivalent for climate research through assimilation of space-borne

radiometer data and ground-based measurements. *Remote Sens. Environ.*, 115, 12, 3517–3529, 2011.
40. Fassnacht, S.R., Brown, K.S.J., Blumberg, E.J., López Moreno, J.I., Covino, T.P., Kappas, M., Huang, Y., Leone, V., Kashipazha, A.H., Distribution of snow depth variability. *Front. Earth Sci.*, 12, 4, 683–692, 2018.
41. Jonas, T., Marty, C., Magnusson, J., Estimating the snow water equivalent from snow depth measurements in the Swiss Alps. *J. Hydrol.*, 378, 1, 161–167, 2009.
42. McCreight, J.L., Small, E.E., Larson, K.M., Snow depth, density, and SWE estimates derived from GPS reflection data: Validation in the western U. S. *Water Resour. Res.*, 50, 8, 6892–6909, 2014, doi:10.1002/2014wr015561.
43. Shi, J. and Dozier, J., Estimation of snow water equivalence using SIR-C/X-SAR. I. Inferring snow density and subsurface properties. *IEEE Trans. Geosci. Remote Sens.*, 38, 6, 2465–2474, 2000.
44. McCreight, J.L. and Small, E.E., Modeling bulk density and snow water equivalent using daily snow depth observations. *Cryosph.*, 8, 2, 521–536, 2014.
45. Painter, T.H., Rittger, K., McKenzie, C., Slaughter, P., Davis, R.E., Dozier, J., Retrieval of subpixel snow covered area, grain size, and albedo from MODIS. *Remote Sens. Environ.*, 113, 4, 868–879, 2009.
46. Tarricone, J., Webb, R., Marshall, H.-P., Nolin, A., Meyer, F.J., Estimating snow accumulation and ablation with L-band interferometric synthetic aperture radar (InSAR). *Cryosph.*, 17, 1997–2019, 2023.
47. López-Moreno, J.I., Fassnacht, S., Heath, J.T., Musselman, K., Revuelto, J., Latron, J., Morán-Tejeda, E., Jonas, T., Small scale spatial variability of snow density and depth over complex alpine terrain: Implications for estimating snow water equivalent. *Adv. Water Resour.*, 55, 40–52, 2013.
48. Sospedra-Alfonso, R., Melton, J., Merryfield, W., Effects of temperature and precipitation on snowpack variability in the Central Rocky Mountains as a function of elevation: Climate Driven Alpine Snowpack. *Geophys. Res. Lett.*, 42, 4429–4438, 2015.
49. Colbeck, S.C., An overview of seasonal snow metamorphism. *Rev. Geophys.*, 20, 1, 45–61, 1982.

Part 3

HISTORICAL AND CONTEMPORARY MISSIONS

8

The National Aeronautics and Space Administration Magellan Mission

Dipan Kumar Das[1], Padmaja Patnaik[1]* and Sudip Kumar Das[2]

[1]*Department of Physics, Centurion University of Technology and Management, Odisha, India*
[2]*Department of Physics, CV Raman University, Bilaspur, Chhattisgarh, India*

Abstract

Between 1989 and 1994, the NASA Magellan Mission revolutionized how geology and topography on Venus were understood and mapped using advanced synthetic aperture radar (SAR) technology to map over 98% of the surface of that planet. A pioneering planetary mission, Magellan, thus afforded unprecedented high-resolution radar images that enable geoscientists to discern and interpret Venus's surface composition, tectonic activities, and volcanic processes. This abstract synthesizes important findings from relevant literature by focusing on Magellan's technological advances and scientific legacy. The Magellan radar system implemented cutting-edge block adaptive quantization technologies to optimize data transmission and image resolution. Through such methods, scientists were able to topographically map volcanic plains, highland regions, and impact craters at impressive resolutions. Automated matching of radar images helped retrieve information related to topographic knowledge needed for the examination of surface processes on Venus. Mission's radar-grammetric techniques revealed vast tectonic deformation, as well as volcanic resurfacing, illuminating the history of a dynamic planet.

Magellan further contributed to identifying distinct volcanic and tectonic features, as in the Alpha Regio and Atla Regio, with radar emissivity data. Stereo data processing techniques developed by Hensley yielded fine-scale topographic maps that helped in explaining the role of lithospheric flexure in corona formation. These analyses showed elevated heat flow that supported hypotheses on Venusian active interior processes. The mission furthered the knowledge in the areas of radar image processing and planetary surface modelling. Techniques, such as fractal

**Corresponding author*: padmaja.patnaik@cutm.ac.in

Pooja Joshi, Ashutosh Bhatt, Neelam Sharma and Kapil Joshi (eds.) Synthetic Aperture Radar for Disaster Management: Bridging Science and Solutions, (207–234) © 2026 Scrivener Publishing LLC

geometry analysis and watershed identification algorithms for radar-based techniques, improved noisy SAR data interpretation. The results indicated that Venus's unique geological textures. The methods also inspired future radar missions like VenSAR and video image stabilization and registration (VISAR), and supported global comparisons of planetary tectonics. Magellan's engineering achievements were also amazing. The monopropellant thruster design and the aero braking techniques presented during the mission became leading tools in advanced orbital insertion strategies in later planetary missions. Similarly, McMahon (1996) indicated data management procedures during the mission as a template for the planetary data system and ensured the availability of long-lived planetary science data. Magellan's discoveries not only changed our perception of Venus but also provided standards for later missions. Many contemporary studies on Venus and its climatic and geological evolution are influenced by the work of this mission, from next-generation radar platforms to contemporary planetary remote sensing with cutting-edge spacecraft, planet, instrument, C-matrix, and events (SPICE) tools for onboard spacecraft geometry. Magellan highlights the power of interdisciplinary study and technological ingenuity in planetary exploration. The contributions of the mission persist in guiding exploration strategies for Venus and other terrestrial planets, thus becoming a cornerstone in planetary science.

Keywords: NASA, planetary exploration, Magellan, Venus, VISAR

8.1 Introduction

Venus has been a focus for scientific research due to its similarity and drastic contrasts with Earth. The early observations made with the telescope pointed toward a thick atmosphere, making numerous missions try to investigate it. Soviet missions, such as Venera and Pioneer Venus by the National Aeronautics and Space Administration (NASA), provided initial peeks into the surface and atmosphere of Venus. These missions revealed a hot surface temperature and a heavy carbon dioxide-rich atmosphere. High-resolution mapping became a necessity as scientists aimed to probe Venus's past geological history and possible habitability in the distant past. Magellan was planned to bridge this fundamental gap based on radar imaging technology. The Magellan mission was ground-breaking in planetary exploration since it was the first spacecraft to chart Venus entirely using synthetic aperture radar (SAR). It revealed hitherto unseen information on Venus's topography and presented the world's first general map of the surface features. The results from Magellan corroborated the existence of vast volcanic plains, impact basins, and tectonic processes on Venus, discounting previous inferences about its geologic evolution. Its accomplishment validated the applicability of vast radar mapping deep within space and had a subsequent effect on subsequent missions. Apart from Venus, Magellan's technology breakthroughs were essential to the development of radar techniques for subsequent planetary missions. Before Magellan, Venus exploration was confined to short flybys and lander missions with limited fragmentary data. The high temperature at the surface and dense atmosphere created huge hurdles to direct observation. Magellan was designed to transcend these limitations by using radar imaging to image through the dense cloud cover of Venus. Launched in 1989, its primary objectives were to map 70% of the photos of the planet's surface in high resolution, study Venus's gravitational field, and understand its geology. The mission far surpassed expectations, actually charting more than 98% of the surface, hugely expanding our understanding of Venus.

Understanding the Earth's and the solar system's origins and evolution, as well as the quantity and makeup of near-earth space resources, are the objectives of NASA's planetary exploration mission. There have been flights to the planets for missions to achieve this, and others are currently being planned or prepared. Now, let us start with the spaceship in orbit (Pioneers and Voyagers). Magellan, Mars Observer, and Galileo are scheduled to debut this decade. It is suggested for the future by the Solar System Exploration Committee; the existing and projected planetary exploration program is explained. These include the extremely difficult but scientifically fruitful sample returns, a set of Mariner Mark IIs with greater ambition, and a series of modest target Observer missions [1]. When space shuttle operations resume in the middle of 1988, a variety of upper stage propulsion technologies for planetary missions and geosynchronous orbit applications will be used on several of the first flights. For the next Space Shuttle missions—the Advanced Communications Technology Satellite, the Tracking and Data Relay Satellite, the Magellan, Galileo, Ulysses, and Mars Observer—some changes are needed to the Inertial Upper Stage (IUS) and the Transfer Orbit Stage. Additionally, the definition and capabilities of the Orbital Maneuvering Vehicle will be covered, which NASA is presently developing to carry out a variety of on-orbit missions and functions in support of orbiting spacecraft [2]. On May 4, 1989, NASA sent the Magellan Venus Radar Mapper spacecraft on an interplanetary transfer orbit to Venus. On August 10, 1990, the probe will go into orbit around Venus. The orbiter is equipped with a multimodal radar system with a wavelength of 12 cm. The majority of Venus's surface was imaged with a resolution of greater than 300 m while in the synthetic aperture mode, with an approach of 120 m across more than half of the planet. If the surface relief is not too great, it can calculate topographic relief in the altimeter mode with a vertical precision of more than 50 m mean across a surface resolution cell around 10 km in diameter. The radar receiver can measure the brightness temperature of surface radio emission in the radiometer mode with a resolution of 2K and an absolute accuracy of 20K [3]. Planetary research has long found Venus to be an intriguing target. Venus was one of the earliest destinations for spacecraft missions at the start of the space era. Venus, our solar system neighbor and Earth's twin sister in terms of size, was supposed to resemble our planet quite a bit. However, the US Mariner, Pioneer Venus, and Magellan missions, as well as the Soviet Venera and Vega spacecraft families, found a completely new and exotic world concealed behind a thick cloud cover during the first phase of Venus satellite investigation from 1962 to 1992. In addition to providing us with

Table 8.1 Different Venus exploration and major findings.

Mission name	Launch year	Key objectives	Major findings/ contributions	Technology advancements
Venera (Soviet Union)	1960s–1980s	Study Venus's surface and atmosphere	Revealed high surface temperatures, dense CO2 atmosphere, and initial surface glimpses	Landers, atmospheric probes
Pioneer Venus (NASA)	1978	Map Venus's atmosphere and surface	Provided early atmospheric and surface data, confirmed extreme conditions	Atmospheric probes, radar altimeter
Magellan (NASA)	1989	Map Venus's surface with high-resolution radar, and study its gravity field	High-resolution radar mapping of 98% of Venus's surface revealed volcanic plains, impact basins, and tectonic features	Synthetic aperture radar (SAR)
Venus Express (ESA)	2005	Study Venus's atmosphere, plasma environment, and surface	Advanced understanding of atmospheric chemistry, dynamics, and surface-atmosphere interactions	Orbiter with advanced spectrometers
Venus Orbiter (Japan - Akatsuki)	2010 (arrived 2015)	Study Venus's atmospheric circulation	Provided data on atmospheric dynamics, cloud structures, and potential volcanic activity	Infrared and ultraviolet cameras

a fundamental understanding of the planet's circumstances, these investigations raised several concerns about the atmospheric and geological evolution, surface-atmosphere interactions, chemistry, structure, dynamics, composition, and the atmosphere's plasma environment. The "morning star" is still a mystery despite having over 20 spacecraft exploring it. However, Venus has been a "forgotten" planet for almost 10 years, with no new missions planned by international space organizations. It is now experiencing a renewed interest in this world: The European Space Agency (ESA) orbiter mission Venus Express has passed the ESA selection stage, the Venus Orbiter mission is being accepted in Japan, and numerous ideas for Venus Discovery are competing for NASA's attention [4]. NASA, which is currently marking its fiftieth anniversary in 2008, is going to send astronauts to the moon by 2020. Yet, one of President Obama's advisory panels has recommended alternative pathways to Mars. NASA has already arrived on Mars with robots, showcasing its technical prowess. The ultimate challenge lies in sending people to Mars. NASA's exploration of space shares similarities with Columbus and Magellan's expeditions. Despite domestic economic and foreign policy challenges, NASA's continued exploration is crucial, as it could potentially save America from potential loss if it did not exist [5]. Table 8.1 summarizes major Venus exploration missions and their technological contributions.

8.2 Mission Design and Development

8.2.1 Conceptualization and Planning

Magellan's roots lay in NASA's strategy to chart Venus at high resolution using radar. In the late 1970s, mission concepts were first conceived. The planning included operating between NASA's Jet Propulsion Laboratory and foreign space agencies. The mission was redesigned several times as technology capabilities and budget concerns evolved. By combining parts from earlier missions such as Voyager and Galileo, NASA maximized cost while maintaining Magellan's scientific viability. The ultimate configuration highlighted radar mapping as its priority, hence a highly specialized mission specifically designed for the exploration of Venus. Beginning in April 1989, three interplanetary spacecraft would be sent into distant space aboard the IUS booster. However, NASA cancelled the Centaur upper stage in 1986 due to the Space Shuttle Challenger disaster and safety issues. As a result, a search for a new upper stage began for the three planetary spacecraft, focusing on selecting the IUS and its scientific significance [6].

8.2.2 Spacecraft Design and Instrumentation

The Magellan spacecraft had a modular design built with known components from previous missions. Its central structure comprised a high-gain antenna for communication with Earth and radar imaging. It had one instrument, the SAR, which was used to map Venus's surface with clear high-resolution detail. An added medium-gain antenna provided stand-by communication, and a propulsion system allowed for the correction of its trajectory. Its power was based on a solar array and battery, designed with efficiency in view for long-running operations. All this made Magellan the most efficient of its kind of planetary explorers.

The needs and characteristics of microelectronics employed in planetary spacecraft are distinct from those of other applications. The electronics must function for 12 years in space without supervision, which puts a high demand on their dependability. In a planetary flyby, energetic particles and cosmic rays may give a large dosage of radiation. The framework for this treatment of planetary spacecraft is the Voyager's 12-year flight to Jupiter, Saturn, Uranus, and Neptune. The presentation will cover developments in microprocessors, detectors, and memory from the Voyager era to the present, including Magellan and Galileo, and will extend to the forthcoming Cassini and Mars Observer programs. Galileo is already on its way to Jupiter, and in September of 1992, the Mars Observer is to be launched. In 1997, the ESA's Huygens Titan probe launched the Cassini mission to Saturn. The original Cassini Project design was to fly the second of the Mariner Mark II spacecraft, modular ones that were adaptable to several new missions beyond the inner Solar System. Cassini had to be modified in 1992 due to new NASA financial limitations. Additionally, the GaAs IC power amplifier component failure issue will be explained. There will be presentations of new scientific instrument detectors and a microelectronics perspective beyond the existing planetary missions [7]. Concern was raised during the first mapping mission of the Magellan spacecraft about the greater temperatures expected for Cycle 2, the first prolonged mapping mission of Venus. A six-month test program was carried out at the NASA Jet Propulsion Laboratory Edwards Facility (JPL-EF) to assess the performance of a 0.2 lbf (0.9N) monopropellant thruster under temperatures ranging from 120 to 150 °C to build confidence in thruster performance at the high temperatures. Daily spaceship reaction wheel desaturations were simulated by firing thrusters in pulse mode once a week. The Magellan spacecraft's thruster performance was tested at temperatures up to 140°C, with consistent performance at 150°C. The test program demonstrated that throughout Cycle 2 flight, the quantity of thruster pulses needed to eliminate momentum during a reaction wheel desaturation remained constant. Although conditions aboard the Magellan spacecraft have not yet reached the severe levels required by the test program, the data support the implementation of a mission rule that Cycle 2 mapping activities should cease if the thrusters exceed 140°C [8].

8.2.3 Radar System: Synthetic Aperture Radar and Its Capabilities

Magellan's SAR system was its most important tool, enabling it to see through Venus's dense clouds and create accurate surface maps. In contrast to optical cameras, SAR imaged the landscape with radio waves, offering topography and composition information. The system had multiple incidence angles and recorded various views of the landscape. The SAR data enabled scientists to map volcanic features, impact craters, and fault lines with unprecedented precision. The radar also yielded altimetry information, which permitted the construction of three-dimensional surface models of Venus. This technology transformed planetary mapping methods.

During its first complete revolution of the planet, the Magellan mission will use S-band, HH-polarized SAR (synthetic aperture radar) to survey 80% of Venus's surface. An altimeter will concurrently collect near-nadir S-band data. During extended-mission cycles, Magellan will be able to collect VV data; it could even be able to collect data at different incidence angles. Based on results from the Mojave Field Experiment, NASA compared predictions made using a second-order Bragg-Rice electromagnetic interaction model. In the California Mojave Desert, targets with low, moderate, and severe roughness were examined by the JPL Airborne Imaging Radar, a full polarimetric SAR operating at the L, C, and P bands. The dielectric constant and target surface roughness were determined at the same time. Later, HH and VV photos were calibrated and rebuilt to mimic data anticipated from the Magellan extended mission. HH and VV returns were predicted using ground observations [9]. The U.S. NASA Magellan (MGN) mission to Venus will employ a data compression approach to lower the SAR data rate. The sole scientific instrument on board the MGN spacecraft is a radar system, utilized for surface radiation measuring, altimetric topography profiling, and surface imaging. If the mission's scientific objectives are implemented simply, the data rate will be higher than what the system bandwidth can handle. To meet the scientific criteria, a data-rate reduction plan that incorporates block adaptive quantization of the SAR data and radar operating in burst mode is chosen [10].

8.2.4 Challenges and Innovations in Spacecraft Engineering

The development of Magellan encountered many engineering challenges, varying from spacecraft weight constraints to thermal management. Venus's hostile environment called for a robust design capable of withstanding high temperatures and high radiation. Engineers came up with innovative solutions, such as the use of passive thermal control systems and the implementation of redundant electronics to enhance reliability. Navigation accuracy was also an issue, with sophisticated trajectory correction maneuvers required. Using recycled components from previous missions helped reduce development while not sacrificing strength. These technological improvements establish standards for upcoming planetary exploration missions.

NASA's Magellan Mission started photographing Venus's surface with a SAR system in 1990. This limited the analysis to two dimensions, with terrain elevation not available. However, authors attempted to assess nontraditional surface elevation measurements using data from the impact crater Danilova, which confirmed a rim height of 1560 m with an uncertainty of ± 200 m [11]. NASA began its activities related to civilian spaceborne radar remote-sensing in 1960 with the development of rocket-based L-band radar. Since then, NASA has examined the depth and scope of the solar system using radar techniques that leverage the range of electromagnetic waves, such as spectral band, bandwidth, waveform coding, polarisation, and interferometric phase. Radar's intrinsic benefits, such its controllable source and direct interaction with the macroscopic characteristics of surfaces and subsurfaces, enable customised measurements to meet particular information goals. Earth remote sensing began in 1978 with Seasat's L-band single polarisation radar, and in 1994, the Shuttle Imaging Radar-C used X-band, C-band, and Lband fully polarimetric synthetic aperture radars to characterise the Earth across a range of disciplines. In just ten days after its 1994 launch, the Shuttle Radar Topography Mission scanned the topography of the planet. Cloudsat, a 94 GHz radar for cloud profiling, and Aquarius, an L-band scatterometer/radiometer for measuring sea surface salinity, are examples of US Earth-observing radar projects. Important solid-earth and cryoscience findings have resulted from the vast time series produced by international civilian radar operations. International Earth science research now relies heavily on frequency-diverse, multi-mode radar flights. Wide swath altimeters for mapping dense ocean topography, global ocean wind, water cycle missions using high frequency two-aperture interferometer techniques, and repeat pass polarimetric interferometry radar missions for mapping surface deformation and biomes are among the future missions that NASA has proposed. The sub-surfaces of the moon, Mars, Titan, and other smaller worlds are being mapped using long-wavelength radars [12].

8.3 Launch and Journey to Venus

8.3.1 Space Shuttle Atlantis-30, Space Transportation System-34, and the Launch Campaign

Magellan launched Space Shuttle Atlantis (STS-30) on May 4, 1989. It was the initial space shuttle launch of a planetary probe. The flight was initially planned on STS-34 but got re-allocated due to re-planning in schedules. Launch utilized an IUS booster to place Magellan on its interplanetary trajectory. In-depth pre-launch preparation ensured spacecraft systems were in functional condition before deployment. The mission demonstrated NASA's capability for integrating planetary flights with shuttle operations. As shown in Figure 8.1, Venus exploration has evolved through successive missions highlighting the importance of radar technology.

8.3.2 Interplanetary Trajectory and Navigation

Upon deployment, Magellan took 15 months to reach Venus using precise trajectory computations for an optimum flight path. The spacecraft carried out a few course correction maneuvers to ensure precise targeting of the

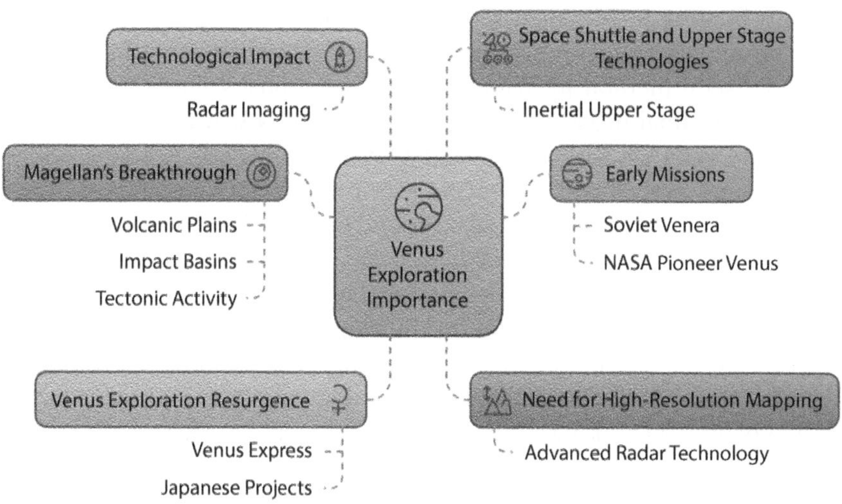

Figure 8.1 Venus exploration importance.

Venusian orbit. Navigation methods included gravity-assist planning and continuous monitoring of the probe's velocity and attitude. Ground controllers at JPL continuously monitored the spacecraft and made course corrections when needed. These maneuvers provided Magellan the means to cost-effectively fly by Venus without expending costly fuel for later orbital missions. The Galileo and Magellan missions at JPL marked the beginning of Mission Operations Assurance (MOA) in the late 1980s. After two Mars missions failed in the late 1990s, it underwent a dramatic evolution. The return of the Stardust sample to Earth marks the present maturity of MOA. The Project Manager and the Office of Safety and Mission Success at JPL and NASA Headquarters received independent risk evaluations from MOA during the sample return phase. In addition, an incompressible test list was created, safety and mission success issues were examined, and JPL Flight Project Practices and Design Principles were examined for any remaining hazards. The results of MOA's growth at JPL and the Space Operations Community are based on these achievements [13].

8.3.3 Aero-Braking and Orbital Insertion at Venus

Magellan was inserted into Venus on August 10, 1990, in an important orbital insertion maneuver. Instead of a pure circular orbit, it was initially inserted into an elliptical orbit, gradually finding a stable mapping altitude. The spacecraft employed aero-braking techniques, using Venus's upper atmosphere to decelerate and trim its trajectory. This technique allowed for fuel conservation and extended mission duration. The success of aero-braking in Magellan set the stage for future planetary missions, proving its effectiveness in orbital maneuvers. Aero-braking has proven to be a useful method in maximizing the mission's scientific return and payload into final orbit. Nevertheless, successful NASA missions such as Magellan, Mars Global Survey, Mars Odyssey, and Mars Reconnaissance Orbiter have proven that embracing aero braking has a number of drawbacks, such as higher operational costs and built-in risk. The complexity of the working procedures and the high level of atmospheric change are the main reasons for the risk of aero-braking. During the aero-braking period, which usually spans several months, this uncertainty necessitates 24-hour activities (orbit determination, pass planning, and risk mitigation), raising the mission's overall cost. Higher on-board autonomy is necessary in the design of an aero braking system to offset this cost rise. Aero-braking can boost the return of missions that are currently in flight, as well as serve as an enabling technology for upcoming missions or a means of increasing the scientific return. In Magellan's case, the spacecraft was sent into an aero braking phase that lowered the orbit's apoapsis height from 8500 km to 540 km after the main scientific mission goals had been met. By improving the precision of the gravity data at high latitudes, this orbital modification allowed for the scientific return of an already successful mission and the improvement of Venus's gravity models. Europe has scheduled a number of Mars missions, such as ExoMars, MarsGen, and MSR, for which the aero-braking can be beneficial. The ESA's Venus Express and Mars Express missions are a starting point to prove aero braking methods for future missions to Mars. Autonomous aero-braking strategies will be tested in these missions, with key elements including flight corridor definition and GNC design. Simulation results will assess system performance and autonomy level [14].

8.4 Radar Mapping of Venus

8.4.1 Radar Imaging Techniques

Magellan's SAR system utilized side-looking radars to capture high-quality pictures of the surface of Venus. The technique allowed for advanced analysis despite thick cloud cover. Different imaging angles were employed for the purpose of depth perception and terrain modeling. Backscatter data were employed for the detection of geological structure and surface composition. The creation of stacks of multi-temporal or multi-incidence angle pictures ("cubes") of co-registered images, matching for change detection, and stereo-parallax measurements for digital elevation excitation may all be automated using computerized image correlation. Traditionally, a skilled stereo operator makes precise parallax measurements. Although they have been studied, automated techniques have not gained widespread acceptance. For instance, radar mapping is necessary to create Digital Elevation Models from stereo radar pictures and to match numerous coverages from successive cycles of image acquisition in NASA's Magellan mission to the planet Venus. According to a normalized cross-correlation measure, there are frequently ±2 pixel discrepancies between the best image-matching technique and the skilled stereo-operator in an experiment to evaluate the machine matches' quality for stereo parallax identification in radar pictures. It was observed that machine matching errors are often lower in scanned aerial photography, and compared to radar pictures, SPOT data produces automated matching with manual matches based on subpixel differences [15]. The Magellan mission, which NASA launched to map the planet Venus, has been a huge success. As a consequence, the whole planet has been covered by an extraordinary collection of radar photographs, many of which are multiple views. Although this data collection was not included in the original NASA program or nominal mission plan, it was shown to permit the extraction of precise topography data with an accuracy of maybe ± 200 m. The methods now employed to acquire this data lend credence to the idea of creating a model of topographic elevation of any region covered by several photos using the Magellan image data collection. In addition to producing topographic data, this will also produce a collection of co-registered and ortho-rectified images that could serve as the foundation for a distinctive and contemporary image database and planet-related information system that planetary scientists of the future may use [16].

8.4.2 Data Collection and Processing

The gathered radar data were highly processed to eliminate distortions and improve clarity. Image mosaics were composed to yield a continuous global map of Venus. Sophisticated algorithms were applied to decipher elevation and texture contrasts. This provided for a thorough Venusian topographic analysis.

Some algorithms for processing radar images, like shape-from-shading, need a lot of computation and time. Additionally, processing photos on several workstations or parallel processing systems could make sense if the data set to be analyzed is enormous. Shape-from-shading, gridding, resampling, stereo matching, and terrain model visualization were all built to run on workstation clusters or parallel computers. the extensive picture data collection from NASA's Magellan mission to Venus, but were also inspired by Austria's MISSIOM effort and the European Union's Center for Warth Observation program, which uses publicly available algorithms to examine remote sensing photographs on distant workstations. A multi-processor method known as Concurrent and Distributed Image Processing was created. Whether the processors are on a single system or dispersed in a cluster, the speedup for image processing jobs rises almost with the number of processors in a linear fashion. It aids users of complicated image processing methods by removing the need for specialized parallel processing hardware and reducing the work required for code porting and maintenance [17]. Magellan, a NASA mission to Venus in the early 1990s, used a microwave radiometer and an S-band (12 cm) synthetic aperture radar to scan almost the whole planet's surface and analyze the topography using radar altimeters. Our knowledge of geomorphology, geology, and geophysical processes that have influenced the development of Venus's surface has been completely transformed by these data. However, the final response to important concerns about the processes and development of Venus's surface has been impeded by the absence of topography of the planet with a greater resolution than that acquired by the Magellan expedition. Volcanism, internal dynamics, surface/atmospheric interactions, and cratering, to mention a few, all leave their mark on the relief of planetary surfaces in subtle to grandiose ways (e.g., Maxwell Montes on Venus) (e.g., quasi-circular depressions on Mars). Major assumptions and uncertainties probably still exist even if a great deal of morphological work and interpretation can be done without topographic data. Approximately 25% of the Earth was surveyed using radar stereo as part of the Magellan project. Using a strict stereo sensor model and automated matching procedures, a recently funded NASA project aims to methodically analyze all of the Magellan stereo data. These findings are contrasted with earlier stereo discoveries, and some potential scientific uses for the stereo data are discussed [18].

8.4.3 Key Contributions of Magellan to Venusian Topography and Surface Composition

Magellan revealed a volcano-dominated planet with tectonic deformations and impact craters. The data show active resurfacing mechanisms, which contradict earlier models of Venus's geology. Surface compositions provided information about Venus' geological history and internal dynamics. The Aglaonice area of Venus is shown on a geological map. This is the first attempt to map all of the available NASA Magellan mission data at the highest resolution yet published. It is mostly based on altimetry data from the same area and geological interpretation of a SAR picture at a resolution of 1:1.5,000,000. The map illustrates the geologic history made visible by mapping at this scale, covering an area of around 1.6 × 106 km^2, or about 1% of the Venusian surface. Five tectonically and texturally separate volcanic plains material units are visible to us instead of a regional plains unit. Nearly two-thirds of the mapped units come from recognizable source locations, indicating that most units on Venus have a known origin if they are mapped at the Magellan data resolution [19]. Radar images of Venus created by the Arecibo observatory during its 57-year lifespan helped us better comprehend the planet's surface features and geologic processes. Before the Magellan expedition, dual-circular polarization photos were first obtained in 1988 using the S-band (2380 MHz) transmitter, which had the best spatial resolution of 1-2 km. In addition to the 1988 observations, these legacy data sets have been preserved by archiving high-resolution photos from observing runs in 2012, 2015, 2017, and 2020 with the NASA

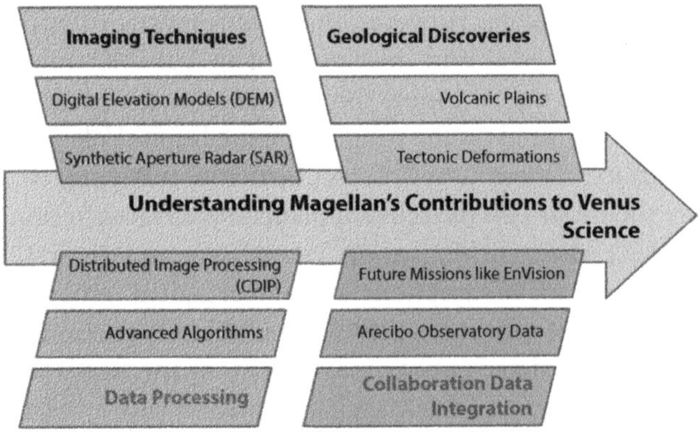

Figure 8.2 Analyzing Magellan's impact on Venus exploration.

Planetary Data System. The Arecibo Venus data collecting and processing techniques, special features of the delay-Doppler imaging, and the determination of relative calibration factors that connect the Magellan data and final multilook maps were all covered. The circular polarization ratio and maybe the Stokes vector components can also be derived from the measurements. These findings are especially pertinent to the EnVision orbiter mission's long-term spin rate monitoring, surface change detection, and S-band polarimetry planning [20]. Figure 8.2 illustrates Magellan's scientific and technological impact on Venus exploration.

8.5 Key Scientific Discoveries

8.5.1 Surface Geology: Volcanism, Cratering, and Tectonics

The Magellan mission provided us with hitherto unknown data concerning Venus's surface geology in the form of giant volcanic plains, gigantic shield volcanoes, and huge lava flows. Radar mapping at high resolution evidenced current volcanicity, and consequently, there was much speculation that Venus remains geologically active. The surface of the planet is crisscrossed with impact craters, but their relatively even distribution indicates that there must have been a global resurfacing episode about 300–500 million years ago. Tectonic lineaments, including rift troughs, fault zones, and mountain ranges, indicate likely global deformation of the lithosphere, although Venus is tectonically distinct from Earth. The findings have demystified our understanding of planetary resurfacing and geological processes outside our own world. The University of Hawaii is now working on many projects that use orbital radar and aerial radars to study the geologic history and eruption patterns of volcanoes on Earth and Venus. The SIR-B and ERS-1 Earth-orbital radars, the Magellan mission to Venus, and a NASA/JPL AIRSAR deployment to Hawaii in August 1990 are the sources of the radar data. Additionally, it was organizing several studies using the JERS-1 and SIR-C/X-SAR orbital radars from 1992 to 1996. Our research is also pertinent to the TOPSAT radar, which NASA is now considering for an Earth probe-class mission at the end of the century. Our science objectives focus on using radar to characterize surface morphology, temporal changes due to volcanic activity, and determine volcano topography to assess eruption characteristics and hazards. While radar studies have potential benefits in volcanology, the operational mode of radar data analysis is still being defined [21]. Impact crater ejecta are a significant source of fine-grained minerals on Venus, and the geographical distribution of craters has been used to estimate the relative age and resurfacing rates of sizable areas. The possible meeting point of tesserae, severely distorted formations that could be the earliest materials on Venus, and distant crater deposits is of special interest. Although the makeup of tesserae is unclear, it is essential to comprehend whether water contributed to crustal development. Tesserae are therefore perfect locations for upcoming landed missions to find potential felsic elements, but because surface landers do not last long, care must be taken to prevent contaminating surface materials from craters. A method using NASA Magellan radar data was developed to detect distal crater ejecta on tessera terrain across Venus. Results showed uneven distribution of fine-grained ejecta, with many tesserae lacking thick mantling material. At Sudenitsa Tesserae, radar signatures of mantling debris were observed, but no source crater was found. The source crater may have been resurfaced by volcanic activity within the past 80 million years. Similar deposits may have built up over time [22]. A geological map of Venus's Alpha Regio (V-32) quadrangle at a scale of 1:2,500,000. The V-32 quadrangle has an area of around 7,600,000 km^2 and stretches from 0° to 25 S, 0° to 30 E. Using full-resolution (up to 75 m/pixel) SAR, altimetry, and stereo-derived topography data from NASA's Magellan mission in ArcGIS 10.5, geological mapping was carried out. There were around 40,000 lineaments mapped. Alpha Regio and Minu-Anni Tessera are the two main regions that contain the oldest unit, Tessera terrain. Two large fracture bands, both aligned around NNW-SSE, and four smaller fracture belts have been found and analyzed. Two unrecognized wrinkle ridge trends have been identified in the northeastern corner of the quadrangle, and 77 geological units were mapped, with plains material divided into 27 units. Estimates of coronae diameters have been extended by hundreds of kilometers [23].

8.5.2 Insights into the Venusian Atmosphere and Interior

Magellan radar observations, together with information obtained from earlier missions, allowed for the refinement of models for Venus's dense atmosphere and internal structure. The thick CO_2-dense atmosphere places enormous surface pressure, almost 92 times that of Earth, and is responsible for the planet's runaway greenhouse effect. Radar soundings made indirect suggestions regarding subsurface layering and suggested potential mantle convection causing Venus's resurfacing processes. Also, gravitational measurement variations suggested crustal thickness and core compositional differences. It is important to study the atmospheric and internal processes of Venus to compare planets and understand runaway greenhouse effects. Five years ago, UCLA scientists were chosen by NASA to establish the Planetary Plasma Interactions Node, which will aid scientists in locating, retrieving, and archiving particles and field data from planetary missions. The PPI Node is equipped with a Radio Astronomy Subnode at GSFC, an Inner Planets Subnode at UCLA, and an Outer Planets Subnode at the University of Iowa, since planetary plasma data is varied and requires expertise in other areas. The PPI Node has made an effort to support the scientific community by offering them high-quality data packages. It has worked with individual scientists and missions to gather and thoroughly document the highest-quality data available. The PPI Node has successfully created 171 CD-ROMs with peer-reviewed data and stored more than 1011 bytes of data. Data archiving rates have steadily increased thanks to this effective technique. Restoring Pioneer Venus data, compiling Galileo Venus archive and radio scientific data from Magellan, and generating CD-ROM data packages for Voyager, Pioneer, and Ulysses data at Jupiter and Saturn are among the ongoing efforts at the node. The Small Bodies Node is also using it to prepare comet data and to facilitate the collision between Jupiter and the comet Shoemaker-Levy 9. Asteroidal data from Galileo will also go into the archives. To handle and read data, users are given software tools that are independent of the machine, operating system, and data format. To minimize development work, scalable solutions have been created that enable independent investigators to handle data on a single CD-ROM. The PPI Node data holdings may be accessed online [24]. At least a dozen enormous coronae on Venus have previously been shown to exhibit signs of lithospheric flexure. Calculating mechanical thickness and surface heat fluxes is made possible by the elastic parameters that are produced by thin plate models fitted to topographic profiles, given an estimated yield strength envelope. However, the low precision of altimetry data from the NASA Magellan mission has made it challenging to investigate most coronae, particularly those with a diameter of less than a few hundred kilometers. Here, flexural signals surrounding 99 coronae spanning around 20% of the surface were sought using stereo-derived topography, which was recently created from SAR images and Magellan altimetry data. Using Cartesian and axisymmetric models, the study calculated elastic thicknesses ranging from 2 to 30 km at 19 coronae. Thermal gradients greater than 24 K km^{-1} are indicated by mechanical thicknesses estimated to be less than 19 km. Surface heat fluxes more than 95 mW m^{-2} indicate that Venus is cooling more quickly than anticipated or that coronae are a significant source to the overall heat budget. According to binomial statistics, flexural signs are less likely to be present in incomplete fracture annuli. Extensional faulting is predicted by stress estimates [25].

8.5.3 Contributions to Planetary Evolution Theories

Magellan's findings have made a significant contribution to theories of planetary evolution, particularly why Venus and Earth, with the same size and composition, should have so dissimilar hostile environments. Lack of plate tectonics on Venus suggests alternative means for heat release, such as periodic global resurfacing. The mission also provided evidence to support the hypothesis that Venus underwent a catastrophic climatic event that made a potentially habitable world one that is uninhabitable in its present form. These discoveries are crucial to understanding thick atmospheres of exoplanets and in forecasting long-term evolution for Earth-like planets. Perhaps the most surprising result of the NASA Magellan mission to Venus was the preservation of more than 970 nearly pristine impact craters that were dispersed almost randomly throughout the planet's surface. Many people believe that the craters are proof of a nearly worldwide catastrophic volcanic resurfacing that occurred between 10 and 100 million years ago, or around 500 million years ago. This perception of Venus is widely accepted in popular science and textbooks and is seldom challenged. In examining data from the Magellan mission, the image of a severely resurfaced Venus first appeared. Using data from Magellan data analysis, the distribution of impact craters and its implications for the resurfacing and evolutionary processes of Venus have been re-examined. The global stratigraphy hypothesis, which suggests catastrophic resurfacing involved lava flows burying pre-flood craters, is not supported by detailed geologic mapping. Geologic mapping shows thin flows cover preflood surfaces, and 8% of the surface preserved in elevated plateaus does not correlate with Venus's oldest surfaces. The SPITTER theory, which incorporates elements of earlier theories, predicts that near-steady-state impact craters would form and be destroyed

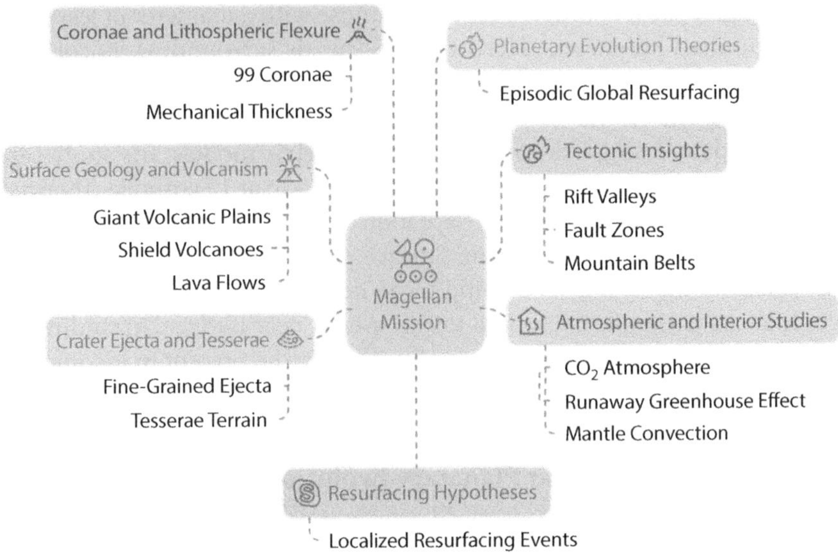

Figure 8.3 Magellan mission: Geological and atmospheric insights.

when the lithosphere is globally thin. Large local regions punctuated by time and space saw the time-integrated formation of multiple crustal plateaus, which destroyed craters. Crater removal processes were rendered inoperable by a secular shift to a thick lithosphere, and craters started to build on the surface. Geologic mapping of Magellan data has revealed a rich and lengthy history of Venus's surface, as opposed to a condensed perspective of the last 500 million years or more [26]. Figure 8.3 summarizes geological and atmospheric insights gained from the Magellan mission.

8.6 Data Analysis and Interpretation

8.6.1 Use of SPICE Tools in Data Handling

The Spacecraft, Planet, Instrument, C-matrix, Events (SPICE) software package was crucial for data processing of Magellan's massive datasets; it delivered accurate positional and orientation information, with the ability to accurately geo-reference radar images. With the inclusion of SPICE kernels, scientists were able to reconstruct the path of the spacecraft, which contributed to the accurate mapping of Venusian terrain. SPICE also supported data fusion of information from previous missions, promoting comparative studies. The successful application of SPICE tools established a benchmark for contemporary planetary exploration, illustrating the significance of strong data-handling structures. For giving scientists the basic observation geometry required to do photogrammetry, create maps, and analyze other types of planetary science data, NASA's "SPICE" supplemental information system is the de facto international standard. Reference frame parameters, underlying time system conversions, field-of-view geometry, and instrument mounting alignment, target body size and shape information, and location and orientation ephemerides of the target body and the robotic spacecraft are all provided by SPICE. To obtain the data and compute derived variables, such as height, illumination angles, instrument viewing latitude/longitude, and more, SPICE comes with a full suite of software known as the SPICE Toolkit. The SPICE system was first developed for the Magellan expedition to Venus and has subsequently expanded to better serve the demands of engineers and scientists. For example, the SPICE Toolkit, which was previously only available in Fortran 77, is now available in Java Native Interface, C, IDL, and MATLAB. Previously only possible through APIs, SPICE computations may now be carried out using a geometry engine's client-server interface. The components are free of cost, licensing, and export limitations, and data visualization is now accessible. Attendees at the ISPRS meeting may find the SPICE system, which is maintained by the Navigation and Ancillary Information Facility, interesting [27].

8.6.2 Techniques for Image Processing and Topographical Analysis

Magellan SAR images were processed using advanced processing techniques to obtain high-resolution topographical maps. Radar backscatter analysis was used to differentiate surface geology and materials. Methods such as radar interferometry and stereo imaging were used to create 3D elevation models, revealing the complex surface of Venus. Image processing methods, including contrast transformation and noise removal, improved feature recognition, making it easier to identify volcanic flows, fault zones, and impact craters. These image processing methods remain rudimentary in planetary remote sensing but are still being used in modern space missions. More than 30 years have passed since the groundbreaking NASA Magellan Mission gave scientists their last close-up look at Venus's surface. To address some of the key problems on Venus's development and present geologic activity, NASA and ESA are now studying two new mission ideas that would return to the planet and gather more precise images and mapping data. To observe the surface through the dense, optically impenetrable atmosphere, both missions would use synthetic aperture radars. The optimization of radar designs for missions aiming to understand Venus's unique characteristics, highlighting the complementary nature of these missions, their design and performance, and their potential to advance Venus and planetary science discussed [28], which is a novel method for identifying patterns in SAR pictures. Several zones displaying polygonal-shaped faults on the planet's surface have been identified through a visual study of the photos from NASA's Magellan mission to Venus. The aim is to provide a way to identify these zones automatically. Conventional edge detectors, such as those based on gradient pictures, are ineffective, and automated image analysis is challenging due to the multiplicative nature of SAR images and their high level of noise. A plan based on a two-scale analysis and an enhanced watershed algorithm was presented. Potential edges in the SAR picture are extracted using the approach provided, and the patterns found are analyzed to determine whether the image comprises a "polygon region." Other SAR or optical pictures, such as the observation of Jupiter's satellite Europa and Mars, can also be used with this approach. Category of EDICS: 2-SEGM (Segmentation) [29].

8.6.3 Integration of Magellan Data into Scientific Research

Magellan's large dataset has been important in shaping Venusian science, and its impact has been extended to most studies of comparative planetology, atmospheric dynamics, and planetary geology. The data from Magellan have been blended with that of other missions, such as Akatsuki and Venus Express, for better models of surface composition and geology. The mission data have also been crucial in developing simulations of Venus's climatic history and potential for past habitability. Even years since its completion, Magellan data continue to form the foundation of Venus research, guiding new scientific inquiry and exploration plans. Remotely sensed datasets gathered by spacecraft offer a great way to demonstrate and explain geology concepts and scientific reasoning to pupils. Synthetic aperture radar photography from NASA's Magellan mission to Venus is a particularly helpful dataset. It acts as a virtual field trip guide to three Venusian field locations, demonstrating the value of this dataset in enhancing geoscience education. Each region exhibits geologic relations that offer optimal conditions for discussing basic geology concepts and engaging in scientific reasoning. The USGS provides free online images of Venus, allowing for easy integration into petrology, geomorphology, structure, field methods, and introductory courses. The extreme surface conditions add a unique twist to geologic studies [30]. Exoplanetary science may influence our study of Venus, and how terrestrial exoplanetary science can influence our understanding of

Venus. The disparities in knowledge seem sharp on the surface. It has been seeing Venus for thousands of years and doing telescopic observations for centuries. When it was established that Venus was a hothouse planet rather than the tropical paradise depicted in science fiction, spacecraft observations started with Mariner 2 in 1962. The level of exploration and understanding of Venus remains below that of Mars, leading to major questions.

Table 8.2 Data analysis and interpretation aspects of the Magellan mission.

Aspect	Techniques/tools	Contributions/ impact	Applications/ significance
Data Handling	SPICE software package (kernels, toolkit)	Accurate geo-referencing, spacecraft trajectory reconstruction, data fusion, and standardized observational geometry	Foundation for modern planetary exploration, improved data analysis and visualization, enhanced comparative studies
Image Processing & Topographical Analysis	Synthetic Aperture Radar imaging, radar backscatter analysis, radar interferometry, stereo imaging, contrast transformation, noise removal, enhanced watershed algorithm	High-resolution topographical maps, surface geology differentiation, 3D elevation models, feature recognition (volcanic flows, faults, craters)	Understanding Venus's geological history, developing advanced radar designs for future missions, and automating pattern recognition in SAR images
Integration into Scientific Research	Combined Magellan data with data from Akatsuki, Venus Express, and other missions; geological simulations, Virtual field trips, and Exoplanetary comparisons	Improved models of surface composition and geology, climatic history simulations, potential habitability studies, geological education, and exoplanetary science	Continued foundation for Venus research, guiding new scientific inquiries and exploration plans, enhancing geoscience education, and aiding in exoplanet research

Exoplanetary science has grown since the discovery of Pegasus 51b in 1995, and multi-million to billion-dollar/euro exoplanet-focused spacecraft missions like JWST and its successors will be flown in the coming decades. Excitement about Venus exploration is again blooming with confirmed and proposed missions from India, Russia, Japan, the European Space Agency, and NASA [31]. Table 8.2 outlines the key data handling and interpretation techniques used in the Magellan mission.

8.7 Future Prospects in Venus Exploration

8.7.1 Follow-Up Missions Inspired by Magellan

Magellan's success paved the way for its successor Venus missions that would take advantage of its findings. NASA's VERITAS and ESA's EnVision missions, due to be launched in the 2030s, will make use of advanced radar technology to explore Venus's surface and geological evolution. They hope to ascertain if Venus experiences volcanism in current times and the way its crust evolves on timescales. Magellan has also influenced atmospheric probes like the Deep Atmosphere Venus Investigation of Noble gases, Chemistry, and Imaging (DAVINCI), which will analyze the chemistry and dynamics of the dense atmosphere on Venus. Such successor missions will reveal more about the planet's history and whether life is present. The only completely steerable radar in the world for high-resolution imaging and ranging of planetary and small-body objects is the Goldstone Solar System Radar facility. For a vast range of solar system objects, these observations offer details on their polar ices, orbits, rotations, and surface properties. The produced data are often used to support scientific research on these objects as well as NASA flight programs, such as the numerous solar system exploration missions conducted over the last three decades. Mars Exploration Rovers, Cassini, Hayabusa, MESSENGER, NEAR, SOHO recovery, Mars Pathfinder, Lunar Prospector, Clementine, Magellan, and Viking are just a few of the space missions that have benefited greatly from the Global Surveyor. Recent accomplishments include the assessment of lunar terrain at high resolution, the characterization and orbit refinement of near-Earth asteroids, and radar scientific outcomes, including the imaging of near-Earth objects and the accurate estimation of the rotation of Mercury [32].

8.7.2 Proposed Technologies for Advanced Exploration

Future exploration of Venus will entail new technologies to endure its hostile environment. New orbiters will possess higher-resolution radar and infrared spectrometers to continue to examine the surface and atmosphere. High-temperature electronics advances would enable landers and rovers to spend more time on the surface, collecting *in situ* geological and atmospheric information in real time. Theoretical aerial platforms, like solar-powered drones or balloon observatories, would be able to penetrate the cloud layers of Venus, which are more Earth-like in pressure and temperature. These advances would revolutionize knowledge of Venus and its place within comparative planetary studies. Using a microwave radiometer and 12 cm S-band synthetic aperture radar, the early 1990s NASA mission Magellan scanned almost the whole surface of Venus and measured the topography using radar altimeters. These findings have fundamentally changed our understanding of the geomorphology, geology, and geophysical processes that have shaped the surface of Venus. However, the conclusive response to important concerns regarding the processes and development of Venus's surface has been impeded by the paucity of imaging and topography of the planet with a greater resolution than that acquired by the Magellan mission. Using a multispectral NIR emissivity mapping sensor and an X-band radar, the Venus Emissivity, Radio Science, InSAR Topography and Spectroscopy (VERITAS) Mission seeks to acquire high-resolution pictures and topography of Venus [33]. Since Magellan thirty years ago, US Venus exploration has relied on flyby and Earth-based observations of missions destined elsewhere, archived data, and international programs like Venus Express (ESA) and Akatsuki (JAXA) until the selection of NASA Discovery missions VERITAS and DAVINCI and participation in ESA's EnVision. These missions will examine several significant scientific issues about the past and present of Venus. Even if these next missions are successful, many significant and unresolved questions about Venus's history, such as how it differs and how it is similar to Earth, will remain. Significant distinctions between Venus's and Earth's planetary evolution require knowledge that the variety of future missions selected cannot supply, but if previous research and technology are successful, an innovative approach within the next 20 years can provide it. To better understand Venus's origin, development, geology, and interactions with the atmosphere, the Venus *In Situ* Transfer and Analysis mission idea seeks to offer precise knowledge about the planet's surface and atmosphere. A long-term, sky-borne laboratory in Venus's upper atmosphere that is kept at ideal temperatures and pressures for prolonged laboratory measurements is what the mission suggests. Samples will be collected by VISTA from many locations on the planet's surface and atmosphere, transported to a state-of-the-art airborne laboratory, and then thoroughly examined. The data gained will provide insights into surface composition, geological

Table 8.3 Key future prospects in Venus exploration inspired by Magellan.

Mission name	Agency	Key objectives	Technologies/ contributions	Relationship to Magellan
VERITAS	NASA	Map Venus's surface to determine its geologic history, and understand why it evolved differently from Earth.	Synthetic Aperture Radar, infrared spectroscopy.	Builds upon Magellan's radar mapping by providing higher-resolution data and additional spectroscopic information.
DAVINCI	NASA	Analyze Venus's atmosphere composition, including noble gases, and image the planet's surface during descent.	Descent probe, atmospheric spectrometers, imaging systems.	Complements Magellan's surface mapping by providing *in situ* atmospheric data and high-resolution descent imagery.
EnVision	ESA (with NASA contribution)	Study Venus's geology and atmosphere, including its interior, surface, and atmospheric processes.	Synthetic Aperture Radar (VenSAR), subsurface radar sounder, and spectrometers.	Advances Magellan's radar mapping with higher resolution and subsurface probing capabilities.
Shukrayaan-1	Indian Space Research Organisation	Study the geological and atmospheric processes of Venus.	Synthetic aperture radar and other instruments.	Will provide further radar mapping data and contribute to the global effort of mapping Venus, building upon the work of Magellan.

provinces, cloud processes, atmospheric aerosols, atmospheric dynamics, and infrequent seismic and volcanic phenomena. Asset rendezvous, sample processing, and a long-lasting laboratory platform are only a few of the major operational and technological obstacles that VISTA must overcome. The flying laboratory, many sampling landers, ascending vehicles, and sample retrievers are all part of the mission concept [34]. Table 8.3 presents future Venus exploration missions inspired by Magellan's legacy.

8.8 Conclusion

The Magellan mission was a ground-breaking achievement in planetary exploration, greatly enhancing our knowledge of Venus. Using radar mapping, Magellan obtained the first high-resolution global images of Venus's surface, discovering a dynamic world of volcanoes, rift valleys, and extensive plains sculpted by geological processes. The mission also gave us information regarding Venus's gravitational field and knowledge regarding the planet's internal structure. Although it was destroyed in 1994 when it was intentionally sunk into Venus's atmosphere, Magellan's legacy continues to reign supreme in planetary science. Its data continues to be the foundation of comparative planetology and informs future missions such as NASA's VERITAS and Europe's upcoming EnVision, whose mission is to continue to dismantle Venus's secrets. Magellan not only proved the possibility of radar mapping but also opened new and novel methods of deep-space exploration; it is therefore one of the most successful planetary missions ever performed by NASA.

AI disclosure statement: The author confirms that no artificial intelligence (AI)-generated content or AI-assisted writing tools were used in the preparation of this chapter (including summaries and the abstract). All text, analysis, and conclusions are the result of the author's original work.

References

1. Mills, R.A. and Bourke, R.D., Planetary Programs, in: *EASCON Record: IEEE Electronics and Aerospace Systems Convention*, pp. 193–8, 1985.
2. Saucier, S., Near term space transportation systems for earth orbit and planetary applications, in: *SAE Technical Papers*, 1987.
3. Saunders, R.S., Pettengill, G.H., Arvidson, R.E., Sjogren, W.L., Johnson, W.T.K., Pieri, L., The Magellan Venus Radar Mapping Mission. *J. Geophys. Res.*, 95, B6, 8339–55, 1990.
4. Titov, D.V., Baines, K.H., Basilevsky, A.T., Chassefiere, E., Chin, G., Crisp, D., *et al.*, Missions to Venus, in: *European Space Agency*, pp. 13–20, (Special Publication) ESA SP, Noordwijk, The Netherlands, 2002.
5. Lambright, W.H., Exploring space: NASA at 50 and beyond. *Public Administration Rev.*, 70, 1, 151–7, 2010.
6. Bangsund, E. and Knutson, R., STS 30, 34 and 44 - a rebirth of the planetary missions, in: *Space Congress Proceedings*, 1988.
7. Draper, R.F., Microelectronics for planetary spacecraft, in: *GaAs IC Symposium Technical Digest 1992*, pp. 11–4, 1992.

8. Costanza, R.A., Freeman, R.J., Bell, T.L., Simulated mission testing and flight performance of a 0.2 lbf magellan monopropellant thruster, in: *AIAA/ASME/SAE/ASEE 28th Joint Propulsion Conference and Exhibit, 1992*, 1992.
9. Wall, S.D. and VanZyl, J.A., Investigation of the relationship of SAR HH and VV backscatter to surface roughness and dielectric constant. *Dig. - Int. Geosci. Remote Sens. Symp. (IGARSS)*, 1, 22–3, 1989.
10. Kwok, R. and Johnson, W.T.K., Block Adaptive Quantization of Magellan SAR Data. *IEEE Trans. Geosci. Remote Sens.*, 27, 4, 375–83, 1989.
11. Leberl, F.W., Maurice, K., Thomas, J.K., Millot, M., Automated radar image matching experiment. *ISPRS J. Photogramm. Remote Sens.*, 49, 3, 19–33, 1994.
12. Elachi, C., The legacy and future of civilian radar missions, in: *CIE International Conference of Radar Proceedings*, 2006.
13. Faris, G.B. and Bryant, L.W., Stardust blazes MOA trail, in: *SpaceOps 2010 Conference*, 2010.
14. Gil-Fernandez, J., Cadenas-Gorgojo, R., Escorial-Olmos, D., Fedele, A., Proving aerobraking with mars express and venus express, in: *61st International Astronautical Congress 2010, IAC 2010*, pp. 1488–96, 2010.
15. Leberl, F.W., Maurice, K., Thomas, J., Kober, W., Radargrammetric measurements from the initial Magellan coverage of planet Venus. *Photogramm. Eng. Remote Sensing*, 57, 12, 1561–70, 1991.
16. Leberl, F.W., How to extract topographic information from magellan radar images, in: *Proceedings of SPIE - The International Society for Optical Engineering*, pp. 33–44, 1993.
17. Goller, A. and Leberl, F., Radar image processing with clusters of computers. *IEEE Aerosp. Electron. Syst. Mag.*, 24, 1, 18–22, 2009.
18. Hensley, S., Mitchell, K., Nunes, D., Shaffer, S., Deen, R., Parcheta, C., *et al.*, Systematic processing of high resolution topography of Venus from Magellan radar stereo data and science applications, in: *Proceedings of the European Conference on Synthetic Aperture Radar, EUSAR*, 2016.
19. Grindod, P.M. and Guest, J.E., 1:1.5,000,000 geological map of the Aglaonice region on Venus. *J. Maps*, 2, 1, 103–17, 2006.
20. Campbell, B.A. and Campbell, D.B., Arecibo Radar Maps of Venus from 1988 to 2020. *Planet. Sci. J.*, 3, 3, 2022.
21. Mouginis-Mark, P.J., Radar studies in physical volcanology, in: *International Geoscience and Remote Sensing Symposium (IGARSS)*, pp. 389–91, 1992.
22. Whitten, J.L. and Campbell, B.A., Recent volcanic resurfacing of Venusian craters. *Geology.*, 44, 7, 519–22, 2016.
23. Bethell, E.M., Ernst, R.E., Samson, C., Geology of the Alpha Regio (V-32) Quadrangle, Venus. *J. Maps*, 15, 2, 474–86, 2019.
24. Walker, R.J., Joy, S.P., King, T.A., Russell, C.T., McPherron, R.L., Kurth, W.S., The planetary plasma interactions node of the Planetary Data System. *Planet. Space Sci.*, 44, 1 SPEC. IS, 1996.

25. O'Rourke, J.G. and Smrekar, S.E., Signatures of Lithospheric Flexure and Elevated Heat Flow in Stereo Topography at Coronae on Venus. *J. Geophys. Res.: Planets*, 123, 2, 369–89, 2018.
26. Hansen, V.L. and Young, D.A., Venus's evolution: A synthesis, in: *Special Paper of the Geological Society of America*, vol. 419, pp. 255–273, 2007.
27. Acton, C., Bachman, N., Semenov, B., Wright, E., Spice tools supporting planetary remote sensing, in: *International Archives of the Photogrammetry, Remote Sensing and Spatial Information Sciences - ISPRS Archives*, pp. 357–9, 2016.
28. Hensley, S., Campbell, B., Perkovic-Martin, D., Wheeler, K., Kiefer, W., Ghail, R., VISAR and VenSAR: Two Proposed Radar Investigations of Venus, in: *IEEE National Radar Conference - Proceedings*, 2020.
29. Moreels, P. and Smrekar, S.E., Watershed identification of polygonal patterns in noisy SAR images. *IEEE Trans. Image Process.*, 12, 7, 740–50, 2003.
30. Lang, N.P., Examples for integrating Venus Magellan imagery into geoscience classes. *J. Geosci. Educ.*, 56, 2, 135–41, 2008.
31. Way, M.J. and Wang, J., Venus topography and boundary conditions in 3D general circulation modeling, in: *Lecture Notes in Geoinformation and Cartography*, pp. 325–335, 2019.
32. Slade, M.A., Benner, L.A.M., Silva, A., Goldstone solar system radar observatory: Earth-based planetary mission support and unique science results. *Proc. IEEE*, 99, 5, 757–69, 2011.
33. Hensley, S., Smrekar, S., Shaffer, S., Paller, M., Figueroa, H., Freeman, A., *et al.*, VISAR: A next generation interferometric radar for venus exploration, in: *Proceedings of the 2015 IEEE 5th Asia-Pacific Conference on Synthetic Aperture Radar, APSAR 2015*, pp. 362–6, 2015.
34. Izenberg, N., Scott, V., Fultz, B., Agrawal, R., Aiazzi, C., Andrade, J., *et al.*, VISTA: Venus *in Situ* Transfer and Analysis Mission Concept, in: *IEEE Aerospace Conference Proceedings*, 2023.

9

Synthetic Aperture Radar in Solar System Exploration

Sachin Srivastava*, Vineet Rawat, Aditya Singh Rana and Abhay Dhasmana

Department of Aerospace Engineering, Uttaranchal University, Dehradun, India

Abstract

Synthetic aperture radar (SAR) technology was initially designed to perform Earth remote sensing and planetary missions; however, it is now evolving into an indispensable system for solar system investigations. This chapter discusses the SAR system application for the study of solar and helio-physical processes, especially its role in investigating the solar influence on interplanetary space and planetary environments. It demonstrates the potential of SAR to deliver high-resolution imaging, sounding, and sensing through various extraterrestrial media. The chapter elaborates on the integration of SAR payload for solar monitoring missions, as well as points out some of the possible design challenges associated with the operation in proximity to the Sun. In this context, it examines the potential for using SAR for mapping coronal mass ejection (CMEs), tracking the interaction of CMEs with the solar wind, and monitoring the solar-forced effects on small bodies, planetary surfaces, and atmospheres. The study includes recent advancements in engineering SAR technology, such as new SAR antennas, power management technologies, signal processing algorithms, and other relevant technologies for missions in the vicinity of the Sun. In addition, the chapter focuses on special SAR configurations such as multi-frequency and polarimetric SAR systems to perform detailed, multi-scale planetary observations. This theoretical knowledge is supported by the analysis of real solar missions and studies of the space weather environment, including the Parker Solar Probe mission, as well as other potential future missions to study the solar and interplanetary environments. This chapter also focuses on emerging trends, such as artificial intelligence-driven data analysis, SAR integration with spectrometry and magnetometry, and the fusion of theoretical models with real-world applications, which makes it highly valuable for researchers, engineers, and mission planners in aerospace and space sciences.

**Corresponding author*: sachinrash2004@gmail.com

Pooja Joshi, Ashutosh Bhatt, Neelam Sharma and Kapil Joshi (eds.) Synthetic Aperture Radar for Disaster Management: Bridging Science and Solutions, (235–264) © 2026 Scrivener Publishing LLC

Keywords: Coronal mass ejection, synthetic aperture radar, P-band, polarimetric SAR, interferometric SAR

9.1 Introduction

Synthetic aperture radar (SAR) technology is considered the backbone of remote sensing, revolutionizing Earth observation, planetary exploration, and deep-space exploration. Unlike optical imaging systems, SAR uses microwave radiation, which allows it to produce high-resolution images even under atmospheric interference, lighting, and surface reflection [1]. Because it can see through clouds, dust, and surface material, SAR is fundamental for planetary missions, particularly those to environments that are unfavorable for traditional remote sensing methods. Synthetic aperture radar has been primarily applied to land and planetary missions, such as mapping the surface of Venus through thick clouds or Martian topography, but now, it is considered a crucial technology for studying solar and heliophysical phenomena. Here, we introduce how SAR can be applied to solar system exploration and show how it improves understanding of solar activity, the heliosphere, and solar-induced changes on planets [2].

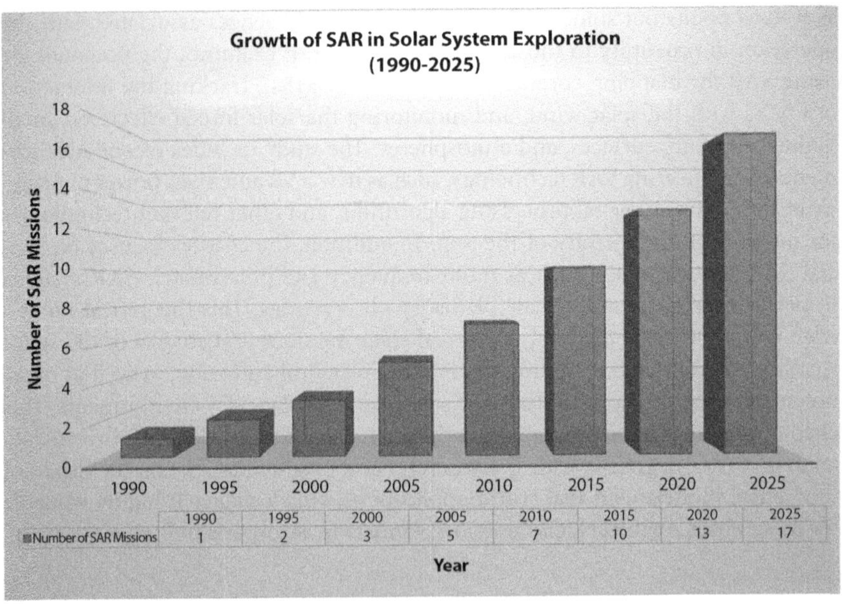

Figure 9.1 The growing relevance of SAR methods for planetary and solar system exploration is highlighted by the graph, which shows the rise in SAR-equipped missions between 1990 and 2025.

Table 9.1 Key SAR missions and their contributions to solar system exploration.

Mission name	Launch year	Target body	SAR frequency band	Key objectives	Major findings
Magellan (NASA)	1989	Venus	S-band	Surface mapping through thick atmosphere	High-resolution Venusian surface maps, volcanic activity evidence
Cassini (NASA/ESA)	1997	Titan (Saturn's moon)	Ku-band	Surface imaging through dense methane atmosphere	Detection of lakes, seas, and river networks on Titan
Mars Reconnaissance Orbiter (MRO)	2005	Mars	SHARAD (20 MHz)	Subsurface ice detection	Discovery of extensive underground water ice layers
Lunar Reconnaissance Orbiter (LRO)	2009	Moon	Mini-RF (S and X-band)	Mapping lunar surface and searching for water ice	Detection of possible water ice in permanently shadowed regions
Sentinel-1 (ESA)	2014	Earth (reference for planetary studies)	C-band	Earth observation, SAR calibration for space missions	Improved SAR imaging techniques for planetary exploration
Parker Solar Probe (NASA)	2018	Sun (heliosphere)	Not a SAR mission, but informs SAR technology	Studying solar wind and corona dynamics	Data helps in designing SAR for solar observations
NISAR (NASA-ISRO)	2024 (Planned)	Earth, with applications for planetary SAR	L-band & S-band	Surface deformation studies, potential for Mars/Lunar applications	Enhanced SAR techniques applicable to planetary research

Figure 9.1 displayed the proof of the growing relevance of SAR methods for planetary and solar system exploration is highlighted by the graph, which shows the rise in SAR-equipped missions between 1990 and 2025. Table 9.1 shows the Key SAR missions with yearwise and their contributions to solar system exploration.

The inclusion of SAR in solar and heliospheric science presents a unique opportunity for studying the dynamic Sun and its impact on the surrounding space environment. Using coronal mass ejection (CME), solar wind propagation observations, and heliospheric structural variations, SAR-derived observations are expected to provide new insights into space weather events and consequences for planetary atmospheres and magnetospheres. Furthermore, the implementation of SAR-enabled spacecraft in the inner solar system poses novel engineering challenges, such as requiring high orbital thermal protection, radiation-hardened electronics, and power-saving power management systems for near-sun operation. In this chapter, the basics of SAR technology, its engineering applications for solar system missions, and the potential SAR-based space exploration directions will be discussed. By reviewing previous missions, present research, and upcoming missions, this work will serve as a standard reference for scientists, engineers, and mission designers interested in utilizing SAR for novel knowledge for heliophysics and interplanetary science.

9.2 Fundamentals of Synthetic Aperture Radar Technology

Synthetic aperture radar technology is a significant remote sensing tool, as it can image planetary surfaces, atmosphere, and even subsurface features with good resolution. Because SAR actively transmits microwave waves and receives their reflections, its performance is not dependent on weather or lighting conditions, as compared to traditional optical imaging systems that use sunlight. A significant method in space exploration, SAR can offer complete images and penetrate various planetary materials. This part discusses the SAR imaging principles, high-resolution imaging capabilities, subsurface penetration techniques, and advantages of multi-frequency and polarimetric SAR configurations.

9.2.1 Principles of Synthetic Aperture Radar Imaging

Synthetic aperture radar is an advanced high-tech imaging radar technology that oscillates a miniature antenna over a path of a similar form and style to those of satellites or space probes to create an immense effective aperture. Compared to orthodox radars, the SAR gets better resolution in this process. The SAR has a basic concept in which microwave pulses are sent towards the target, and reflected signals are detected while Doppler shifts due to platform motion are used for the creation of high-resolution images. By combining a great number of pulses over time, SAR can produce highly accurate images of the detailed surfaces. One of the significant advantages of SAR is its capability to operate in the absence of illumination sources. Thus, it can be utilized to study planets and moons with several long periods of darkness, dense atmospheric cover, or fierce weather. Another advantage is that SAR signals can penetrate through barriers that block optical sensors, such as haze, dust storms, and clouds. Furthermore, surface roughness, texture, and dielectric characteristics can be detected using SAR, which offers important information about the climate and geology of planets.

There are two main kinds of resolution types utilized in SAR imaging. Range resolution depends on the length of the transmitted radar pulse, which increases with the signal's bandwidth. Conversely, azimuth resolution is achieved using the synthetic aperture method, which, through the relative motion of the radar platform, simulates a far larger antenna to enhance image clarity. Because of this feature, SAR is extremely useful for documenting the minute details of planetary surfaces and their changes dynamically with time.

9.2.2 High-Resolution Imaging and Surface Characterization

Synthetic aperture radar is an innovative imaging radar that scans with the help of moving a small antenna in a trajectory, resembling a space satellite or spaceship; a huge effective area is thus covered. Due to this technique, SAR achieves high resolution as to the traditional method of radar systems. The basic concept of SAR is to send microwave pulses in the direction of a target, receive the backscattered signals, and then use the Doppler changes brought on by the platform's motion to create a high-resolution image. Synthetic aperture radar creates remarkably accurate, detailed images of surfaces by combining many radar pulses over time. Such characterizations of surfaces are achieved from SAR by analyzing the returned backscattered signals, with information on roughness, materials, and the dielectric; the intensity in the reflected signals helps in various terrain types as rocky surfaces or icy plains/volcanic deposits; in this regard, radar backscatters can classify dunes on Mars, as well as formations of ice at Europa and volcanic lava flows of Venus; and another important application in SAR surface characterization is called Interferometric SAR (InSAR) whereby images of one area are produced from different aspects and the changes in phase as the returned signal is compared accurately to measure any deformations about the surface-the technique is well suited for planetary tectonics, changes in subsurface, or changes in any planetary surface within time [6].

9.2.3 Subsurface Penetration Capabilities

Among other specializations is the ability of SAR technology to pierce surfaces on planets in search of what exists beneath such features. Radar waves of more extensive wavelengths are capable of reflecting subsurface geological formations or passing through dry, frozen, or half-transparent surfaces, thus achieving their target, determining penetration depth, as influenced by the radar frequency, surface makeup, and the existence of water and metallic ores, among other elements [7]. Lower frequency bands of P-band radar and other SAR signals can penetrate deeper into subterranean structures such as buried impact craters, ice layers, and lava tubes. However, higher frequency signals like X-band radar provide greater resolution but penetrate over a much shorter range. A combination of different frequency bands provides scientists with a more complete picture of subsurface structures at multiple depths.

Water ice deposits have now been located easily with the assistance of subsurface imaging, as offered by SAR. Subsurface ice, massive in extent and volume, has been discovered at different places across Mars. According to the National Aeronautics and Space Administration (NASA), ice beneath Mars was detected for the first time using NASA's Mars Reconnaissance Orbiter shallow radar (SHARAD). Subsurface ice deposits have also been discovered in permanently shadowed craters by radar scans of the Moon, which can be used as important resources for future lunar exploration missions [8].

9.2.4 Multi-Frequency and Polarimetric Synthetic Aperture Radar

Since this technology utilizes the various frequencies' wavelengths to evaluate various depths and material categories, multi-frequency SAR can thus probe planetary surfaces to a greater extent. Greater SAR frequencies provide extensive detail of features on the surface, while lower frequency signals penetrate further through the surface. Blending different frequencies offers scientists a better dataset encompassing surface and underground features. Another cutting-edge method for improved data acquisition is polarimetric SAR (PolSAR), which monitors the interaction of radar waves with planetary surfaces in terms of polarization properties. Additional information regarding the material content, moisture levels, and surface roughness can be obtained by transmitting and receiving signals in horizontal and vertical polarization states. The procedure is effective when finding mineral constituents, changes in surface quality, or differences in geological formation. Multi-frequency and polarimetric SARs are also applications for major uses, since they study how radiation from the sun affects the surfaces of the planets. Cosmic rays and solar wind continue to bombard the surface compositions on planetary bodies with continuous radiation, giving them distinctive space weathering attributes. Ultimately, scientists will use it to understand more about the long-term impact of solar activity on the planets' surface variations [9].

9.3 Design Considerations for Synthetic Aperture Radar in Solar System Exploration

Solar system exploration using SAR technologies poses unique engineering challenges that are even more complicated by the extreme nature of the space environment. While traditional SAR systems used for Earth observation are designed for relatively benign threat levels, deep-space SAR systems must be capable of withstanding extreme radiation levels, temperature variations, and power constraints. Moreover, the huge distances involved lead to communication delays and constraints on data transfer, necessitating advanced data-compression techniques and self-operation. For SAR technology to perform adequately in deep-space missions, the fundamental design challenges and innovative remedies are analyzed here.

9.3.1 Challenges in Operating Synthetic Aperture Radar in Harsh Space Environments

The lower atmosphere and near-orbital conditions of Earth are quite different from the environment in which SAR systems built for solar system exploration must operate. The following issues need to be resolved:

Variability in Extreme Temperatures: There is extreme variability in space temperatures. Where the Sun is nearby, temperatures reach as high as 600°C (1112°F). Temperatures can approach nearly zero, -270°C (-454°F), in deep space, where there is little sunlight. Additionally, landers or planetary orbiters carrying SAR payloads must endure rapid heating and cooling cycles as they transition from areas receiving sunlight to areas in shadow. These thermal changes result in expansion and contraction of the materials, which wear them out, and make the SAR antennas get out of alignment, and can even cause problems with structural failure. Advanced thermal protection technologies such as active cooling systems, multi-layer insulation, and heat-resistant materials would be needed to overcome these impacts.

High Radiation Exposure: One of the most important elements of SAR systems in deep space is high radiation exposure. Cosmic rays and solar storms target missions operating close to the Sun or in regions with high magnetospheres, such as Jupiter and Saturn, with high-energy particle attacks. Long-term exposure can impair sensor performance, memory storage, and semiconductor devices. Because of these reasons, radiation-hardened electronics and lead, polyethylene, and reinforced composite materials used as shields are required for SAR systems. Data processing also employs error-correcting algorithms to reduce radiation-induced errors and prevent failures in the system.

Limitations in Communication and Data Transmission: High-bandwidth communication lines are required to send high-resolution photos back to Earth. Due to the large distances, deep-space missions have limited data transfer speeds and lengthy communication delays. For example, depending on planetary alignment, it can take anywhere from 4 to 24 minutes for a signal from a Mars spacecraft to reach Earth. These constraints would have to be bypassed with onboard data processing and compression within SAR systems. With the utilization of artificial intelligence (AI) and machine learning (ML), SAR payloads are capable of self-identifying relevant features automatically, extracting important information, and eliminating redundancy during transmission.

Inhomogeneities in Atmospheric and Gravitational Environments: Each celestial object's gravitational regime impacts the function of a SAR in different ways. For instance, missions to high-gravity planets, such as Jupiter, could potentially be subjected to strong orbital perturbations, whereas SARs placed on low-gravity moons, such as Titan, could suffer from unstable orbits caused by tidal forces. Moreover, the quality of SAR images is decreased by signal attenuation and scattering because of dense planetary atmospheres such as Venus and Titan. The SAR frequencies should be chosen in the most optimal way to provide the necessary signal penetration while minimizing distortion to these problems.

9.3.2 Radiation and Thermal Stress Considerations

Strong thermal and radiation management techniques are needed for SAR missions to endure harsh space environments.

9.3.2.1 *Thermal Control Techniques*

Since they are exposed to extreme temperature fluctuations, SAR systems within the solar system must be in thermal equilibrium. Thermal control techniques are used to protect SAR electronics from extreme cold in deep space and from intense heat near the Sun [10]. Highly reflective thin film material multi-layer insulation is used to minimize thermal radiation absorption and emission. Heat pipes and radiators remove excess heat from electronic components, maintaining them within a safe operating temperature range. For high thermal fluctuation missions, phase-change materials dynamically absorb and release heat, stabilizing temperature fluctuations. Deployable sunshields are also used to reflect solar radiation, especially for SAR instruments close to the Sun, like those on solar observation satellites. These thermal control systems work together to allow SAR sensors and electronics to operate optimally in various space environments.

9.3.2.2 *Radiation Hardening Techniques*

High-energy solar storm radiation, cosmic rays, and planetary magnetospheres seriously threaten SAR electronics in space. To reduce damage caused by radiation, radiation-hardened microelectronics are utilized, utilizing special semiconductor materials that are not degraded. Shielding materials like lead, aluminum, and polyethylene are incorporated into spacecraft structures to absorb or deflect high-energy particles. Figure 9.2 shows the classification/flow chart of the hardening techniques. Three independent electronic circuits conduct the same function simultaneously using triple modular redundancy, a fault-tolerance method; if one circuit is

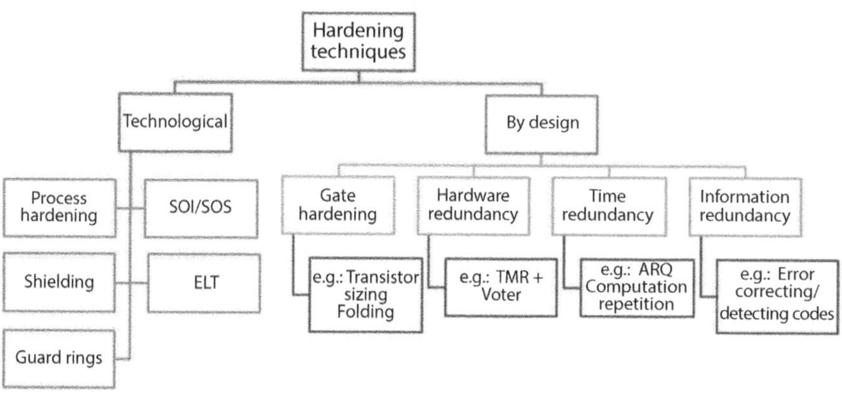

Figure 9.2 Classification of the hardening techniques [11].

damaged due to radiation exposure, the majority output is utilized to ensure system integrity. Error correction methods are also required because they enable SAR processors to identify and repair radiation-induced bit-flip faults in computation and memory storage. Combined with other methods, these ensure that SAR systems can resist long exposure to deep space's harsh radiation environment.

9.3.2.3 Power Management Under Extreme Conditions

Synthetic aperture radar instruments are power-hungry to function efficiently, but power generation is difficult in space due to intermittent solar intensity and deep-space darkness. High-efficiency solar panels with radiation-hardened photovoltaic cells are used in missions near the Sun that can operate irrespective of solar radiation damage. Radioisotope thermoelectric generators are used for missions beyond Mars, where sunlight is inadequate, to supply continuous power by harnessing heat from radioactive decay to produce electricity. Supercapacitors and lithium-ion batteries are used as power reservoirs, ensuring energy supply during high-intensity SAR operations. Adaptive power scheduling is offered by SAR instruments to optimize efficiency, where power consumption is automatically managed by the system according to operational demand and optimizing battery life and energy use [12].

9.3.3 Antenna and Signal Processing Innovations

Facilitate improved SAR performance in extraterrestrial conditions, technological innovation in antenna technology, and signal processing is needed.

Antenna Designs for Intelligent SAR: SAR effectiveness is highly reliant on the antenna design, which influences image quality, signal strength, and compatibility with operation under deep-space conditions. Phased-array antennas are most widely employed in spaceborne SAR because they can be folded compactly before launch and expanded in orbit, allowing large aperture sizes necessary for high-resolution imaging. Metamaterial-based antennas employ artificially engineered materials to maximize gain, minimize mass, and maximize energy efficiency. Electronically steerable antennas reduce wear and potential failure points by steering the beam direction by phases, without mechanical movement. These new antenna technologies allow SAR to collect high-fidelity data on a broad range of mission objectives, from heliospheric plasma studies to mapping planetary surfaces [13].

Advances in Signal Processing for Spaceborne SAR: To facilitate solutions to issues with long-range data acquisition, signal distortion, and bandwidth constraints, deep-space SAR missions require sophisticated signal processing algorithms. Onboard data compression methods, such as wavelet and Fourier-based algorithms, compress the SAR imaging size, thus facilitating efficient transmission of data over long interplanetary distances. Furthermore, autonomous feature extraction based on AI allows SAR sensors to pre-process images onboard, stripping irrelevant information and highlighting items of interest before transmission to Earth [14]. In addition, employing different wavelengths and polarizations to detect different material compositions, multi-frequency and polarimetric imaging techniques improve SAR's capability to explore different planetary surfaces and subsurface structures. All these signal processing advances ensure that SAR is an extremely useful remote sensing tool for exploring deep space.

Overcoming Doppler and Motion-Induced Challenges: Planetary rotation, orbital flight, and spacecraft speed-induced Doppler shifts must be compensated for in SAR systems in space since they warp imaging data. Through dynamic adjustment of radar signal characteristics to relative motion between the SAR platform and target surface, Doppler compensation algorithms eliminate frequency shifts. Despite high spacecraft speeds, SAR instruments utilize synthetic aperture refinement methods in low-orbit flight operations to enhance image resolution by integrating signal acquisition over time [15]. Motion correction algorithms also stabilize SAR imaging and provide accurate geolocation of acquired data by compensating for changes in spacecraft trajectory. Through these developments, SAR can sustain operation in dynamic orbital and interplanetary environments with high imaging accuracy.

9.3.4 Power Management Strategies for Deep Space Synthetic Aperture Radar Operations

One of the strongest limitations for SAR missions is the availability of power.

Power Generation and Storage: SAR missions pose serious concerns regarding power availability, necessitating high-efficiency sources of energy to enable operations for longer durations. High-efficiency solar array photovoltaic cells are the main source of power for inner solar system missions. These panels are specially designed to withstand heat cycling and harsh sun radiation. However, through the process of converting heat from decaying radioactive isotopes into power, radioisotope thermoelectric generators offer a stable alternative in deep-space missions where sunlight is limited. New energy storage devices, including lithium-ion batteries and supercapacitors, store excess energy and discharge when required to offer stable power delivery during SAR imaging's peak power requirements. Synthetic aperture radar devices can be operated stably in a variety of space conditions owing to these power solutions [16].

Power Optimization Techniques: Power optimization techniques with high efficiency are the key to maximizing energy usage since SAR imaging is a high-power requirement. Duty-cycled SAR operations minimize unnecessary power usage and maximize mission life by activating SAR sensors only when required. Power distribution is dynamically allocated by adaptive energy distribution systems, which deliver energy to essential components based on operating real-time requirements [17]. In addition, by adjusting radar pulse lengths and transmission frequencies, energy-efficient signal transmission techniques minimize power drain and ensure SAR data collection remains effective without high power consumption. Even in areas with limited energy resources, these techniques enable SAR missions to remain sustainable.

Heat Dissipation Strategies for Power-Intensive SAR Systems: The high dissipation power of SAR systems produces significant heat generation, where optimal thermal dissipation methods must be employed to prevent overheating. Radiative panels cool passively by emitting the excess heat out to space using a vacuum atmosphere for thermal control. Phase-changing materials are used in the absorption and emission of heat and to stabilize the oscillations of the temperature of SAR payloads. For miniaturized SAR systems, the heat sinks must be custom-designed to provide increased thermal conductivity and uniform heat distribution. A localized hot spot is not formed, and the overheating of the critical components is avoided. As a result, the SAR instruments will be able to operate at safe temperatures, and their efficiency and longevity will be ensured by this thermal management in the deep-space environment [18].

9.4 Integration of Synthetic Aperture Radar in Solar and Heliospheric Research

Synthetic aperture radar has been an important Earth observation and planetary exploration instrument since the 1960s. Its high (up to a few meters once processed) resolution imaging, surface penetration, and multi-frequency band operation make it suitable for solar and heliosphere studies. It requires advanced observation techniques and systems that can work in harsh environments to monitor solar activity and its impact on the interplanetary medium. Synthetic aperture radar provides unique advantages in observing solar dynamic phenomena, CMEs, solar wind interpenetrations, and resultant effects on planetary systems. This chapter discusses the use of SAR for observing solar activity and energy transport through the heliosphere, as well as space weather phenomena.

9.4.1 Synthetic Aperture Radar Payloads for Solar Observation Missions

Integrating Synthetic Aperture Radar (SAR) into solar observation missions opens a new way of monitoring the solar dynamic processes. Previously, solar observations used optical, ultraviolet, X-ray, and radio telescopes to study activities in the solar surface, coronal phenomena, and space weather. The use of SAR incorporates high-resolution imaging, deep penetration into the surfaces of celestial bodies, and insensitivity to external lighting. Synthetic aperture radars would find applications in monitoring solar surface phenomena, plasma motion, and magnetic field interaction. The SAR payloads on solar probes must be considered under harsh environmental conditions. The intense radiation and temperatures of the Sun require advanced thermal protection and radiation-hardened electronics to protect SAR instruments. High-frequency antennas must be deployable, compactly storable, and efficient in deep space operations. Additionally, power management remains a challenge since SAR systems are high-power consumers that must be efficiently captured from solar panels without risking overheating [19]. Upcoming missions would put SAR-equipped spacecraft in high-inclination orbits to study the Sun's poles, which are underexplored. These missions would supplement other solar observing missions, such as Parker Solar Probe (PSP) and Solar Orbiter, and would offer more insight into solar activity through high-resolution SAR imagery.

9.4.2 Monitoring Solar Dynamics Using Synthetic Aperture Radar

Solar dynamics are long-term variations in the Sun's structure and activity, including solar flares, prominences, and magnetic reconnection events. Synthetic aperture radar technology can observe these phenomena at higher resolution than ever by imaging structures on the solar surface continuously at high spatial resolution and observing their temporal evolution. Unlike conventional imaging methods, SAR is capable of penetrating stormy solar regions and providing data on plasma flow, density variations, and subsurface solar activity. For the study of different layers of the Sun, multi-frequency SAR architectures may be considered with *in-situ* records of solar corona and chromosphere dynamics. The use of various SAR polarizations provides more information on the structure of the solar magnetic fields that cause the dynamic processes of the Sun. Observing these dynamic processes together with the advanced capabilities of SAR and the improvement of space weather prediction will provide a better understanding of the effects of solar activity on the heliosphere and planetary environments. New missions that incorporate SAR with magnetometer and spectroscopic sensors have the promise to create a large number of measurements that allow for the study of energy transfer mechanisms in the Sun's atmosphere as the cause of the heliosphere and interplanetary space influence [20].

9.4.3 Mapping Coronal Mass Ejections and Solar Wind Interactions

Coronal mass ejections are enormous bursts of solar plasma and magnetic field output from the Sun's corona into interplanetary space. They have significant effects on space weather, influencing satellite operations, communications systems, and planetary magnetospheres. Synthetic aperture radar can potentially image CME structure and propagation at high resolution, illuminating their origin, speed, and interaction with the solar wind [21]. Employing polarimetric and multi-frequency SAR, scientists can analyze the density and magnetism of CMEs during their passage through space. The ability of SAR to track these processes near-real time can help refine forecasting models of space weather so that adequate preparations can be undertaken against threats of radiation and geomagnetic storms. Additionally, SAR would also be used to study CME interaction with planetary magnetospheres, particularly those on Venus and Mars, because they do not have a strong magnetic field. Consequently, their atmosphere becomes too sensitive to solar activity. Utilizing SAR for interplanetary missions scheduled for the future would enable monitoring and prediction of CME impacts on both natural and artificial space environments more effectively.

9.4.4 Synthetic Aperture Radar in the Study of Heliosphere Transport Phenomena

The heliosphere is the vast region of space influenced by the solar wind and magnetic field of the Sun, well beyond the orbit of Pluto. Understanding how energy, particles, and magnetic fields transport themselves through the heliosphere is crucial for space exploration and forecasting space weather. Synthetic aperture radar can become a significant contribution to studies of these transport phenomena, providing high-resolution diagnostics of plasma flow, shock waves, and turbulence in the heliospheric plasma. The ability of SAR penetration and fine-scale resolution in space plasmas is what makes it a valuable tool for observing solar wind variations and their interaction with interstellar space. Imaging heliospheric currents and shock fronts could permit SAR to observe regions of turbulence where charged particles and cosmic rays are accelerated. The SAR data paired with magnetometric and spectroscopic data may also provide the ability to comprehend solar activity's impact on the outer heliosphere and heliopause, where the interstellar medium comes into contact with the solar wind. Later flights with SAR hardware might provide immediate observation of such dynamic events, yielding new insight into the physical processes of the Sun's action at a distance.

9.5 Synthetic Aperture Radar for Planetary Surface and Atmospheric Studies

Synthetic aperture radar has emerged as an instrument of profound importance to the study of planetary surface and atmospheric dynamics, with the capabilities of high-resolution imaging through clouds, atmospheric dust, and even subsurface layers. When applied to solar system exploration, SAR allows for detailed observation of planetary terrain, geological configuration, and solar-powered atmospheric activity. One of the foremost areas of concern is the exploration of solar-forced changes at planetary surfaces where solar wind, coronal mass ejections, and extreme ultraviolet radiation couple with planetary crusts and exospheres, resulting in erosion, ionization, and other transformative processes. SAR is equally useful in studying planetary atmospheres by observing cloud structures, storm dynamics, and atmospheric composition modification due to solar heating and magnetic interactions. Additionally, SAR is also important in investigating dust transport and space weathering on airless or thin-atmosphere bodies like the Moon and Mars through analyzing surface roughness, albedo changes, and regolith redistribution. With the aid of multi-frequency and polarimetric SAR methods, researchers can achieve a complete overview of how solar-powered processes condition planetary environments throughout the solar system [22].

9.5.1 Monitoring Solar-Forced Changes on Planetary Surfaces

The Sun is important in determining planetary surface characteristics through solar radiation, interactions of charged particles, and effects of space weather, resulting in long-term surface composition, structure, and thermal property changes. One of the most important processes affected by solar activity is space weathering, in which prolonged exposure to solar wind, cosmic rays, and micrometeoroid impacts changes the optical and dielectric characteristics of surface materials, as seen on the Moon and Mercury. Synthetic aperture radar technology is priceless in the detection of these changes through the examination of radar backscatter changes, giving information on regolith evolution [23]. Also, the wide variations in temperature resulting from solar heating and cooling cycles cause thermal stress on planetary surfaces. This leads to rock fracturing and regolith formation, a process that can be detected by InSAR, which accurately quantifies surface deformations.

On ice-capped planets like Mars, solar-powered sublimation-deposition cycles dominate atmospheric and climate processes. Synthetic aperture radar becomes a primary method to monitor ice-thickness changes and subsurface layering, as shown by the Mars Reconnaissance Orbiter's SHARAD instrument. Additionally, solar-forced tectonic and volcanic processes, especially on geologically active planets, can be traced with SAR for ground deformation and subsurface structural changes. By performing high-resolution imaging, subsurface penetration, and long-term monitoring, SAR is essential in investigating solar-forced variations on planetary surfaces, providing essential information about planetary evolution and climate processes throughout the solar system.

9.5.2 Synthetic Aperture Radar Applications in Studying Planetary Atmospheres

Synthetic aperture radar technology is a highly effective means of investigating planetary atmospheres, possessing distinct advantages over optical and infrared sensors, especially in penetrating cloud cover and yielding high-resolution images under diverse atmospheric conditions. Atmospheric processes on planets like Venus, Mars, Titan, and the gas giants are solar radiation-driven, affecting weather patterns, cloud dynamics, wind speeds, and ionospheric interactions. Synthetic aperture radar, when used in combination with other remote sensing methods, delivers valuable information about these atmospheric processes. The following are the major uses of SAR in planetary atmospheric studies, detailed pointwise as below: Figure 9.3 is displayed the synthetic aperture radar applications in studying planetary atmospheres.

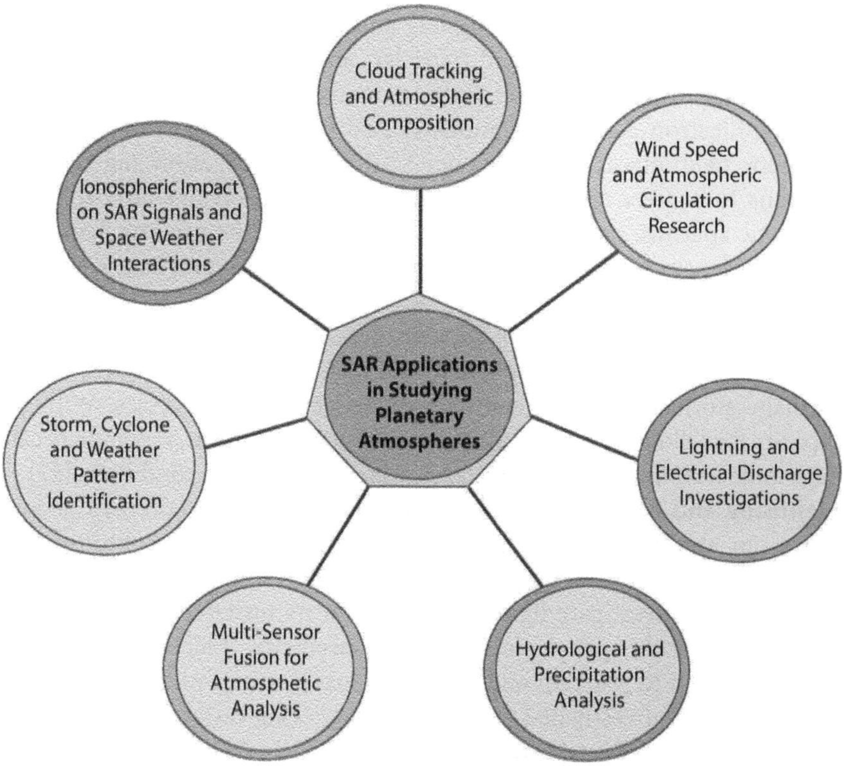

Figure 9.3 Synthetic aperture radar applications in studying planetary atmospheres.

Cloud Tracking and Atmospheric Composition: SAR can penetrate dense clouds and deliver high-resolution surface imaging even in atmospherically dense planets like Venus and Titan. At Venus, SAR from missions such as Magellan has uncovered the hidden surface beneath its permanently sulfuric acid cloud cover, supporting atmospheric opacity and dynamic research. The capability of SAR to function at various frequency bands provides sensitivity to changes in cloud density and composition [24].

Wind Speed and Atmospheric Circulation Research: Doppler SAR methods can measure planetary wind speeds by detecting frequency changes in radar waves due to traveling atmospheric particles. In Titan, Cassini SAR data demonstrated surface structures formed by powerful methane winds, and from them, insights were gained about atmospheric circulation processes. Monitoring temporal changes in wind speed helps to know about climate dynamics and seasonal changes in planetary atmospheres [25].

Ionospheric Impact on SAR Signals and Space Weather Interactions: Planetary ionospheres, charged particles, and solar radiation all impact radar signal distortions as an indirect method to examine ionospheric activity. Electron density variations caused by solar storms or planetary magnetic field interactions influence SAR wave propagation and serve as a means to analyze space weather influences on planetary atmospheres. Synthetic aperture radar ionospheric reflections can be used to chart ionospheric density variations, especially on terrestrial-type planets such as Earth and Mars [26].

Storm, Cyclone, and Weather Pattern Identification: SAR has the capacity to identify broad weather phenomena, such as Mars' dust storms, Titan's methane storms, and gas giant-like Jupiter's cyclonic processes. With its capacity for following the real-time development of a storm, researchers can determine its influences on planetary surface and climatic conditions [27].

Lightning and Electrical Discharge Investigations: Certain SAR geometries can also observe electromagnetic disturbances with lightning and electrical discharges within planetary atmospheres. For Jupiter, SAR may be applied to investigate lightning storms in its unstable atmosphere in combination with optical and radio measurements. Electrical discharge features related to Martian and Venusian dust storms and cloud developments might be investigated *via* SAR backscattering variations [28].

Hydrological and Precipitation Analysis: SAR can be employed to detect evidence of precipitation and liquid buildup in planetary atmospheres, especially on active hydrological cycle bodies like Titan. On Earth, SAR has been widely applied to track rainfall and flood regimes, a method that could be extended to the examination of extraterrestrial liquid methane and ammonia precipitation [29]. Radar reflectivity variations in regions of the atmosphere can be signs of changing moisture content and precipitation amounts.

Multi-Sensor Fusion for Atmospheric Analysis: SAR, in conjunction with spectrometry and magnetometry, offers a more complete view of planetary atmospheric states. With the integration of thermal and optical sensors, cross-validation of the data enhances the accuracy of atmospheric analysis. Data analysis through AI could be used in future missions to amplify SAR's ability to predict weather and climate on distant planets [30].

9.5.3 Synthetic Aperture Radar-Based Investigation of Dust Transport and Space Weathering

Two major processes that control the formation of planetary body surfaces are dust transport and space weathering, which are dominated by solar radiation, solar wind, and micrometeoroid impact. The SAR technology is particularly crucial in studying these processes because it offers high-resolution imagery and penetration into the subsurface, enabling one to study patterns of dust movement, surface alteration, and regolith development over time. On Mars, to consider one example, dust storms are the preeminent atmospheric process, periodically redrawing the surface and changing thermal and electric properties. Synthetic aperture radar can monitor these movements of dust, charting their erosion and deposition footprints, and observations of surface roughness and dielectric constant variations due to long-term exposure to solar radiation. Likewise, on airless bodies such as the Moon and Mercury, for which the space weathering is solar wind and micrometeoroid impact-driven, SAR allows regolith maturity to be discerned and scientists to determine the impact of solar-driven processes on radar reflectivity and mineralogy [31].

Outside planetary surfaces, SAR may be used to examine dust motion in interplanetary space, especially in CME-disturbed and solar wind variable zones. Electromagnetically excited charged dust grains play a critical role in structuring planetary magnetospheres and exospheres and are also factors of space weather that will affect future exploration missions. Utilizing polarimetric and multi-frequency SAR configurations, scientists can investigate the scattering properties of dusty regions, ascertain particle size distributions, and their role in planetary evolution. By combining SAR data with other observing methods, such as spectrometry and magnetometry, our ability to model dust transport processes, predict their long-term impact on planetary surfaces, and quantify the effect of solar activity on space weathering processes is enhanced. Future SAR-capable missions can provide a deeper understanding of these interactions, enabling advances in planetary science and interplanetary exploration.

9.6 Case Studies of Synthetic Aperture Radar in Solar System Exploration

Synthetic aperture radar has been at the core of planetary exploration, but its solar system science applications, specifically to investigate solar and heliophysical phenomena, are only beginning to be realized. Synthetic aperture radar technology can be used in space missions to enhance research on the Sun, its influence on the interplanetary medium, and solar radiation and solar wind-affected planetary surfaces. Here, we present distinguished case studies exemplifying the application of SAR in space exploration, starting with the PSP mission to observe the Sun's corona and the dynamics of solar wind. Further, we present conceptual missions that will utilize SAR technology to observe solar and interplanetary conditions. Lastly, we present a comparative overview of SAR missions in space science, examining their technological progress, setbacks, and opportunities for solar system exploration.

9.6.1 Parker Solar Probe and Its Observational Capabilities

The NASA PSP, which was launched in 2018, Figure 9.4 shows the schematic view of PSP instruments with their components is among the most sophisticated missions dedicated to understanding the outer corona of the Sun and its interaction with the adjacent interplanetary medium. Though the probe does not have an onboard SAR system, its strategy for high-resolution imaging and electromagnetic field sensing serves as the backbone for the incorporation of SAR on future solar missions. The PSP uses an array of instruments like the Wide-Field Imager and FIELDS Experiment to observe dynamic solar structures, including CMEs and solar wind turbulence. Synthetic aperture radar technology has the potential for such missions in the form of solar plasma structures mapped at high resolution, electromagnetic field variations observed at fine scales, and tracking solar-driven space

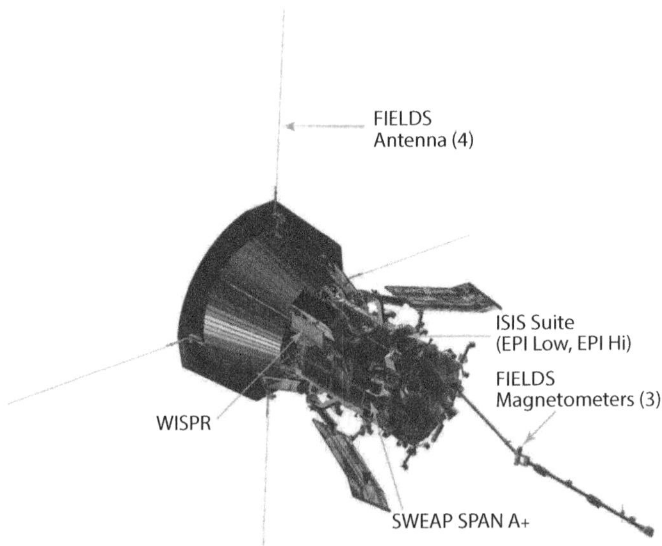

Figure 9.4 Schematic view of PSP instruments [Source: Wikipedia].

weather impacts. A SAR system designed for such a harsh environment would need to be high-power efficient, radiation-shielded, and capable of real-time data processing to offer greater insights into solar activity [32].

As PSP heads towards the Sun, the data it offers creates an essential basis for SAR-fitted spacecraft design for a mission that might be conducted in harsh thermal and radiation environments. Incorporating SAR in these missions would boost the imaging potential of 3D mapping of CMEs, real-time analysis of solar wind speed by Doppler, and observations of sub-surface solar structure. Next-generation ideas based on PSP could include SAR-fitted small satellites acting in coordination with probes, such as PSP, to perform multi-frequency radar imaging of the heliosphere to enhance space weather forecasting and planning for deep-space travel [33].

9.6.2 Conceptual Missions Utilizing Synthetic Aperture Radar for Solar and Interplanetary Studies

Conceptual SAR missions for solar and interplanetary studies aim at pushing radar-based observational methods beyond planetary surfaces to active solar and space plasma environments. In contrast to traditional SAR utilized for Earth and planetary imaging, these missions would utilize ultra-sensitive, space-borne radar systems with the capability of observing plasma structures, magnetized turbulence, and solar ejecta propagation. Conceptualized missions, such as Solar Radar Explorer (SRE), would send SAR arrays to solar orbit to chart the interaction of solar plasma, magnetic fields, and interplanetary medium. Another idea is interferometric SAR methods used in heliospheric research, with a fleet of small SAR satellites creating a synthetic aperture large enough to track solar ejecta and shock waves through the solar system ultra-precisely [34].

An essential design problem with such missions involves supporting SAR operations during harsh solar radiation and thermal gradients. Subsequent missions would consider using adaptive, light antennas, onboard ML-based processing, and efficient energy radar topologies to remain durable in conditions of solar proximity. Synthetic aperture radar technology can further be employed with the aid of spectrometry, magnetometry, and plasma diagnostics to provide detailed insight into the solar-induced phenomenon of space weather effects on planetary atmospheres, as well as heliosphere structure. The concept studies exemplify how SAR will transform real-time monitoring of the Sun's control over the solar system [35].

9.6.3 Comparative Analysis of Synthetic Aperture Radar Missions in Space Science

Synthetic aperture radar has been widely applied in planetary exploration, and missions like NASA's Magellan (Venus), Figure 9.5 depicts the pictorial view of NASA's Magellan. The European Space Agency's Mars Express, and NASA-ISRO's NISAR have highlighted the capability of SAR to image through planetary surfaces and delineate geological structures. However, using SAR in solar and interplanetary science has a distinct set of challenges and opportunities over conventional planetary SAR applications. Planetary SAR systems mostly work at microwave frequencies (L-band, X-band, C-band) for subsurface and surface imaging, while solar-oriented SAR would need to work at higher frequencies and tailored polarization methods to meaningfully interact with the dynamic plasma environment of the Sun [36].

In the context of comparing the application of SAR in planetary vs. solar studies, major differences occur in signal processing techniques, antenna design, and mission constraints. Planetary SAR missions use reflected radar signals from solid surfaces, whereas solar SAR would have to interpret reflections from highly ionized, turbulent plasma environments. Furthermore, planetary SAR systems are intended for image-stable surfaces, while solar SAR would be geared towards monitoring swift, high-energy phenomena such as CMEs, solar flares, and magnetized plasma waves. Based on a comparison of historical and planned missions, SAR incorporation into heliophysics is not only possible but could yield revolutionary improvements in space weather prediction and deep-space travel.

Figure 9.5 NASA's Magellan [Source: NASA Science].

With the use of AI-based data analysis and multi-sensor fusion, upcoming SAR missions can close the gap between planetary and solar radar use and provide a new era of space-based SAR studies [37].

9.7 Future Trends and Emerging Technologies

As the technology of SAR develops, so too are expected to be revolutionary improvements in solar system exploration. Emerging trends revolve around advanced data processing abilities, enhanced synergies with complementary observation methods, and more effective, miniaturized SARs for deep-space exploration. Machine learning-based analysis of data is transforming SAR use by facilitating automatic feature extraction, pattern detection, and real-time decision-making to guide spacecraft control. The conjunction of SAR with magnetometry and spectrometry provides a multi-modal approach to studying planetary surfaces, atmosphere, and heliophysical interactions with higher accuracy. Additionally, advances in the technology of miniaturized and autonomous SAR are facilitating low-cost, small-form missions that can operate in hostile space environments. In the coming years, the strategy for future SAR missions suggests the application of AI, multi-sensor fusion, and next-generation power management methods to ensure solar system exploration becomes more efficient and scientifically fruitful.

9.7.1 Artificial Intelligence-Driven Data Analysis in Synthetic Aperture Radar Applications

The complementary combination of AI and ML in SAR technologies is revolutionizing data interpretation and processing, especially when it comes to the exploration of the solar system. While SAR data processing was formerly synonymous with sophisticated signal processing techniques, AI techniques are now responsible for facilitating feature extraction, classification, and anomaly detection within gigantic databases in real-time. This greatly enhances efficiency as well as accuracy, particularly invaluable in deep space missions involving humongous amounts of observation data. Deep learning techniques refine SAR performance through enhanced image clarity, noise elimination, and repairing damaged or incomplete images. These make data stronger and maintain their scientific worth despite receiving incomplete and degraded signals from remote spacecraft. Artificial intelligence is key to detecting solar activity patterns such as solar flares and CMEs through observing SAR-based images. It can image

CME structure, speed, and propagation to help predict space weather and its effects on planetary atmospheres. In addition, AI-driven SAR applications enable planetary surface monitoring through the generation of high-resolution change maps. The maps reveal solar radiation-produced changes, i.e., regolith movement and atmospheric variation, giving insight into solar domination of planetary bodies. In autonomous space missions, AI reduces the need for significant quantities of data transmission and acquisition through pre-processing and filtering onboard, which is important information. It improves deep-space operations efficiency as well as mitigates dependence on ground processing to make faster decision-making and improve mission flexibility [38].

9.7.2 Integration of Synthetic Aperture Radar with Spectrometry and Magnetometry

Integrating SAR with spectrometry and magnetometry provides a multimodal platform for observing solar system phenomena. Synthetic aperture radar imaging is optimally suited for surface and subsurface mapping, while spectrometry is applicable in material composition analysis, and magnetometry provides data on magnetic field variations. This integration is particularly effective for observing planetary bodies affected by solar radiation and space weather [39]. For example, SAR detects planetary surface alteration caused by solar wind interaction, while spectrometry establishes the elemental composition of altered regions, and magnetometry establishes the impact on planetary magnetic fields. The combined technique enhances the investigation of planetary atmospheres, space weather coupling, and solar-driven geological processes, leading to enhanced understanding of the solar system.

9.7.3 Advancements in Miniaturized and Autonomous Synthetic Aperture Radar Systems

Miniaturization of SAR technology is enabling the attachment of small power-sipping SAR instruments to mini-satellites, CubeSats, and deep-space probes. Thanks to these breakthroughs, cheap, mass-producible SAR missions can be placed in orbit in huge constellations to monitor solar system activity constantly. Deep-space exploration also relies increasingly on autonomous SAR systems where decision-making is required on a real-time basis and under ever-changing situations. Self-tuning SAR sensors are capable of dynamically adjusting imaging parameters in response to changes observed, enhancing data acquisition efficiency. Additionally, power-efficient SAR architectures with enhanced power management methodologies are being researched to provide maximum performance in constrained energy environments, such as Sun proximity missions or interplanetary deep-space missions [40].

9.7.4 Roadmap for Future Synthetic Aperture Radar Missions in Solar System Exploration

The future of SAR in the exploration of the solar system revolves around growing autonomy, data fusion, and higher observational capability. The AI-supported SAR processing will enhance data accuracy to facilitate automated detection of solar events, planetary changes, and interplanetary interaction. Future missions will have SAR systems on board next-generation spacecraft, including lunar and Martian orbiters, Venus probes, and interstellar flights. The innovation of next-generation SAR antennas with improved frequency bands and adaptive polarimetric array configurations will further open the horizon for SAR applications. Moreover, combined missions of SAR with optical, infrared, and radio telescopes will provide a complete picture of solar-planetary interactions. With advancing technology, SAR will become a backbone in shaping space exploration, enabling planetary defense, space weather observation, and deep-space exploration.

9.8 Conclusion

The use of SAR in exploring the solar system is an important milestone for our capacity to explore the Sun, planets, and interplanetary space at a level previously unattained. Utilizing high-resolution images, penetration in subsurface information, and innovative signal processing tools, SAR has established its role in planetary and Earth sciences. As this technology finds application outside its conventional areas, its applicability in solar dynamics research, CMEs, and heliospheric interactions is increasingly obvious. The ability of SAR to observe solar-caused planetary surface deformation and atmospheric interaction also makes it more applicable to space exploration. But operating SAR systems in the harsh environment of space poses substantial engineering challenges, such as resistance to radiation, thermal management, and power efficiency. These challenges must be met by ongoing developments in antenna design, signal processing, and onboard data handling methods to achieve efficient operation under high-radiation and deep-space conditions.

In the future, the direction of SAR in solar system exploration will be dictated by new technologies, such as AI-assisted data processing, sensor integration, and miniaturized autonomous SAR payloads. Synthetic aperture radar combined with spectrometry and magnetometry will enable a fuller understanding of planetary and heliophysical phenomena and strengthen constructive collaboration between multiple observation methods. Artificial intelligence-driven automation will transform SAR data processing, allowing real-time feature detection and adaptive mission planning. Moreover, the creation of smaller, power-efficient SAR systems will unlock new possibilities for deep-space missions, making cost-effective, scalable deployments throughout the solar system possible. The future SAR mission roadmap emphasizes interdisciplinary collaboration, integrating aerospace engineering, planetary science, and AI to advance the frontiers of space research. With the advancement of SAR technology, its contribution to demystifying the solar system will continue to grow, with new scientific breakthroughs and the development of our understanding of the Sun's role in planetary conditions.

AI disclosure statement: The author acknowledges the use of AI-based tools for minor language editing or grammatical correction. However, all intellectual contributions, analyses, and conclusions presented in this chapter are entirely the author's own.

Bibliography

1. Slade, M.A., Benner, L.A., Silva, A., Goldstone solar system radar observatory: Earth-based planetary mission support and unique science results. *Proc. IEEE*, *99*, 5, 757–769, 2010.
2. Dell'Agnello, S., Boni, A., Cantone, C., Ciocci, E., Martini, M., Patrizi, G., Maiello, M., Next-generation laser retroreflectors for GNSS, solar system exploration, geodesy, gravitational physics and earth observation, in: *International Conference on Space Optics—ICSO 2014*, vol. 10563, SPIE, pp. 347–355, 2018, January.
3. Rincon, R., Carter, L., Lu, D., Du Toit, C., Perrine, M., Hollibaugh-Baker, D.M., Neish, C.D., Space Exploration Synthetic Aperture Radar (SESAR), in: *IGARSS 2019-2019 IEEE International Geoscience and Remote Sensing Symposium*, IEEE, pp. 8320–8323, 2019, July.
4. Herman, D.H., Niehoff, J.C., Spadoni, D.J., An alternative approach to solar system exploration planning. *Acta Astronaut.*, *8*, 11-12, 1371–1388, 1981.
5. Prats-Iraola, P., Scheiber, R., Rodriguez-Cassola, M., Mittermayer, J., Wollstadt, S., De Zan, F., Moreira, A., On the processing of very high

resolution spaceborne SAR data. *IEEE Trans. Geosci. Remote Sens., 52*, 10, 6003–6016, 2014.

6. Ciarletti, V., A variety of radars designed to explore the hidden structures and properties of the Solar System's planets and bodies. *C.R. Phys., 17*, 9, 966–975, 2016.
7. Elachi, C., Space imaging radar in planetary exploration and earth observation. *AIAA J., 39*, 4, 553–563, 2001.
8. Alberti, G., Biccari, D., Bortone, B., Caramiello, C., Catallo, C., Croce, A., Seu, R., The Ground System of the SHAllow RADar (SHARAD) Experiment. *Memorie della Soc. Astronomica Ital. Supplementi, 11*, 57, 2007.
9. Xue, F., Wang, X., Xu, F., Wang, Y., Polarimetric SAR interferometry: A tutorial for analyzing system parameters. *IEEE Geosci. Remote Sens. Mag., 8*, 2, 83–107, 2020.
10. Domenech, I.I., Rincon, R., Carter, L., Perrine, M., Lu, D., Banting, R., Nehmetallah, G., Systems integration and environmental testing activities performed on the Space Exploration Synthetic Aperture Radar (SESAR), in: *Sensors and Systems for Space Applications XIV*, vol. 11755, SPIE, pp. 42–56, 2021, April.
11. Toro, D.G., Arzel, M., Seguin, F., Jézéquel, M., Soft error detection and correction technique for radiation hardening based on C-element and BICS. *IEEE Trans. Circuits Syst. II: Express Br.*, 61, 12, 952–956, 2014.
12. Al-Amri, F., Maatallah, T.S., Al-Amri, O.F., Ali, S., Ali, S., Ateeq, I.S., Kayed, T.S., Innovative technique for achieving uniform temperatures across solar panels using heat pipes and liquid immersion cooling in the harsh climate in the Kingdom of Saudi Arabia. *Alexandria Eng. J., 61*, 2, 1413–1424, 2022.
13. Kim, Y. and Jordan, R.L., Spaceborne SAR Antennas for. *Spaceborne Antennas Planet. Explor.*, 305, 2006.
14. Shimada, M., *Imaging from spaceborne and airborne SARs, calibration, and applications*, CRC press, 2018.
15. Guo, H., Wang, Z., Li, Z., Zhang, Q., Liu, L., Tang, F., Zhang, Z., Wide-swath Ocean Current Measurement Based on MIMO Along-track Interferometry SAR, in: *IEEE Trans. Geosci. Remote Sens.*, 2024.
16. Elhadidy, M.A. and Shaahid, S.M., Parametric study of hybrid (wind+ solar+ diesel) power generating systems. *Renewable Energy, 21*, 2, 129–139, 2000.
17. Mohandes, M.A., Modeling global solar radiation using Particle Swarm Optimization (PSO). *Sol. Energy, 86*, 11, 3137–3145, 2012.
18. Sarkar, S.N., Alsop, D.C., Madhuranthakam, A.J., Busse, R.F., Robson, P.M., Rofsky, N.M., Hackney, D.B., Brain MR imaging at ultra-low radiofrequency power. *Radiology, 259*, 2, 550–557, 2011.
19. Paterna, S., Advanced Methodologies for Planning and Scheduling Payload Operations for Planetary Exploration Missions, 2022.
20. Shimada, M., Tadono, T., Rosenqvist, A., Advanced Land Observing Satellite (ALOS) and monitoring global environmental change. *Proc. IEEE, 98*, 5, 780–799, 2009.

21. Feng, L., Gan, W., Liu, S., Wang, H., Li, H., Xu, L., Ying, B., Space weather related to solar eruptions with the ASO-S mission. *Front. Phys.*, *8*, 45, 2020.
22. Kowalski, M.P., Barstow, M., Bruhweiler, F., Cruddace, R., Dupree, A., Holberg, J., Naval Research Lab Washington DC, Million Degree Plasmas in Extreme Ultraviolet (EUV) Astrophysics. White Paper in Response to Astro2010 Science Call, 2010.
23. Pyrina, M., North Atlantic Ocean decadal variability over the past millennium from climate model simulations and proxy based reconstructions.
24. Glaze, L.S., Wilson, C.F., Zasova, L.V., Nakamura, M., Limaye, S., Future of Venus research and exploration. *Space Sci. Rev.*, *214*, 1–37, 2018.
25. Palaszewski, B.A., Atmospheric Mining in the Outer Solar System: Resource Capturing, Exploration, and Exploitation, in: *49th AIAA/ASME/SAE/ASEE Joint Propulsion Conference*, p. 3765, 2013.
26. Hapgood, M., Linking Space Weather Science to Impacts—The View from the Earth, in: *Extreme events in Geospace*, pp. 3–34, Elsevier, 2018.
27. Liu, Y., Racah, E., Correa, J., Khosrowshahi, A., Lavers, D., Kunkel, K., Collins, W., Application of deep convolutional neural networks for detecting extreme weather in climate datasets. *arXiv preprint arXiv:1605.01156*, 2016.
28. James, M.R., Wilson, L., Lane, S.J., Gilbert, J.S., Mather, T.A., Harrison, R.G., Martin, R.S., Electrical charging of volcanic plumes. *Space Sci. Rev.*, *137*, 399–418, 2008.
29. Rink, K., Şen, O.O., Hannemann, M., Ködel, U., Nixdorf, E., Weber, U., Kolditz, O., An environmental exploration system for visual scenario analysis of regional hydro-meteorological systems. *Comput. Graphics*, *103*, 192–200, 2022.
30. Priyanka, E.B., Thangavel, S., Mohanasundaram, R., Anand, R., Solar powered integrated multi sensors to monitor inland lake water quality using statistical data fusion technique with Kalman filter. *Sci. Rep.*, *14*, 1, 25202, 2024.
31. Valla, P.G., Lowick, S.E., Herman, F., Champagnac, J.D., Steer, P., Guralnik, B., Exploring IRSL50 fading variability in bedrock feldspars and implications for OSL thermochronometry. *Quat. Geochronol.*, *36*, 55–66, 2016.
32. Raouafi, N.E., Matteini, L., Squire, J., Badman, S.T., Velli, M., Klein, K.G., Woolley, T., Parker solar probe: Four years of discoveries at solar cycle minimum. *Space Sci. Rev.*, *219*, 1, 8, 2023.
33. Hadid, L.Z. and Witasse, O., Parker Solar Probe, in: *Encyclopedia of Astrobiology*, pp. 2253–2255, Springer Berlin Heidelberg, Berlin, Heidelberg, 2023.
34. Piotrowski, W.L., Use of Space Station for Earth and Planetary Exploration, in: *Optical Engineering for Cold Environments*, vol. 414, pp. 162–178, SPIE, 1983, September.

35. Young, L.A. and Aiken, E.W., Vertical lift planetary aerial vehicles: Three planetary bodies and four conceptual design cases, in: *27th European Rotorcraft Forum*, Moscow, Russia, 2001, September.
36. Zhang, K., Guo, H., Jiang, D., Han, C., Analysis of Geometric Characteristics and Coverage for Moon-Based/Spaceborne Bistatic SAR Earth Observation. *Remote Sens.*, *15*, 8, 2151, 2023.
37. Jia, Y., Zhang, Z., Qin, L., Ma, T., Lv, B., Fu, Z., Zou, Y., Research of lunar water-ice and exploration for china's future lunar water-ice exploration. *Space Sci. Technol.*, *3*, 0026, 2023.
38. Ben-Itzhak, S., The Use of Artificial Intelligence in Outer Space Capabilities. *SAIS Rev. Int. Aff.*, *44*, 2, 5–30, 2024.
39. Bennett, J.S., Vyhnalek, B.E., Greenall, H., Bridge, E.M., Gotardo, F., Forstner, S., Bowen, W.P., Precision magnetometers for aerospace applications: A review. *Sensors*, *21*, 16, 5568, 2021.
40. D'Amico, S., Pavone, M., Saraf, S., Alhussien, A., Al-Saud, T., Buchman, S., Farhat, C., Miniaturized autonomous distributed space system for future science and exploration, in: *International Workshop on Satellite Constellations and Formation Flying*, pp. 1–20, International Astronautical Federation's Astrodynamics Committee, Delft, The Netherlands, 2015, June.

Part 4

CHALLENGES, POLICIES, AND FUTURE DIRECTIONS

10

Applications of Synthetic Aperture Radar in Disaster Management

Gagandeep Singh

National Institute of Disaster Management, Ministry of Home Affairs, New Delhi, India

Abstract

Synthetic aperture radar (SAR) technology has emerged as a valuable asset for disaster risk management by providing advanced imaging capabilities that surpass traditional optical remote sensing approaches. It holds clear advantages over optical systems through its ability to collect data regardless of weather conditions or light levels, offering persistent monitoring possibilities. This comprehensive review examines the diverse applications of SAR across all phases of disaster governance. Recent innovations in SAR instrumentation now facilitate near-real-time, high-definition geospatial intelligence. Analytical techniques, including coherence analysis, deformation time-series mapping, and fused SAR-optical datasets, demonstrate great potential for flood surveillance, landslide detection, critical infrastructure oversight, and glacial lake monitoring. Real-world cases from India, Nepal, and the United States illustrate how SAR data support impact-based forecasting, emergency response coordination, and post-catastrophe recovery in diverse geohazard contexts. However, challenges persist that hamper SAR's full integration into global disaster management strategies, particularly in low- to middle-income nations. The open sharing of SAR archives, cross-sector collaboration platforms, and locally contextualized capacity-building initiatives could help realize SAR's promise. When paired with artificial intelligence, light detection and ranging, global navigation satellite systems, and web-based platforms, SAR may even transform hazard monitoring, resilience planning, and climate change adaptation for increasingly complex risk landscapes.

***Keywords*:** Synthetic aperture radar (SAR), remote sensing, hazard monitoring, disaster risk management, artificial intelligence

Email: gagandeep.nidm@nic.in

Pooja Joshi, Ashutosh Bhatt, Neelam Sharma and Kapil Joshi (eds.) Synthetic Aperture Radar for Disaster Management: Bridging Science and Solutions, (267–304) © 2026 Scrivener Publishing LLC

10.1 Introduction: Unveiling the Potential of Synthetic Aperture Radar in Disaster Management

Communities across the globe are prone to higher risks from natural hazards due to altered climate patterns and changes within geophysical situations [1]. Increased risks need monitoring systems that provide accurate, detailed information within the correct time durations for support of enhanced decision-making processes. Synthetic aperture radar (SAR) provides systems with a necessary base, offering unique advantages that exceed the capabilities of earlier optical remote sensing systems. It is distinguished by its microwave frequency operation, different from optical sensors, as it permits nocturnal as well as adverse-weather imagery [2]. This capability offers the greatest value during crises, as rapid and reliable intelligence is of prime necessity.

Disasters such as landslides, floods, cyclones, earthquakes, and wildfires are increasingly becoming frequent and often extreme around the world [3, 4]. Some of these events occur when cloud coverage becomes extended or water becomes excessive, or when there is total darkness, causing low optical imagery efficiency [5]. The specific capability of SAR makes it capable of penetrating harsh environments, hence providing reliable and precise imaging information regularly [6]. The capability of SAR for day and nighttime target monitoring regardless of any condition makes it highly performative for observing operational tasks from the production of flood maps, building damage surveys, surface deformation identification, and vegetation health analysis. Disaster risk managers now integrate SAR information into full multi-hazard evaluation programs as well as early-warning systems and employ it for post-disaster recovery analysis [7]. The research methods using SAR technology have progressed significantly as they transitioned from being general support for qualitative research for conducting quantitative geospatial analysis [8–16]. High-end SAR methods such as differential interferometric SAR (D-InSAR), as well as persistent scatterer SAR (PS-InSAR) and polarimetric SAR (PolSAR) increase opportunities for high-resolution spatial analysis with temporal accuracy [17–20].

Technological improvements like SAR-optical fusion and AI-supported platforms, viz., Google Earth Engine (GEE) [21], Sentinel Application Platform (SNAP) Toolbox, have simplified processing and broadened access, allowing even non-specialists to utilize SAR for real-time disaster monitoring.

This chapter captures SAR's comprehensive role in disaster risk management from scientific foundations to operational best practices while highlighting how cutting-edge methodologies and real-world examples demonstrate SAR's effectiveness in building resilience and responding to evolving hazards.

10.2 Evolution of Platforms: From Inception to the New Millennium

The development of SAR has witnessed a remarkable trajectory, shaped by technological innovation, military imperatives, and growing civilian applications. This development has progressed from Cold War-era experiments to real-time global disaster observation systems.

Early Developments (1950s–1970s):
Synthetic aperture radar began as an experimental military technology during the Cold War, with airborne systems developed in the US and the Soviet Union, aimed at overcoming limitations of optical imaging under cloud cover or at night [22–24]. Goodyear's experimental SAR in 1951 marked one of the earliest demonstrations [25–27]. The first spaceborne radar, Seasat-A (1978) by the National Aeronautics and Space Administration (NASA), launched SAR into orbit with an L-band sensor. Although operational for only 106 days, Seasat provided the first-ever global radar dataset, setting a precedent for future missions [28].

First Operational Satellites (1990s):
The ERS-1 (1991) and ERS-2 (1995) missions, operated by the European Space Agency (ESA), introduced spaceborne C-band SAR, offering continuous and reliable imagery that revolutionized oceanographic research, land surface monitoring, and natural hazard assessment across the globe [29–31]. Japan's JERS-1 (1992) satellite mission deployed L-band SAR, becoming one of the earliest Earth observation systems targeted to forest structure mapping and biomass estimation, demonstrating sensitivity to canopy volume and vegetation moisture [32–34]. Canada's RADARSAT-1 (1995) mission used C-band radar offering multiple beam modes and flexible incidence angles widely used in ice mapping and disaster relief applications [35–37].

Second Generation High-Resolution Era (2000s):

- Envisat ASAR (2002–2012): ESA's follow-up with enhanced swath and modes expanded SAR capabilities for environmental monitoring and hazard [38].
- ALOS PALSAR (2006): Japan's L-band system became central to topographic mapping, vegetation analysis, and landslide detection [16].
- TerraSAR-X (2007, Germany): This mission introduced very high-resolution (up to 1 m) X-band imagery for urban infrastructure analysis and detailed change detection [39].
- COSMO-SkyMed (2007, Italy): A four-satellite constellation enabling dual-use civil-military monitoring, widely applied in earthquake and flood risk assessments [15].

Open Access and Sentinel Era (2010s):

- RADARSAT-2 (2007): The commercial successor of RADARSAT-1, RADARSAT-2 provided improved resolution and full polarimetric capabilities, making it highly suitable for ground deformation, ice monitoring, and oil spill detection [18].
- Sentinel-1A (2014) and Sentinel-1B (2016): As part of ESA's Copernicus Program, these satellites deliver free and open C-band SAR data, with 6–12-day revisit intervals, enabling cutting-edge research in flood mapping, land subsidence, and glacier motion monitoring [40, 41]
- ALOS-2 (2014): Japan's advanced L-band SAR mission featuring ScanSAR and Stripmap modes, ALOS-2 is widely used in landslide risk assessment, especially in vegetated and mountainous regions due to its deeper penetration and stability in InSAR applications [18].

Recent and Emerging Platforms (2020s):

- SAOCOM 1A/1B (Argentina, 2018/2020): These L-band SAR satellites, developed by Argentina's CONAE, complement Italy's COSMO-SkyMed X-band system as part of the SIASGE constellation, enabling comprehensive, multi-frequency Earth observation for soil moisture and hazard mapping [42].

- ICEYE and Capella Space (2019–ongoing): These NewSpace commercial providers have launched micro-SAR satellite constellations offering sub-meter resolution and daily revisit rates, proving valuable in disaster monitoring, infrastructure surveillance, and environmental change detection [42].
- NovaSAR-S (UK, 2018): Developed as the first operational S-band SAR satellite, NovaSAR-S was designed for flood monitoring, forestry, and agricultural applications, with a focus on cost-effective Earth observation [42].
- NISAR (NASA-ISRO SAR Mission, expected 2025): The first dual-frequency Land S-band SAR mission, NISAR, will enable precise monitoring of soil moisture, biomass, tectonic deformation, and disaster impact assessment [42].
- TanDEM-X & TerraSAR-X Tandem: These German radar satellites operate in close formation to generate global 3D elevation maps with meter-scale vertical accuracy, transforming terrain modeling and digital elevation model (DEM) generation.

These developments collectively mark a transition from experimental systems to global SAR constellations with near-real-time capabilities, supporting interdisciplinary applications in disaster risk reduction, hydrology, ecology, and infrastructure monitoring.

10.3 Foundations of Synthetic Aperture Radar Data Analysis: A Practical Guide for Disaster Practitioners

The power of SAR in disaster risk management depends not only on its imaging capabilities but also on correct data processing and interpretation. This section outlines key steps in SAR preprocessing, advanced analytical techniques, and fusion strategies.

10.3.1 Synthetic Aperture Radar Data Preprocessing

Preprocessing operations are needed for the evaluation process to ensure accurate interpretations and quantifiable analysis outputs. Preprocessing operations entail radiometric calibration, followed by speckle filtering, geometric correction, and terrain normalization as routine procedures

[43–45]. Synthetic aperture radar imagery is readied for disaster management applications through the implementation of preprocessing activities that transform data.

Synthetic aperture radar signals (digital numbers) undergo radiometric calibration to produce backscatter coefficients such as σ°, β°, or γ°, which enables comparison of different acquisitions and sensors [46]. Changes in the calibrated backscatter value enable important detection of temporal shifts and enable accurate disaster mapping classifications.

Synthetic aperture radar imagery contains multiplicative noise. Filters such as Lee, Frost, and Gamma-MAP are used to enhance image clarity, though excessive filtering can obscure fine details in complex urban or mountainous terrains [47]. The interpretation benefits from speckle filtering. However, excessive filtering produces spatial detail loss, affecting urban and mountainous areas that contain landslides and flash flood hazard zones [48–50].

The process of geocoding and orthorectification corrects geometric distortions that originate from sensor motion and topographic relief. The process ensures accurate spatial correlation between SAR data and other geospatial data when integrated. The Range-Doppler and Terrain Correction algorithms are standard methods that operate through the SNAP toolbox or ESA's Sentinel-1 toolbox for performing this task [51].

10.3.2 Advanced Analytical Techniques in Synthetic Aperture Radar Processing

Synthetic aperture radar's power lies in its ability to detect subtle ground and surface changes. Several advanced techniques have been developed to extract quantitative information from multi-temporal or polarimetric SAR datasets.

10.3.2.1 Interferometric Synthetic Aperture Radar

Interferometric synthetic aperture radar exploits the phase difference between two or more SAR images acquired over the same area at different times to measure ground displacement. It is particularly effective for: Earthquake-induced deformation [52], volcanic activity [53], and analysis of landslides and subsidence [15].

However, InSAR is sensitive to temporal and spatial decorrelation, especially in vegetated or snow-covered regions, which is why coherent processing chains like PS-InSAR [17] and small baseline subset (SBAS)

[54] techniques were introduced. These methods are suitable for long-term monitoring in urban environments and infrastructure assessments post-disaster.

10.3.2.2 Polarimetric Synthetic Aperture Radar

Polarimetric synthetic aperture radar measures the amplitude and phase of backscattered signals in different polarization combinations (HH, HV, VH, VV), providing insights into surface geometry, moisture content, and vegetation structure [55–57]. In disaster studies, PolSAR is effective for:

- Flood mapping under vegetation canopy, due to differential scattering characteristics [58]
- Damage classification in urban zones post-cyclones or earthquakes [59].

Advanced decomposition techniques like Freeman–Durden and Cloude–Pottier help disaggregate scattering mechanisms (surface, double-bounce, volume) to improve interpretability [60–62].

10.3.2.3 Time-Series and Change Detection Analysis

Change detection methods compare SAR images before and after a disaster to map the spatial extent and severity of changes. These include:

- Amplitude-based change detection, using difference or ratio images, offers a straightforward method for identifying large-scale changes such as flooding or deforestation, though it can be sensitive to noise and terrain effects [63].
- Coherence-based approaches, which utilize interferometric coherence loss to detect terrain disruption or structural damage, have been successfully applied to earthquakes and urban collapses [64].
- Time-series analytics, such as trend analysis of pre-failure slope movement, are critical for landslide prediction in high-risk areas using sensors like Sentinel-1 and ALOS-2 [65].

Multi-temporal data stacks from Sentinel-1, ALOS-2, or TerraSAR-X enable robust temporal assessments, providing high revisit frequency for ongoing monitoring of environmental hazards [66].

10.3.3 Synthetic Aperture Radar Data Fusion and Integration

To compensate for SAR's limitations particularly in spectral discrimination data fusion approaches are increasingly adopted. These involve integrating SAR with optical, light detection and ranging (LiDAR), or ancillary geospatial data for more comprehensive disaster analysis.

- Synthetic aperture radar–Optical Fusion: Used for flood mapping, where SAR detects water beneath clouds and optical sensors provide land cover context. Techniques such as PCA-based fusion and wavelet transforms have been applied for this purpose [6].
- Synthetic aperture radar–DEM Integration: DEMs (e.g., SRTM, ALOS-PALSAR) are used for terrain correction and hydrological modeling. Topographic features derived from DEMs have also been used as conditioning factors in flash flood and landslide susceptibility modeling [67].
- Synthetic aperture radar–AI Integration: Machine learning algorithms support vector machines, random forest, convolutional neural networks are increasingly used to classify flooded areas, detect damage, and assess structural integrity using SAR time series. Google Earth Engine and cloud platforms facilitate scalable processing of large SAR archives.

A thorough understanding of SAR data processing is essential for effectively utilizing it within disaster management. Each analytical step from fundamental preprocessing methods to sophisticated interferometric and machine-learning techniques enhances our ability to interpret SAR imagery meaningfully. Practitioners aiming to benefit from SAR must therefore possess not only technical expertise but also the skill to interpret SAR-derived information within the context of physical disaster processes. Numerous studies, such as those addressing flash flood susceptibility and vulnerability zonation in mountainous regions [68, 69], illustrate how SAR becomes most valuable when integrated into broader, multidisciplinary approaches. By combining remote sensing data, geospatial analysis, and informed decision-making, SAR significantly contributes to building resilience against disasters.

10.4 Synthetic Aperture Radar Applications Across the Disaster Management Lifecycle

10.4.1 Preparedness: Building Risk Knowledge and Monitoring Precursor Signals

The preparedness stage is concerned with identifying hazard-prone areas, as well as putting in place efficient early-warning systems, both of which receive a great advantage from SAR-based analysis. Synthetic aperture radar is an essential tool for making detailed mapping of previously hazard-prone areas, as well as continuous observation of precursors that could lead to hazardous phenomena, most notably for geohazards like landslides, subsidence, and glacier movement represented in Figure 10.1 [65, 70, 71].

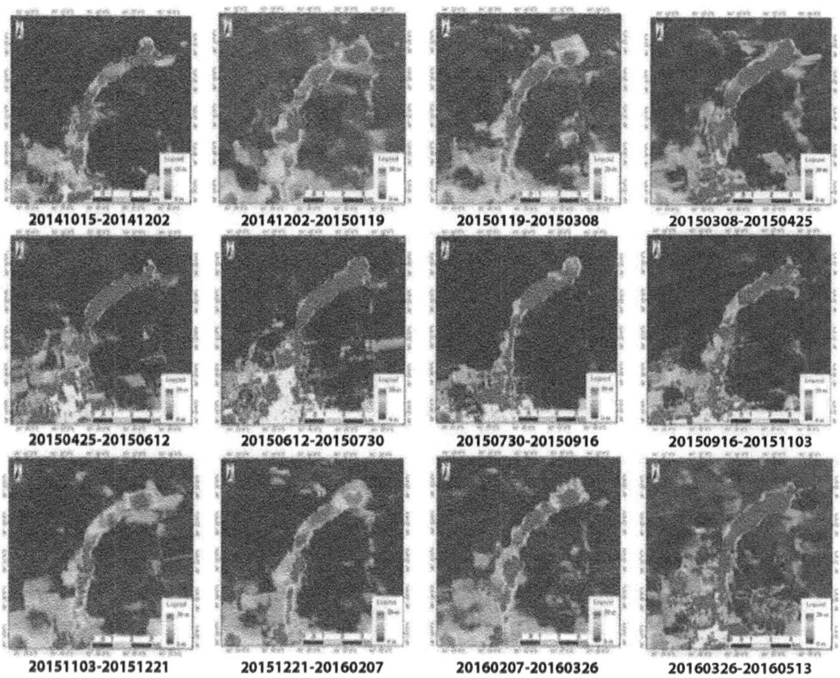

Figure 10.1 Movement distribution ranging from 0 (blue) to 20 (red) meters of the Yiga Glacier between October 15, 2014, estimated using Sentinel-1 images [70].

Hazard and Exposure Mapping

The application of high-resolution SAR imagery facilitates the production of precise baseline hazard maps, identifying clearly areas prone to flooding, unstable terrain subject to landslides, and urban areas of subsidence [67, 69]. These layers, when integrated with GIS data, enhance zoning and contingency planning as shown in Figure 10.2.

Monitoring Slope Stability

Persistent Scatterer InSAR detects slow movements over time, making it invaluable in mountainous terrain. In the Indian Himalayas, seasonal and tectonic deformation trends have been identified through SAR time series [72]. Figure 10.3 shows the PSI scatterer map representing the line-of-sight (LOS) velocity for the Kumaun region in the Indian Himalayas.

Monitoring Hydrometeorological Trends

Synthetic aperture radar measures water levels in rivers and wetlands, supporting flood forecasting. When fused with hydrological models, these insights improve flood early warnings [73].

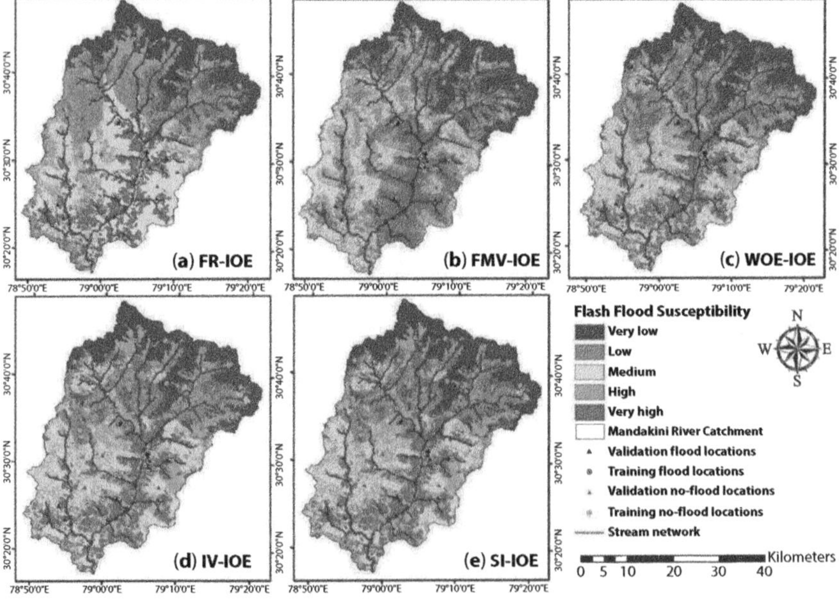

Figure 10.2 Flash flood susceptibility maps for Mandakini River Basin in India generated using multiple remote sensing-based datasets, including the SRTM DEM [67, 69].

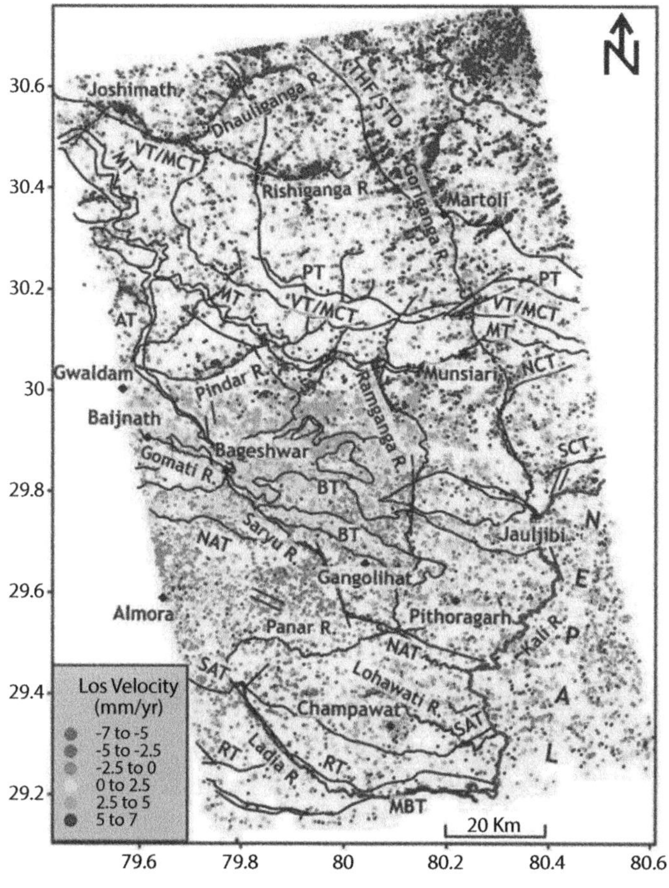

Figure 10.3 Line-of-sight velocity (mm/yr) of Kumaun Himalaya [72].

10.4.2 Response: Rapid Assessment and Real-Time Monitoring

During a disaster, timely, cloud-penetrating SAR data support emergency decisions on evacuation, aid, and infrastructure evaluation.

Flood Inundation Mapping
Satellites like Sentinel-1 and ALOS-2 can map flooded areas hours after onset even under cloud cover using backscatter contrast between water and land [58]. Time-sensitive flood maps have been produced using Sentinel-1 during the 2018 Kerala flood event in India, informing the rescue missions [74, 75]

Damage Detection in Urban Areas

Urban damage mapping after earthquakes or cyclones can be efficiently carried out using coherence analysis and amplitude change detection techniques in SAR. Studies have shown that coherence loss between pre- and post-event SAR pairs correlates strongly with building collapses and surface disruptions [76, 77]. Figure 10.4 shows the results of the coherence loss approach developed to identify and classify the severity of damage in Beirut city, in addition to the damage results reported by advanced rapid image analysis [aria.jpl.nasa.gov], overlaid on the post-event optical satellite image.

Wildfire and Cyclone Impact Monitoring

Though SAR is not the primary tool for active fire detection, it is highly effective in assessing post-fire vegetation degradation and cyclone-induced structural damage, especially through polarimetric decomposition and backscatter analysis [14].

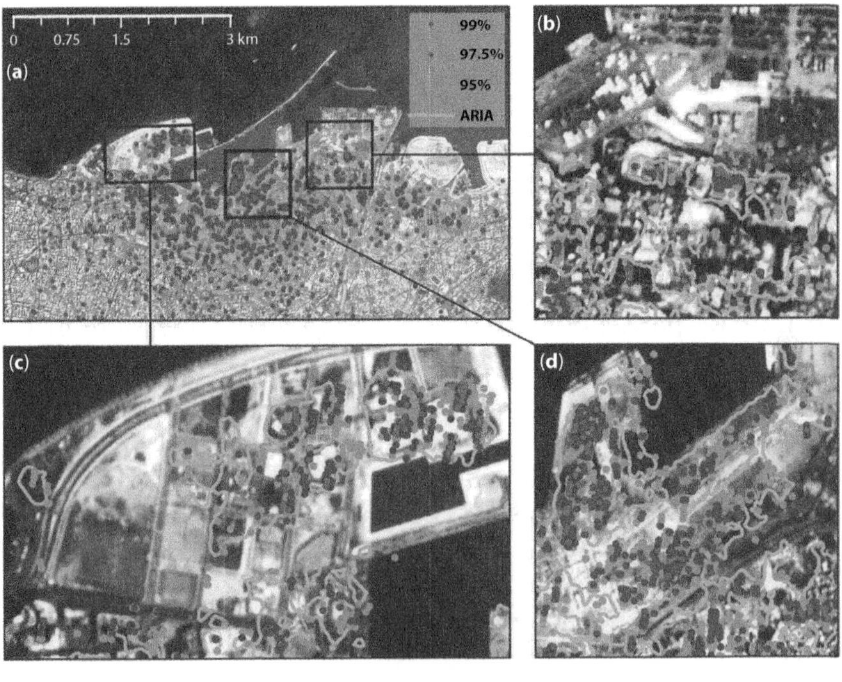

Figure 10.4 (a) Classification of severity of damages to buildings in Beirut city; (b–d) enlargement of damage sections for better illustration [77].

10.4.3 Recovery and Reconstruction: Supporting Long-Term Rehabilitation

The recovery and reconstruction phase focuses on restoring normalcy while building resilience to future disasters. Synthetic aperture radar supports post-disaster recovery through consistent monitoring of rehabilitation progress and landscape evolution [78, 79].

Infrastructure Monitoring
Both PS-InSAR and SBAS-InSAR track deformation in critical infrastructure such as dams, bridges, and roads. In Kathmandu, SAR data were used to understand ground changes post Nepal's 2015 earthquake as represented in Figure 10.5 [80].

10.4.4 Mitigation: Enhancing Resilience Through Informed Planning

Mitigation efforts use SAR to monitor environmental change and inform long-term land-use decisions.

Urban Expansion Monitoring
Synthetic aperture radar's high revisit frequency detects urban area expansion in floodplains and unstable slopes [81].

River Morphodynamics and Coastal Monitoring
Synthetic aperture radar tracks river shifts, coastal erosion, and delta changes helping inform shoreline defense planning. Figure 10.6 shows the flood dynamics in the Mekong Delta [82].

Soil Moisture and Hydrological Monitoring
Synthetic aperture radar can monitor temporal variations in soil moisture to support water management, although it does not measure absolute values directly [83].

The function of SAR has expanded intensely from simple flood-mapping applications to sophisticated ground-motion analysis and detailed infrastructure evaluation. Synthetic aperture radar observations are being used to support early-warning systems and also assist with resilient reconstruction. In the years ahead, expected developments in SAR sensor technology, near-real-time processing techniques, and interface with artificial intelligence techniques will further enhance its contribution toward constructing safer, disaster-resilient communities.

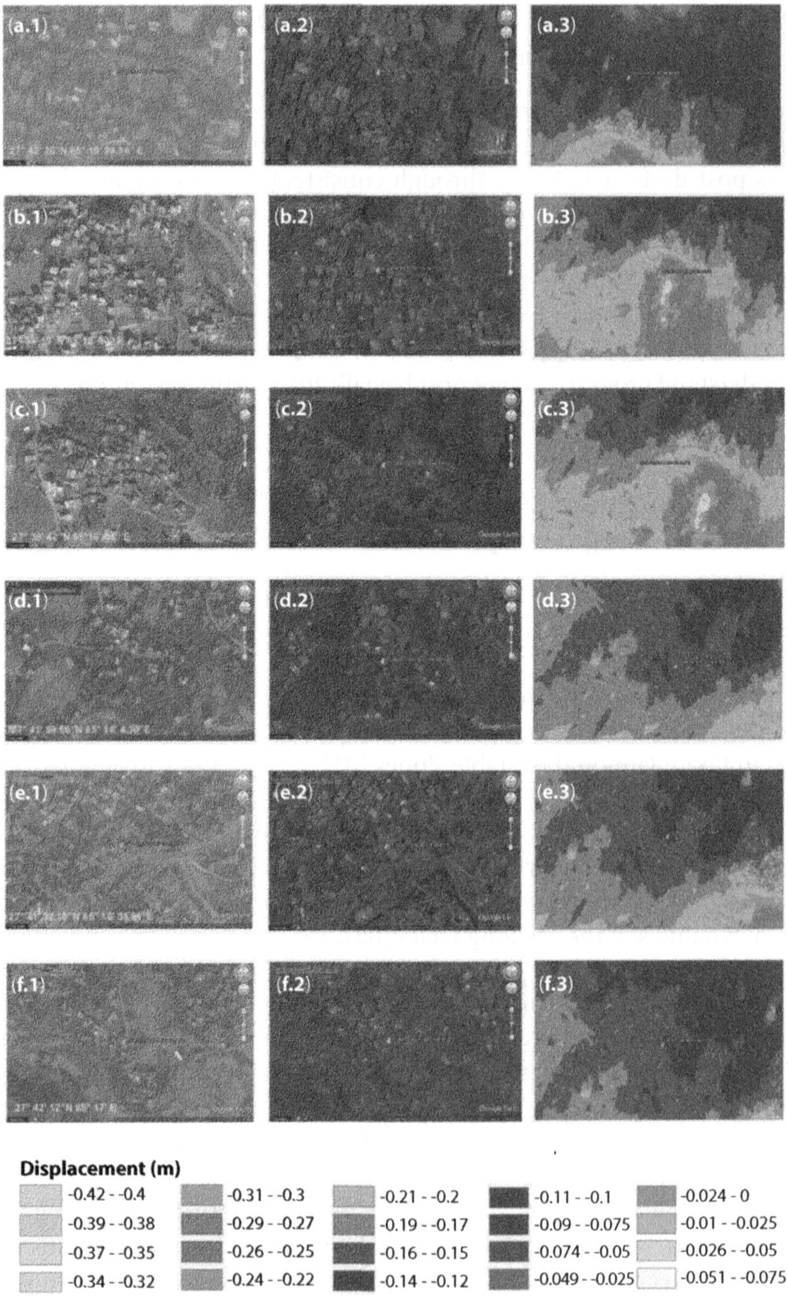

Figure 10.5 (a.1 to f.1) Pre-earthquake Google Earth images; (a.2 to f.2) post-earthquake scenario; (a.3 to f.3) land subsidence generated using SAR interferometry at the corresponding locations [80].

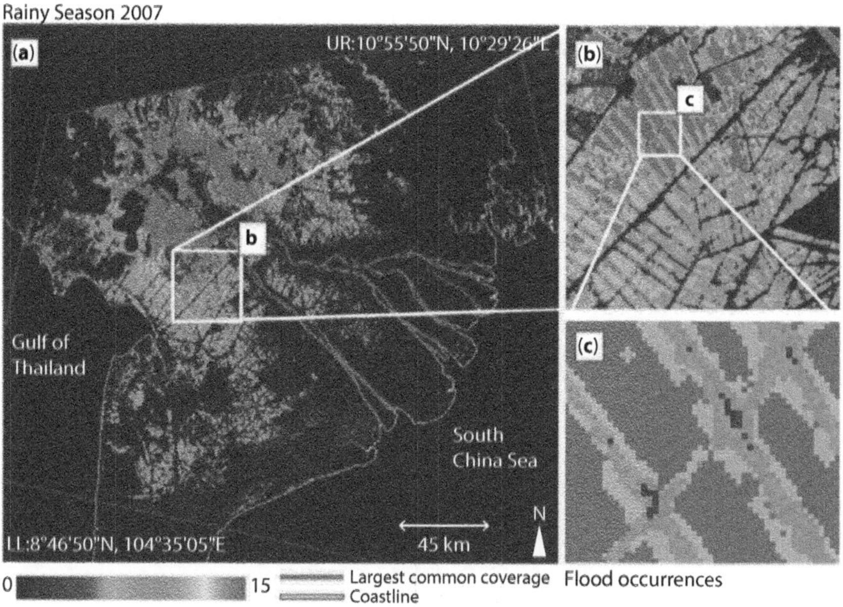

Figure 10.6 Flood dynamics in the Mekong Delta for the 2007 rainy season, based on 15 Envisat Advanced SAR Wide Swath Mode derived water masks [82].

10.5 Case Studies of Synthetic Aperture Radar in Action During Catastrophic Events

Assessing the efficiency of SAR for disaster risk management involves its functioning under actual circumstances. Case studies bring in-depth, evidence-based information on real-world applications, which complement the detailed analysis of SAR capabilities for disaster observation, emergency response, and post-disaster rehabilitation. The current section highlights well-chosen and well-documented international instances where SAR contributed significantly to the handling of diverse geohazards, as well as hydrometeorological hazards.

10.5.1 Nepal Earthquake (2015): Ground Deformation and Damage Mapping

The Mw 7.3 earthquake of Gorkha, affecting the central region of Nepal on April 25, 2015, caused more than 9,000 deaths, as well as widespread structural damage. Satellite SAR methods were employed soon after the earthquake for evaluating surface deformation, as well as urban damage [84–86]. Sentinel-1A SAR imagery from ESA was employed to produce differential interferograms (D-InSAR) shown in Figure 10.7 that identified surface displacements of as much as 2 meters within the Kathmandu

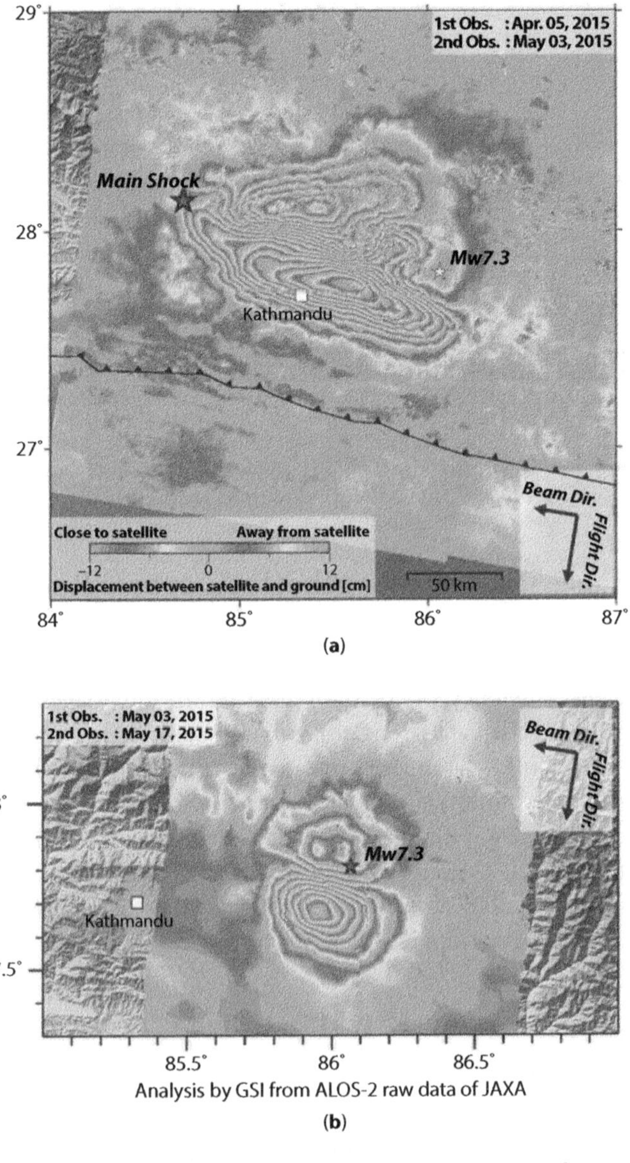

Figure 10.7 (a) Concatenated interferograms generated using SAR interferometry from ScanSAR data acquired before the earthquake; (b) Interferograms showing the crustal deformation associated with the earthquake [87].

Valley [87]. TerraSAR-X and COSMO-SkyMed data helped assess urban structural damage using coherence loss analysis [88]. Synthetic aperture radar insights mapped earthquake rupture zones and identified secondary risks like landslides, informing humanitarian response and reconstruction.

10.5.2 Kerala Floods (India, 2018): Rapid Flood Extent Mapping

Using both optical (Sentinel-2) and SAR (Sentinel-1) data, remote sensing techniques were used to detect changes immediately during the August 2018 flood disasters in Kerala, India. Researchers combined these datasets using GEE to use techniques like SAR backscatter thresholding and the normalized difference water index to differentiate permanent water bodies from flood-induced inundation zones [89]. Tiwari *et al.*, 2020 [74] used SAR data in GEE for mapping flood inundation in August from 2015 to 2018, as shown in Figure 10.8. The resulting real-time maps can support rescue logistics during peak flooding and can be integrated into government and NGO response workflows.

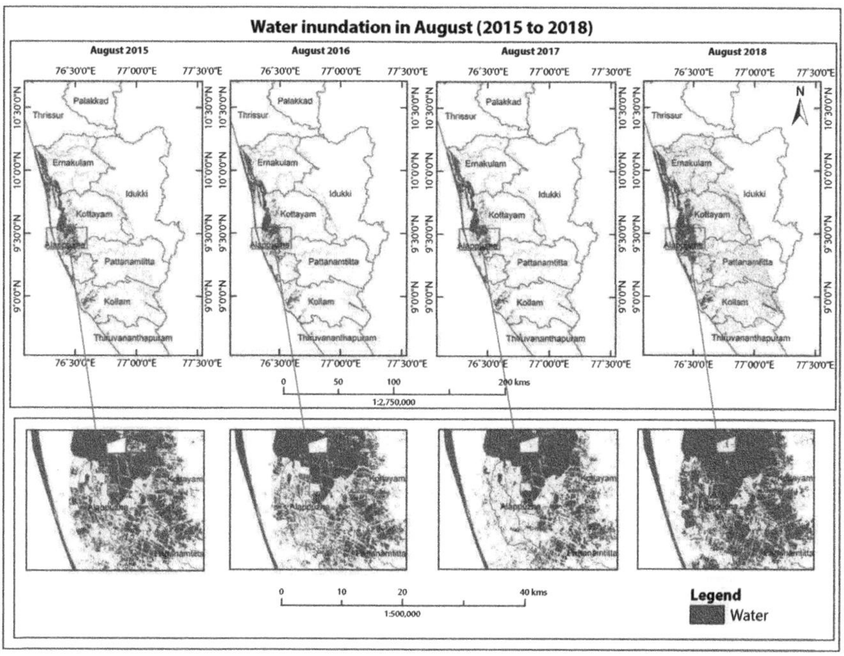

Figure 10.8 Water inundation map of Kerala floods from August 2015–2018 [74].

10.5.3 Cyclone-Induced Damage Assessment and Coastal Inundation

The use of SAR has greatly improved coastal flooding monitoring and cyclone-induced damage assessment in India. For example, after storms like Amphan and Yaas (shown in Figure 10.9), Sentinel-1 and ALOS-2 SAR datasets have been utilized in several coastal Indian states, such as West Bengal and Odisha, to evaluate the level of flooding, damage to infrastructure, and shoreline retreat [90, 91] These studies used InSAR, coherence

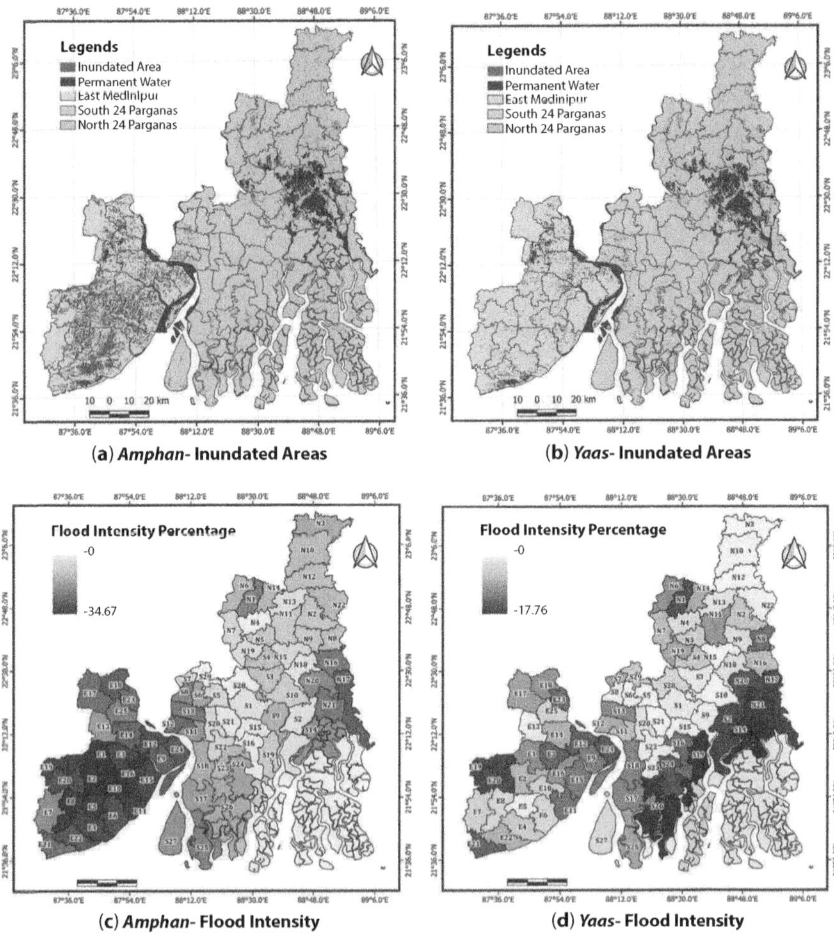

Figure 10.9 The extent of the inundated area was estimated from Sentinel-1 imagery acquired during the occurrence of the tropical cyclones Amphan (between May 20 and 5 June 5, 2020) and Yaas (between May 26 and June 10, 2021) in GEE [90].

analysis, and amplitude-based flood mapping to give practical insights into land subsidence, storm surge impacts, and inundation dynamics, especially when cloud cover prevented optical data [92, 93]. Near real-time mapping is now possible thanks to Google Earth Engine integration. This helps disaster response organizations with focused intervention and recovery.

10.5.4 Wayanad Landslide (India, 2024)

The 2024 Wayanad landslides, triggered by intense monsoonal rainfall across the Western Ghats of Kerala, prompted extensive use of SAR-based remote sensing, particularly Sentinel-1 InSAR time series, for mapping slope deformation, detecting early warning signals, and supporting

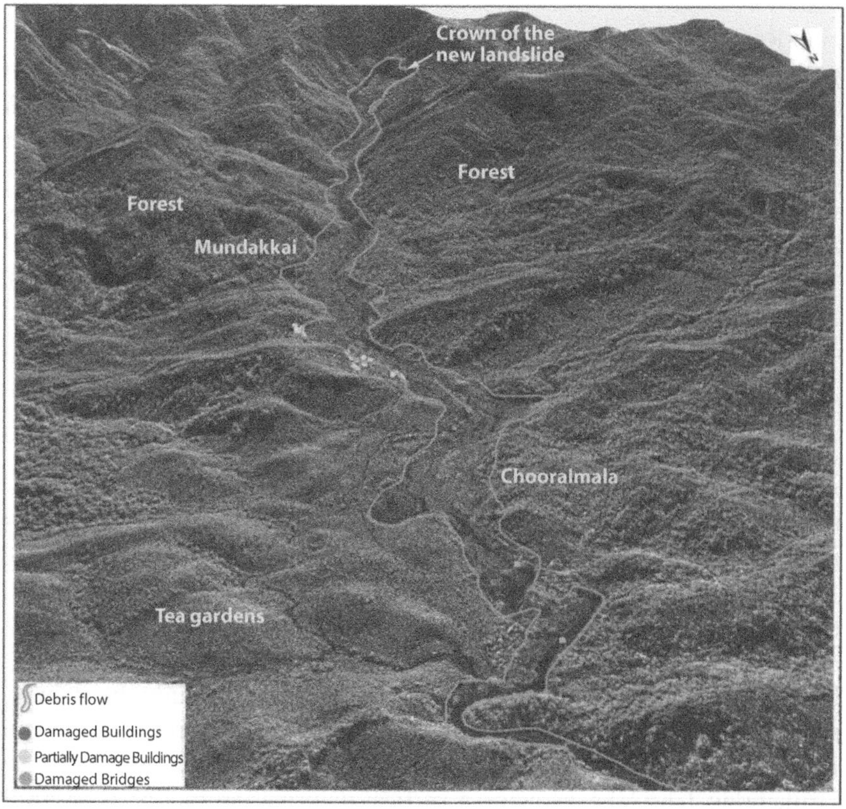

Figure 10.10 Map of damaged buildings in the Chooralmala debris flow. Damaged buildings were identified using PlanetScope, RISAT and Pleiades imagery and overlain on the RISAT image of 2 August 2024 and 31 July 2024 [94].

post-disaster assessment. Researchers employed PSI and SBAS to pinpoint evolving instability patterns (as shown in Figure 10.10), identifying critical precursors to the debris flows that devastated key regions [94]. The analysis established the precursor conditions, viz., slow ground movement detected for a year (up to 11mm per year), steep slopes and weakened rock formations, and intense rainfall (6% of annual rainfall in hours) as the final trigger. Further, the debris flow characteristics were also ascertained, indicating an extremely high velocity of 43 m/s, having massive destructive power. The flow transformed the landscape dramatically, wherein the river width expanded from 30 m to 154 m and carried massive boulders, trees, and soil downstream.

10.5.5 Hurricane Harvey (USA, 2017): Urban Flood Monitoring

Hurricane Harvey caused unprecedented flooding in the Houston metropolitan area in August 2017, resulting in over 100 deaths and $125 billion in damage. National Oceanic and Atmospheric Administration and NASA used data from the L-band ALOS-2 and X-band TerraSAR-X missions for change detection analysis. Amplitude-based and coherence-based methods were applied to detect urban inundation and infrastructure impact zones, as shown in Figure 10.11 [95]. Synthetic aperture radar data supplemented aerial reconnaissance and were integrated into FEMA's response protocols. The ability to detect water under cloud-covered urban areas enabled rapid estimation of affected populations and prioritization of relief efforts.

10.5.6 Glacial Lake Outburst Floods Threat in Lunana (Bhutanese Himalayas): Glacial Monitoring

Glacial lake outburst floods (GLOFs) are among the most dangerous hazards in high-mountain Asia. The Lunana region in the Bhutanese Himalayas has been identified as a high-risk GLOF-prone region due to the presence of five valley glaciers, as shown in Figure 10.12. Sentinel-1 SAR time series images were used to monitor the lake outburst susceptibility. Furthermore, the PSI approach was employed to study the stability of moraines and slopes surrounding the glacial lakes. The results shown in Figure 10.13 clearly reflected the instability of the terminal moraine of the Drukchung Tsho lake, showing a vertical deformation rate of -2.52 mm/year. Similarly, the lateral moraines of two other lakes, namely Thorthomi

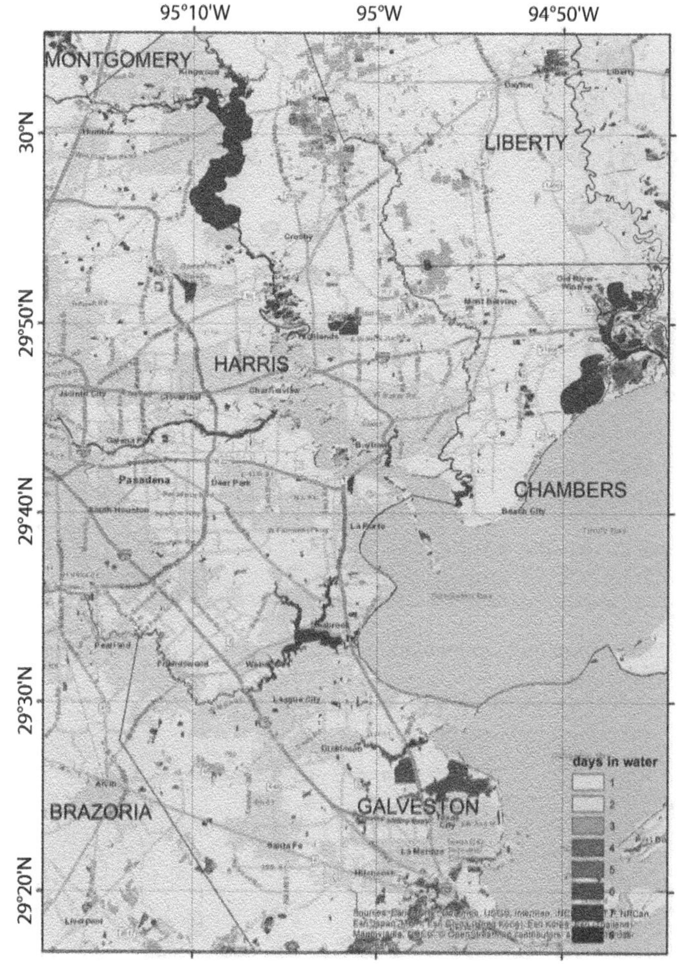

Figure 10.11 The observed times of inundated areas during Sentinel-1 revisits [95].

Tsho and Lugge Tsho, were found unstable with a projected vertical deformation rate of -2.47 mm/year and -2.73 mm/year, respectively [96].

10.5.7 Beirut Port Explosion (Lebanon, 2020): Urban Damage Analysis

Synthetic Aperture Radar Data & Technique

The Sentinel-1A SAR images were utilized to analyze urban damage using coherence change analysis during the August 2020 Beirut Port Explosion. Widespread structural damage was suggested by a notable decrease in

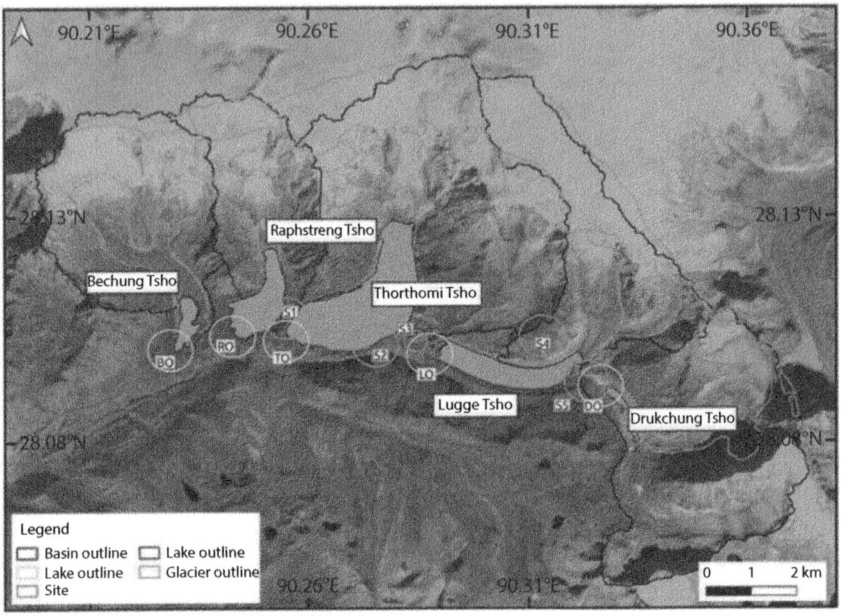

Figure 10.12 Geographical representation of five glacial lakes in the Luanna region of the Bhutanese Himalayas.

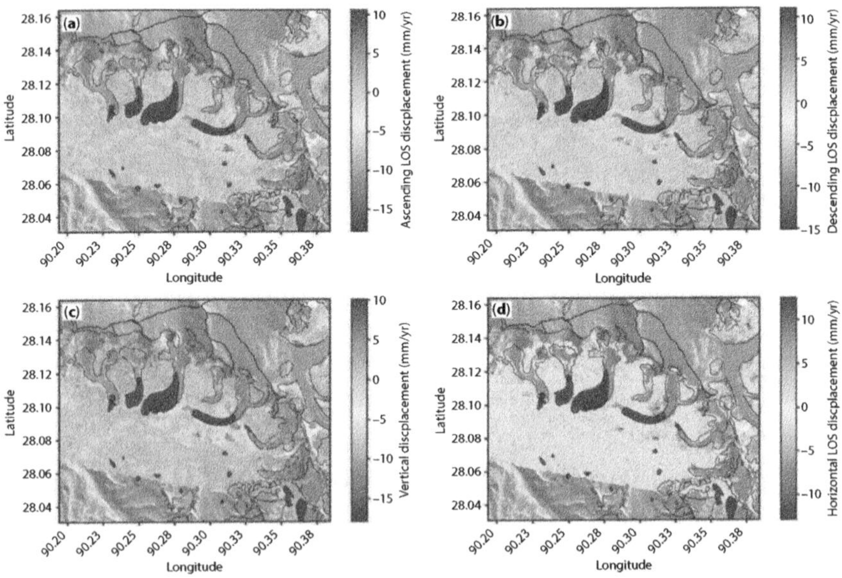

Figure 10.13 Deformation rate maps, Luanna, the Bhutanese Himalayas. (a) Ascending LOS displacement (mm/yr), (b) descending LOS displacement (mm/yr), (c) projected vertical displacement (mm/yr), and (d) projected horizontal displacement (mm/yr) [image source- [96]].

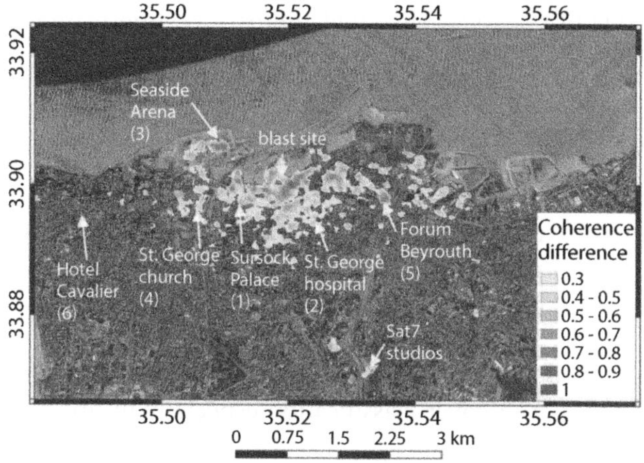

Figure 10.14 Low-pass filtered mean coherence loss map. The color indicates the coherence difference, describing the coherence loss between scenes before and after the explosion [97].

coherence within 1.5 kilometers of the bomb location. The coherence loss maps shown in Figure 10.14, demonstrated the dependability of SAR for catastrophe mapping in crowded urban settings, as they were very compatible with damage confirmed by drone-based assessments [77, 97]. Recovery planning and the effective distribution of humanitarian supplies were greatly aided by the timely delivery of damage assessment maps created from SAR data to UN agencies and the Lebanese government [77, 98]. Following the explosion, access to drone and aircraft surveillance was restricted due to the politically sensitive context, underscoring the vital role that satellite SAR plays in delivering quick, dependable, and scalable situational awareness. One of the earliest geospatial tools utilized to rank emergency operations and infrastructure assessment were the coherence loss maps produced with Sentinel-1 imagery [99].

Together, these case studies highlight the real-world applications of SAR technology in a variety of disaster scenarios, such as urban disasters, coastal floods, and geophysical hazards in mountainous areas.

10.6 Integrating Synthetic Aperture Radar with Other Technologies for Enhanced Disaster Monitoring

In an increasingly complex disaster landscape, SAR's utility is maximized when fused with other geospatial and physical datasets. This section highlights integrated approaches that improve accuracy, timeliness, and decision-making.

10.6.1 Synthetic Aperture Radar and Optical Remote Sensing Integration

Synthetic aperture radar offers structural and topographic insights, while optical imagery provides spectral detail. Together, they significantly enhance disaster monitoring.

- **Enhanced Feature Discrimination**: SAR-optical fusion improves land cover classification especially in flood zones [6, 100].
- **Cloud Penetration & Continuity**: SAR ensures data continuity even when clouds obstruct optical imagery.
- **Case Example**: During the 2018 Kerala floods, combined Sentinel-1 and Sentinel-2 imagery enhanced mapping accuracy [74].

10.6.2 Integration with Light Detection and Ranging and Digital Elevation Model Data

Topographic data enhances SAR's geometric accuracy and enables terrain analysis.

- **Topographic Correction**: DEMs (e.g., SRTM, ALOS) are essential for terrain correction and flood simulation [51].
- **Landslide Monitoring**: LiDAR's precision, fused with InSAR, reveals slope instability with greater clarity.

10.6.3 Integration with Meteorological and Hydrological Data

Synthetic aperture radar complements environmental observations for modeling and forecasting.

- **Hydrological Modeling**: SAR flood maps combined with rainfall, discharge, and soil moisture data improve forecasting [101].
- **Meteorological Integration**: Rainfall and temperature patterns help contextualize SAR-based landslide risk [73].

- **Case Example**: During Hurricane Harvey, SAR data combined with meteorological inputs enabled real-time urban inundation maps [95].

10.6.4 Synthetic Aperture Radar and Global Navigation Satellite System: Ground Truthing and Calibration

A global navigation satellite system (GNSS) provides precise ground displacement data for validating SAR results.

- **Co-Seismic Validation**: GNSS confirms InSAR-detected crustal movements [102].
- **Tectonic Analysis**: SAR-GNSS fusion is critical for fault modeling in seismic zones.

10.6.5 Synthetic Aperture Radar and Hydrological/Hydraulic Models

Hydrological tools benefit from SAR's spatial water data.

- **Flood Modeling:** SAR-derived extents enhance simulation tools like HEC-RAS [103]
- **Flash Flood Zonation:** SAR's geomorphic inputs improve rainfall-runoff hazard models [104].

10.6.6 Leveraging Artificial Intelligence and Machine Learning

Artificial intelligence tools accelerate and scale SAR-based disaster assessments.

- **Feature Extraction:** Convolutional neural networks and other deep learning models automate flood detection and damage classification.
- **Predictive Analytics:** Ensemble models forecast flood spread using SAR, optical, and meteorological inputs [95].

10.6.7 Synthetic Aperture Radar with Unmanned Aerial Vehicles and Drone-Based Observations

Drones provide localized validation and high-resolution insights.

- **Ground Truthing:** Unmanned aerial vehicle imagery verifies SAR damage maps used during Beirut explosion response [77].
- **3D Modeling:** Structure-from-motion complements SAR-derived displacement data.

10.6.8 Web-Based Integration and Decision Support Systems

Online tools deliver SAR-derived insights to stakeholders in real-time.

- **Platforms:** Copernicus EMS, UNOSAT Live Maps, and ISRO Bhuvan combine SAR with exposure layers for fast-response visualization.

10.6.9 Challenges and Future Directions

Despite progress, integration faces several barriers:

- **Data Harmonization:** Differences in spatial resolution, temporal frequency, and sensor geometry among SAR, optical, and LiDAR datasets require sophisticated co-registration and calibration methods to ensure data consistency [6].
- **Computational Load:** Multi-sensor data fusion and AI-driven analyses involve significant computational resources, often necessitating cloud-based platforms and high-performance computing solutions.
- **Standardization:** Developing standardized protocols for data exchange and integration across different platforms remains an ongoing challenge for the global disaster management community.

Future research should focus on developing robust data fusion algorithms, optimizing cloud-based processing workflows, and enhancing interoperability among disparate datasets to further harness the synergy between SAR and complementary technologies.

10.7 Opportunities and Challenges in Operationalizing Synthetic Aperture Radar for Disaster Management

Despite SAR's proven capabilities, mainstreaming its use in disaster workflows still faces technical, institutional, and policy-related barriers. This section outlines the most pressing challenges and emerging opportunities.

10.7.1 Bridging the Skill Gap: Training Disaster Managers to Use Synthetic Aperture Radar Tools

Despite the growing availability of SAR data, its practical utility remains underexploited in many disaster management agencies. This is primarily due to a skill gap between data availability and end-user capacity. Unlike optical imagery, SAR interpretation requires understanding radar physics, backscatter characteristics, and complex image artifacts, such as speckle noise and layover effects [46]. Most disaster managers, especially at subnational levels, lack formal training in these technical aspects.

Capacity-building efforts often remain limited to academic institutions or remote sensing research centres, with little trickle-down to the operational agencies responsible for disaster planning and emergency response. Bridging this gap requires targeted and hands-on training programs, tailored for disaster professionals, not just geospatial scientists. Organizations such as the UN-SPIDER, CEOS Working Group on Disasters, and regional space agencies (e.g., ISRO, JAXA) have begun offering thematic workshops and e-learning modules focused on flood mapping, landslide monitoring, and deformation analysis using SAR.

In India, for example, the National Remote Sensing Centre has developed capacity-building tools like Bhuvan, which offer user-friendly flood mapping applications using Sentinel-1 data. However, institutionalizing SAR literacy across disaster management authorities from district-level officers to national planners remains a long-term but essential goal.

10.7.2 Challenges in Data Acquisition and Processing, Especially in Developing Countries

One of the most persistent barriers to operationalizing SAR is the asymmetry in data access, processing infrastructure, and technical know-how, especially between developed and developing countries.

Although Sentinel-1 C-band data is freely accessible under the ESA's Copernicus Programme, high-resolution data from commercial missions such as TerraSAR-X, RADARSAT-2, and COSMO-SkyMed are often priced beyond the reach of local disaster management institutions. Even when data is available, bandwidth limitations, storage issues, and limited computational infrastructure pose significant constraints to downloading and processing SAR datasets. Moreover, processing advanced techniques like D-InSAR or SBAS-InSAR requires significant computational capacity and proficiency in tools such as SNAP, ISCE, or StaMPS many of which are not widely adopted at the field level.

Promising solutions include platforms like GEE, ESA's Geohazards Exploitation Platform, and UN Open Mapping Hubs, which offer cloud-based processing and easier access.

10.7.3 Policy and Collaboration Imperatives: Global Partnerships for Open-Access Synthetic Aperture Radar Datasets

Maximizing SAR's impact requires policy and institutional support, particularly around open access, interoperability, and global collaboration:

- The International Charter on Space and Major Disasters provides free SAR data during crises.
- Collaborative platforms like Sentinel Asia, CEOS Working Group on Disasters, and the Global Flood Partnership (GFP) promote SAR sharing for humanitarian use [105].

However, many of these mechanisms are event-triggered and short-term. They rarely support long-term recovery or resilience planning.

Future-ready actions include:

- Establishing national policies that recognize SAR as a core disaster monitoring tool.
- Encouraging data-sharing agreements between public and private SAR operators.
- Supporting research-to-policy pipelines for local SAR adoption.

One promising development is the upcoming NISAR mission. Its dual-frequency (L+S band) and open-access policy is expected to dramatically improve data availability, especially in the Global South.

10.8 Conclusion and Future Research Directions

Over the past two decades, SAR has evolved from a specialized scientific tool into a mainstream technology supporting comprehensive disaster management. Its all-weather, day-and-night imaging capability makes it ideal for disaster response, recovery, and resilience planning across diverse geographies.

When combined with optical imagery, LiDAR, DEM, GNSS, and AI, SAR becomes an indispensable tool for multi-hazard analysis and early warning systems. Case studies from Nepal, India, the USA, and Lebanon have demonstrated how SAR transforms complex disaster dynamics into actionable intelligence.

Looking ahead, future research should aim to:

- Enhance automation and real-time analytics through AI and cloud-based platforms, to shorten response times and reduce processing bottlenecks.
- Expand SAR applications to emerging risks, including GLOFs, coastal erosion, and wildfire dynamics.
- Promote inclusive access to SAR data by advancing open-data policies, localized training programs, and user-friendly processing tools, especially in developing countries.
- Strengthen cross-sectoral integration of SAR with decision-making frameworks across urban planning, climate adaptation, and infrastructure resilience.

By continuing to advance the technological, institutional, and collaborative dimensions of SAR deployment, the remote sensing community can ensure that SAR remains at the forefront of building safer, more resilient societies in the face of growing disaster risks. However, despite the remarkable progress, several challenges remain. The skill gap, especially among operational disaster managers, the limited access to high-resolution SAR data, and institutional silos in data governance continue to restrict the full integration of SAR into decision-making workflows, particularly in the Global South. At the same time, there is enormous potential for SAR to shape the future of climate resilience, urban disaster mitigation, and ecosystem-based adaptation.

10.8.1 Key Takeaways

- Synthetic aperture radar offers unmatched capabilities for real-time and retrospective disaster analysis, including flood extent mapping, infrastructure damage detection, terrain deformation, and glacial lake monitoring.
- Cross-sensor integration (e.g., SAR-optical-AI) with other geospatial, physical, and socio-economic data sources enables cross-sectoral decision support, essential for modern disaster governance frameworks.
- Synthetic aperture radar's mainstreaming in operational systems requires training, policy backing, and open-access initiatives.

10.8.2 Future Research Directions

To further advance the use of SAR in disaster risk reduction and climate adaptation, the following research frontiers are recommended:

- Development of Near-Real-Time SAR Analytics: As climate-related hazards become more frequent and dynamic, there is a pressing need for automated SAR analytics capable of producing near-real-time outputs for emergency responders. Research into fast D-InSAR algorithms, coherence change detection, and AI-driven classification models must be prioritized, with an emphasis on user-friendly interfaces.
- Synthetic Aperture Radar for Compound and Cascading Hazards: Most existing SAR applications are hazard-specific. Future research should focus on compound-hazard detection, such as earthquake-triggered landslides or cyclone-induced flooding and erosion. This will require multi-sensor fusion, ML models, and event-sequence mapping capabilities [106].
- Urban SAR Applications in the Context of Smart Cities: With increasing urbanization, SAR can play a key role in monitoring subsidence, infrastructure stress, and urban heat islands. Research should aim to embed SAR-derived indicators into urban resilience dashboards and municipal planning tools.
- Synthetic Aperture Radar for Climate Change Monitoring and Adaptation Planning: SAR's ability to monitor glaciers, wetlands, permafrost, and vegetation dynamics is underutilized in climate adaptation planning. Future studies should

assess how long-term SAR time series can support loss and damage accounting, ecosystem restoration, and resilience financing mechanisms under the UNFCCC framework.

- Locally Adapted SAR Training Frameworks: Research is also needed on how best to localize SAR training both linguistically and pedagogically for disaster managers, community-based organizations, and first responders. Co-designing training modules with end-users will ensure uptake and application.

The power of SAR lies not only in its technical precision but in its potential to inform life-saving decisions, guide resilient reconstruction, and enable sustainable development in the face of recurring disasters. As the frequency and severity of hazards intensify under a changing climate, the need for robust, timely, and scalable Earth observation systems has never been greater. Synthetic aperture radar, when integrated within an inclusive and interoperable disaster management ecosystem, offers a pathway toward data-informed resilience for communities worldwide.

AI disclosure statement: The author acknowledges the use of AI-based tools for minor language editing or grammatical correction. However, all intellectual contributions, analyses, and conclusions presented in this chapter are entirely the author's own.

References

1. IPCC, Climate Change 2023: Synthesis Report, Contribution of Working Groups I, II and III to the Sixth Assessment Report of the Intergovernmental Panel on Climate Change, Core Writing Team, H. Lee and J. Romero (Eds.), pp. 35–115, IPCC, Geneva, Switzerland, 2023.
2. Moreira, A., Prats-Iraola, P., Younis, M., Krieger, G., Hajnsek, I., Papathanassiou, K.P., A tutorial on synthetic aperture radar. *IEEE Geosci. Remote Sens. Mag.*, 1, 1, 6–43, 2013.
3. Summers, J.K., Lamper, A., McMillion, C., Harwell, L.C., Observed changes in the frequency, intensity, and spatial patterns of nine natural hazards in the United States from 2000 to 2019. *Sustainability*, 14, 7, 4158, 2022.
4. Tin, D., Cheng, L., Le, D., Hata, R., Ciottone, G., Natural disasters: A comprehensive study using EMDAT database 1995–2022. *Public Health*, 226, 255–260, 2024.
5. Robinson, T.R., Rosser, N., Walters, R.J., The spatial and temporal influence of cloud cover on satellite-based emergency mapping of earthquake disasters. *Sci. Rep.*, 9, 1, 12455, 2019.

6. Schumann, G. and Di Baldassarre, G., The direct use of radar satellites for event-specific flood risk mapping. *Remote Sens. Lett.*, 1, 2, 75–84, 2010.
7. Ge, P., Gokon, H., Meguro, K., A review on synthetic aperture radar-based building damage assessment in disasters. *Remote Sens. Environ.*, 240, 111693, 2020.
8. Tsatsaris, A., Kalogeropoulos, K., Stathopoulos, N., Geoinformation technologies in support of environmental hazards monitoring under climate change: An extensive review. *ISPRS Int. J. Geo-Inf.*, 10, 2, 94, 2021.
9. Li, T., Cui, L., Xu, Z., Hu, R., Joshi, P.K., Song, X., Tang, L., Quantitative analysis of the research trends and areas in grassland remote sensing: A scientometrics analysis of Web of Science from 1980 to 2020. *Remote Sens.*, 13, 7, 1279, 2021.
10. Dritsas, E. and Trigka, M., Remote sensing and geospatial analysis in the big data era: A survey. *Remote Sens.*, 17, 3, 550, 2025.
11. Plank, S., Rapid damage assessment by means of multi-temporal SAR—A comprehensive review and outlook to Sentinel-1. *Remote Sens.*, 6, 6, 4870–4906, 2014.
12. Tapete, D. and Cigna, F., Trends and perspectives of space-borne SAR remote sensing for archaeological landscape and cultural heritage applications. *J. Archaeol. Sci.: Rep.*, 14, 716–726, 2017.
13. Ivanov, A.Y. and Zatyagalova, V.V., A GIS approach to mapping oil spills in a marine environment. *Int. J. Remote Sens.*, 29, 22, 6295–6315, 2008.
14. Bishop, M.P., James, L.A., Shroder, J.F., Walsh, S.J., Geospatial technologies and digital geomorphological mapping: Concepts, issues and research. *Geomorphology*, 137, 1, 5–26, 2012.
15. Wasowski, J. and Colesanti, C., Investigating landslides with space-borne synthetic aperture radar (SAR) interferometry. *Eng. Geol.*, 88, 3–4, 200–218, 2006.
16. Metternicht, G., Hurni, L., Gogu, R., Remote sensing of landslides: An analysis of the potential contribution to geo-spatial systems for hazard assessment in mountainous environments. *Remote Sens. Environ.*, 96, 1, 56–68, 2005.
17. Ferretti, A., Prati, C., Rocca, F., Permanent scatterers in SAR interferometry. *IEEE Trans. Geosci. Remote Sens.*, 39, 1, 8–20, 2002.
18. Chen, Z., Montpetit, B., Banks, S., White, L., Behnamian, A., InSAR monitoring of Arctic landfast sea ice deformation using L-band ALOS-2, C-band RADARSAT-2 and Sentinel-1. *Remote Sens.*, 13, 22, 4570, 2021.
19. Du, Y., Feng, G., Li, Z., Peng, X., Zhu, J., Ren, Z., Effects of external digital elevation model inaccuracy on StaMPS-PS processing: A case study in Shenzhen, China. *Remote Sens.*, 9, 11, 1115, 2017.
20. Besoya, M., Govil, H., Bhaumik, P., A review on surface deformation evaluation using multitemporal SAR interferometry techniques. *Spatial Inf. Res.*, 29, 267–280, 2021.
21. Singh, G. and Pandey, A., Mapping Punjab flood using multi-temporal open-access synthetic aperture radar data in Google earth engine, in: *Hydrological Extremes: River Hydraulics and Irrigation Water Management*, pp. 75–85, 2021a.

22. Delaney, W.P. and Ward, W.W., Radar development at Lincoln Laboratory: An overview of the first fifty years. *Lincoln Lab. J.*, 12, 2, 147–166, 2000.
23. Chernyak, V.S. and Immoreev, I.Y., A brief history of radar. *IEEE Aerosp. Electron. Syst. Mag.*, 24, 9, 21–27, 2009.
24. Inggs, M., Synthetic aperture radar during the 50 years of the Aerospace and Electronic Systems Society. *IEEE Aerosp. Electron. Syst. Mag.*, 37, 9, 6–21, 2022.
25. Kim, Y., Stuhr, F., van Zyl, J., Rosen, P., Freeman, A., Johnson, W., Shen, Y., Achievements in the Development of Space-borne Imaging Radar, 1999.
26. Ouchi, K., Recent trend and advance of synthetic aperture radar with selected topics. *Remote Sens.*, 5, 2, 716–807, 2013.
27. Lasswell, S.W., History of SAR at Lockheed Martin (previously Goodyear Aerospace), in: *Radar Sensor Technology IX*, vol. 5788, SPIE, 16 May 2005. https://doi.org/10.1117/12.603927.
28. Curlander, J.C. and McDonough, R.N., *Synthetic Aperture Radar: Systems and Signal Processing*, Wiley, 1991. https://books.google.co.in/books/about/Synthetic_Aperture_Radar.html?id=nM98QgAACAAJ.
29. Attema, E. and Desnos, Y.L., Synthetic aperture radar in Europe: ERS, Envisat, and beyond. *Johns Hopkins APL Tech. Dig.*, 21, 1, 55–63, 2000.
30. Johannessen, J.A., Portabella, M., Vachon, P.W., *Satellite oceanography from the ERS synthetic aperture radar and radar altimeter: A brief review*, European Space Agency Review, 2013. https://www.researchgate.net/publication/265207463_Satellite_oceanography_from_the_ERS_synthetic_aperture_radar_and_radar_altimeter_A_brief_review.
31. Crapolicchio, R., De Chiara, G., Floury, N., ERS-2 scatterometer: Mission performances and current reprocessing achievements. *IEEE Trans. Geosci. Remote Sens.*, 50, 5, 1722–1731, 2012.
32. Lucas, R., Clewley, D., Rosenqvist, A., Global forest monitoring with synthetic aperture radar (SAR) data, in: *Global Forest Monitoring from Earth Observation*, pp. 175–202, CRC Press, 2012. https://www.taylorfrancis.com/chapters/oa-edit/10.1201/b13040-15/global-forest-monitoring-synthetic-aperture-radar-sar-data-richard-lucas-daniel-clewley-ake-rosenqvist-josef-kellndorfer-wayne-walker-dirk-hoekman-masanobu-shimada-humberto-navarro-de-mesquita-jr.
33. Rauste, Y., Multi-temporal JERS SAR data in boreal forest biomass mapping. *Remote Sens. Environ.*, 97, 3, 263–275, 2005.
34. Santoro, M., Eriksson, L.E.B., Askne, J.I.H., Smith, G., Assessment of stand-wise stem volume retrieval in boreal forest from JERS-1 L-band SAR backscatter. *Int. J. Remote Sens.*, 27, 6, 1285–1305, 2006.
35. Singhroy, V. and Charbonneau, F.J., RADARSAT: Science and applications. *Phys. Can.*, 70, 4, 227–232, 2014.
36. Samsonov, S.V., Czarnogorska, M., White, D.J., Selecting optimal RADARSAT Constellation Mission beams for monitoring ground deformation in Alberta's oil sands. *Can. J. Remote Sens.*, 41, 5, 408–422, 2015.

37. Ramsay, B., Flett, D., Andersen, H.S., Gill, R., Preparation for the operational use of RADARSAT-2 for ice monitoring. *Can. J. Remote Sens.*, 30, 3, 379–389, 2004.
38. Raucoules, D. and De Michele, M., SAR high-resolution image sub-pixel correlation for landslide monitoring: Application to Salazie (La Réunion Island) and La Valette (Alps), in: *IGARSS Conference Proceedings*, 2012.
39. Voormansik, K., Praks, J., Antropov, O., Mõttus, M., Flood mapping with TerraSAR-X in forested regions in Estonia. *IEEE J. Sel. Top. Appl. Earth Obs. Remote Sens.*, 6, 3, 1841–1851, 2013.
40. Karamvasis, K., Time series methods for Synthetic Aperture Radar data to monitor ground deformation, soil moisture and floodwater. Doctoral dissertation, National Technical University of Athens, 2023.
41. Aswathi, J., Kumar, R.B.B., Oommen, T., Bouali, E.H., InSAR as a tool for monitoring hydropower projects: A review. *Energy Geosci.*, 3, 2, 307–324, 2022.
42. Paek, S.W., Balasubramanian, S., Kim, S., de Weck, O., Small-satellite synthetic aperture radar for continuous global biospheric monitoring: A review. *Remote Sens.*, 12, 16, 2546, 2020.
43. Filipponi, F., Sentinel-1 GRD preprocessing workflow. *Proceedings*, 18, 1, 11, 2019.
44. Mullissa, A., Vollrath, A., Odongo-Braun, C., Slagter, B., *et al.*, Sentinel-1 SAR backscatter analysis-ready data preparation in Google Earth Engine. *Remote Sens.*, 13, 10, 1954, 2021.
45. Flores-Anderson, A.I., Parache, H.B., Martin-Arias, V., *et al.*, Evaluating SAR radiometric terrain correction products: Analysis-ready data for users. *Remote Sens.*, 15, 21, 5110, 2023.
46. Small, D., Flattening gamma: Radiometric terrain correction for SAR imagery. *IEEE Trans. Geosci. Remote Sens.*, 49, 8, 3081–3093, 2011.
47. Argenti, F. and Alparone, L., Speckle removal from SAR images in the wavelet domain. *IEEE Trans. Geosci. Remote Sens.*, 40, 11, 2363–2374, 2002.
48. Pan, B. and Shi, X., Fusing ascending and descending time-series SAR images with dual-polarized pixel attention UNet for landslide recognition. *Remote Sens.*, 15, 23, 5619, 2023.
49. Deijns, A.A.J., Dewitte, O., Thiery, W., Timing landslide and flash flood events from SAR satellite: A regionally applicable methodology illustrated in African cloud-covered tropical environments. *Nat. Hazards Earth Syst. Sci.*, 22, 11, 3679–3694, 2022.
50. Ngo, P.T.T., Hoang, N.D., Pradhan, B., Nguyen, Q.K., Tran, X.T., A novel hybrid swarm optimized multilayer neural network for spatial prediction of flash floods in tropical areas using Sentinel-1 SAR imagery and geospatial data. *Sensors*, 18, 11, 3704, 2018.
51. Wessel, B., TanDEM-X Ground Segment DEM Products Specification Document. DLR TanDEM-X Science Team, Public Document TD-GS-PS-0021, 2016.

52. Massonnet, D., Rossi, M., Carmona, C., Adragna, F., Peltzer, G., Feigl, K., Rabaute, T., The displacement field of the Landers earthquake mapped by radar interferometry. *Nature*, 364, 6433, 138–142, 1993.
53. Burgmann, R., Rosen, P.A., Fielding, E.J., Synthetic aperture radar interferometry to measure Earth's surface topography and its deformation. *Annu. Rev. Earth Planet. Sci.*, 28, 169–209, 2000.
54. Berardino, P., Fornaro, G., Lanari, R., Sansosti, E., A new algorithm for surface deformation monitoring based on small baseline differential SAR interferograms. *IEEE Trans. Geosci. Remote Sens.*, 40, 11, 2375–2383, 2002.
55. Ullmann, T., Schmitt, A., Roth, A., Duffe, J., Dech, S., Land cover characterization and classification of Arctic tundra environments by means of polarized synthetic aperture X- and C-band radar (PolSAR) and Landsat 8. *Remote Sens.*, 6, 9, 8565–8593, 2014.
56. Chandola, S., Polarimetric SAR interferometry for forest aboveground biomass estimation. University of Twente, 2014.
57. Mathieu, R., Naidoo, L., Cho, M.A., Leblon, B., Main, R., Toward structural assessment of semi-arid African savannahs and woodlands: The potential of multitemporal polarimetric RADARSAT-2 fine beam images. *Remote Sens. Environ.*, 134, 24–36, 2013.
58. Martinis, S., Kersten, J., Twele, A., A fully automated TerraSAR-X based flood service. *ISPRS J. Photogramm. Remote Sens.*, 104, 203–212, 2015.
59. Kim, M., Park, S.E., Lee, S.J., Detection of damaged buildings using temporal SAR data with different observation modes. *Remote Sens.*, 15, 2, 308, 2023.
60. Cloude, S.R. and Pottier, E., A review of target decomposition theorems in radar polarimetry. *IEEE Trans. Geosci. Remote Sens.*, 34, 2, 498–518, 1996.
61. Freeman, A. and Durden, S.L., A three-component scattering model for polarimetric SAR data. *IEEE Trans. Geosci. Remote Sens.*, 36, 3, 963–973, 1998.
62. Carreiras, J., Higginbottom, T., Godlee, J.L., Harrison, S., Determinants of L-band backscatter in dry tropical ecosystems: Implications for biomass mapping. *SSRN Electron. J.*, 2024.
63. Crosetto, M., Monserrat, O., Cuevas-González, M., Devanthéry, N., Crippa, B., Persistent scatterer interferometry: A review. *ISPRS J. Photogramm. Remote Sens.*, 115, 78–89, 2016.
64. Zhao, J., Chen, Z., Niu, Y., Zhang, S., Yan, P., Building damage detection of the 2023 Ms 6.2 Jishishan earthquake using multi-temporal dual-polarization ALOS-2/PALSAR-2 data. *J. Geomatics Inf. Sci. Wuhan Univ.*, 50, 1, 36–48, 2025.
65. Mondini, A.C., Guzzetti, F., Chang, K.T., Monserrat, O., Martha, T.R., Manconi, A., Landslide failures detection and mapping using Synthetic Aperture Radar: Past, present and future. *Earth Sci. Rev.*, 216, 103574, 2021.
66. Fattahi, H. and Amelung, F., InSAR bias and uncertainty due to the systematic ionosphere effect. *Geophys. Res. Lett.*, 43, 9, 4198–4205, 2015.

67. Singh, G. and Pandey, A., Hybrid ensemble modeling for flash flood potential assessment and susceptibility analysis of a Himalayan river catchment. *Geocarto Int.*, 37, 25, 9132–9159, 2022b.
68. Singh, G. and Pandey, A., Flash flood vulnerability assessment and zonation through an integrated approach in the Upper Ganga Basin of the Northwest Himalayan region in Uttarakhand. *Int. J. Disaster Risk Reduct.*, 66, 102573, 2021b.
69. Singh, G. and Pandey, A., Identification of flash flood susceptible zones in a highly complex topography and altitude dependent climatically sensitive Himalayan River Basin, in: *EGU General Assembly Conference Abstracts*, pp. EGU22–153, 2022a.
70. Sousa, J.J., Liu, G., Fan, J., Perski, Z., Steger, S., Bai, S., Wei, L., Geohazards monitoring and assessment using multi-source Earth observation techniques. *Remote Sens.*, 13, 21, 4269, 2021.
71. Cigna, F., Observing geohazards from space: Applications of satellite radar for landslide, subsidence and uplift monitoring. *Geosciences*, 8, 2, 59, 2018.
72. Kothyari, G.C., Dumka, R.K., Luirei, K., Kandregula, R.S., Taloor, A.K., Malik, K., Bhan, U., Estimation of active surface deformation using PSInSAR technique of the Central Himalayan region. *Geocarto Int.*, 39, 1, 2302415, 2024.
73. Chini, M., Hostache, R., Pelich, R., Matgen, P., Pulvirenti, L., Pierdicca, N., Probabilistic urban flood mapping using SAR data, in: *IGARSS 2019-2019 IEEE International Geoscience and Remote Sensing Symposium*, IEEE, pp. 4643–4645, 2019, July.
74. Tiwari, V., Kumar, V., Matin, M.A., Thapa, A., Ellenburg, W.L., Gupta, N., Thapa, S., Flood inundation mapping-Kerala 2018; Harnessing the power of SAR, automatic threshold detection method and Google Earth Engine. *PLoS One*, 15, 8, e0237324, 2020.
75. Katiyar, V., Tamkuan, N., Nagai, M., Near-real-time flood mapping using off-the-shelf models with SAR imagery and deep learning. *Remote Sens.*, 13, 12, 2334, 2021. https://doi.org/10.3390/rs13122334.
76. Sun, W., Shi, L., Yang, J., Li, P., Building collapse assessment in urban areas using texture information from postevent SAR data. *IEEE J. Sel. Top. Appl. Earth Obs. Remote Sens.*, 9, 8, 3792–3808, 2016.
77. ElGharbawi, T. and Zarzoura, F., Damage detection using SAR coherence statistical analysis: Application to Beirut, Lebanon. *ISPRS J. Photogramm. Remote Sens.*, 173, 190–202, 2021.
78. Ghaffarian, S., Rezaie Farhadabad, A., Kerle, N., Post-disaster recovery monitoring with Google Earth Engine. *Appl. Sci.*, 10, 13, 4574, 2020.
79. Sheykhmousa, M., Understanding post-disaster recovery through assessment of land cover and land use changes using remote sensing. University of Twente, 2018.
80. Kumar, A., Lal, P., Prasad, A., Tripathy, P., Saikia, P., Analyzing urban damage and surface deformation based hazard-risk in Kathmandu city occurred

during Nepal earthquake (2015) using SAR interferometry. *Adv. Space Res.*, 70, 12, 3892–3904, 2022.

81. Marconcini, M., Metz, A., Esch, T., Zeidler, J., Global urban growth monitoring by means of SAR data, in: *2014 IEEE Geoscience and Remote Sensing Symposium*, IEEE, pp. 1477–1480, 2014, July.
82. Kuenzer, C., Guo, H., Huth, J., Leinenkugel, P., Li, X., Dech, S., Flood mapping and flood dynamics of the Mekong Delta: ENVISAT-ASAR-WSM based time series analyses. *Remote Sens.*, 3, 7, 1075–1096, 2013.
83. Moran, M.S., Hymer, D.C., Qi, J., Sano, E.E., Soil moisture evaluation using multi-temporal synthetic aperture radar (SAR) in semiarid rangeland. *Agric. For. Meteorol.*, 105, 1–3, 69–80, 2000.
84. Watanabe, M., Thapa, R.B., Ohsumi, T., Fujiwara, S., Yarai, H., Detection of damaged urban areas using interferometric SAR coherence change with PALSAR-2 after the 2015 Gorkha earthquake in Nepal. *Earth Planets Space*, 68, 1, 1–12, 2016. https://doi.org/10.1186/s40623-016-0513-2.
85. Liu, W. and Yamazaki, F., Damage detection in the 2015 Nepal earthquake using ALOS-2 satellite SAR imagery, in: *Tenth Pacific Conference on Earthquake Engineering*, pp. 6–8, 2015, November.
86. Nagai, H., Watanabe, M., Tomii, N., Tadono, T., Suzuki, S., Multiple remote-sensing assessment of the catastrophic collapse in Langtang Valley induced by the 2015 Gorkha earthquake. *Nat. Hazards Earth Syst. Sci.*, 17, 11, 1907–1921, 2017.
87. Kobayashi, T., Morishita, Y., Yarai, H., Detailed crustal deformation and fault rupture of the 2015 Gorkha Earthquake, Nepal, revealed from ALOS-2 SAR data. *Earth Planets Space*, 67, 201, 2015.
88. Feng, G., Li, Z., Shan, X., Zhang, L., Zhang, G., Zhu, J., Geodetic model of the 2015 April 25 M w 7.8 Gorkha Nepal Earthquake and M w 7.3 aftershock estimated from InSAR and GPS data. *Geophys. J. Int.*, 203, 2, 896–900, 2015.
89. Singh, V.S.K., Rao, Y.S., Bhatt, C.M., Change detection-based flood mapping using multi-temporal Earth Observation satellite images: 2018 flood event of Kerala, India. *Eur. J. Remote Sens.*, 54, 1, 94–107, 2021. https://doi.org/10.1080/22797254.2020.1867901.
90. Mondal, M., Nanda, M.K., Peña-Arancibia, J.L., Assessment of inundation extent due to super cyclones Amphan and Yaas using Sentinel-1 SAR imagery in Google Earth Engine. Theoretical and Applied Climatology, 2024.
91. Kumar, R., Rani, S., Maharana, P., Assessing the impacts of Amphan cyclone over West Bengal, India: A multi-sensor approach. *Environ. Monit. Assess.*, 193, 578, 2021.
92. Tripathy, P. and Malladi, T., Global Flood Mapper: A novel Google Earth Engine application for rapid flood mapping using Sentinel-1 SAR. *Nat. Hazard.*, 114, 707–727, 2022.
93. Reshma, K.N., Agrawal, R., Ramakrishnan, R., Land subsidence studies in the Godavari Delta regions of the East coast of India using ALOS and Sentinel 1 data. *Ecol. Indic.*, 153, 110219, 2023.

94. Roy, P., Jain, N., Martha, T.R., Jalan, P.R., Das, I.C., A rockslide-induced debris flow caused the catastrophic 2024 Wayanad disaster in Kerala, India. Landslides, 2025. https://doi.org/10.1007/s10346-025-02484-0.
95. Shen, X., Anagnostou, E.N., Allen, G.H., Brakenridge, G.R., Kettner, A.J., Near-real-time non-obstructed flood inundation mapping using synthetic aperture radar. *Remote Sens. Environ.*, 221, 302–315, 2019.
96. Wangchuk, S., Bolch, T., Robson, B.A., Monitoring glacial lake outburst flood susceptibility using Sentinel-1 SAR data, Google Earth Engine, and persistent scatterer interferometry. *Remote Sens. Environ.*, 271, 112910, 2022.
97. Pilger, C., Gaebler, P., Hupe, P., Kalia, A.C., Schneider, F.M., Yield estimation of the 2020 Beirut explosion using open access waveform and remote sensing data. *Sci. Rep.*, 11, 13645, 2021. https://doi.org/10.1038/s41598-021-93690-y.
98. Amhaz, A., The Role and Challenges of Earth Observation in the Beirut Explosion Humanitarian Response, in: *75th International Astronautical Congress (IAC)*, Milan, Italy, 14-18 October 2024, 2024.
99. Mushtaha, A.W., Alaloul, W.S., Baarimah, A.O., Musarat, M.A., Alzubi, K.M., Khan, A.M., A decision-making framework for prioritizing reconstruction projects in post-disaster recovery. *Results Eng.*, 25, 103693, 2025.
100. Martha, T.R., Kerle, N., Jetten, V., van Westen, C.J., Kumar, K.V., Characterising spectral, spatial and temporal information in optical and SAR data for landslide mapping using object-oriented methods. *Remote Sens. Environ.*, 236, 111456, 2010.
101. Riazi, M., Khosravi, K., Shahedi, K., Ahmad, S., Jun, C., Bateni, S.M., Kazakis, N., Enhancing flood susceptibility modeling using multi-temporal SAR images, CHIRPS data, and hybrid machine learning algorithms. *Sci. Total Environ.*, 871, 162066, 2023.
102. Ohzono, M., Takahashi, H., Ohta, Y., Ohno, K., Transient inflation event at Kussharo caldera, eastern Hokkaido, Japan in 2021–2023 captured by multiple GNSS observation networks. *Earth Planets Space*, 77, 1, 41, 2025.
103. Matgen, P., Schumann, G., Henry, J.B., Hoffmann, L., Pfister, L., Integration of SAR-derived river inundation areas, high-precision topographic data and a river flow model toward near real-time flood management. *Int. J. Appl. Earth Obs. Geoinf.*, 9, 3, 247–263, 2007.
104. Di Mauro, C., Hostache, R., Matgen, P., Pelich, R., Chini, M., Van Leeuwen, P.J., Blöschl, G., Assimilation of probabilistic flood maps from SAR data into a coupled hydrologic–hydraulic forecasting model: a proof of concept. *Hydrol. Earth Syst. Sci.*, 25, 7, 4081–4097, 2021.
105. Tralli, D.M., Blom, R.G., Zlotnicki, V., Donnellan, A., Evans, D.L., Satellite remote sensing of earthquake, volcano, flood, landslide and coastal inundation hazards. *ISPRS J. Photogramm. Remote Sens.*, 59, 4, 185–198, 2005.
106. Tilloy, A., Malamud, B.D., Winter, H., Joly-Laugel, A., A review of quantification methodologies for multi-hazard interrelationships. *Earth Sci. Rev.*, 196, 102881, 2019.

11

Sentinel-1 SAR Insights into Arctic Glacier Behavior: Dynamics of Vestre Brøggerbreen and Climate Implications

Namitha K. Narayan, Dhanush S., Dilsa Nasar and Geetha Priya M.*

Centre for Incubation, Innovation, Research and Consultancy, Jyothy Institute of Technology, Bengaluru, India

Abstract

This research underscores the surface dynamics and ice flow behavior of Vestre Brøggerbreen, a glacier in Svalbard, over the period 2018 to 2024, utilizing synthetic aperture radar (SAR) data from Sentinel-1 satellites. The study examines SAR-derived surface displacement data and finds notable seasonal variations. The summer months, especially July and August, have the highest surface velocities, which correspond to when surface melt is at its highest. The accumulation zone demonstrated greater velocities than the Ablation Zones, highlighting the impact of regional hydrology and topography. To evaluate the precision of SAR-derived melt extent estimates, the study also incorporates field validation based on unmanned aerial vehicles (UAVs), which was carried out in September 2019. High-resolution imagery from the UAV surveys was conducted at a flight altitude of 50 meters, and validated the surface melt dynamics seen in the SAR data. Particularly in areas where cloud cover and other environmental conditions frequently make it difficult to collect optical data, these UAV surveys were crucial in improving the accuracy of satellite-based observations. This study demonstrates the effectiveness of SAR data in monitoring glaciers, especially in isolated areas like Svalbard, where inclement weather frequently prevents the use of optical remote sensing. The results demonstrate that SAR-based monitoring can offer a continuous and reliable method for tracking glacier velocity and melt dynamics over extended periods, making it a valuable tool for studying glacier behavior in response to climate change.

**Corresponding author*: geetha.sri82@gmail.com

Pooja Joshi, Ashutosh Bhatt, Neelam Sharma and Kapil Joshi (eds.) Synthetic Aperture Radar for Disaster Management: Bridging Science and Solutions, (305–336) © 2026 Scrivener Publishing LLC

Keywords: Arctic glaciers, synthetic aperture radar (SAR), ablation season dynamics, surface melt extent, ice flow velocity, climate-glacier interactions

11.1 Introduction

Glaciers are big, clumped masses of ice that slowly move. It is important to the Earth's cryosphere because they have significant impacts on temperature and sea levels [1]. The massive ice formations affect ocean circulation and global climatic patterns while also serving as vital freshwater reserves. Glaciers are essential indicators of climate change owing to their sensitivity to climatic variations. The Arctic is one of the places most vulnerable to climate change. Research indicates that the Arctic is experiencing its most significant warming episode in 50,000 years [2]. This phenomenon, also known as Arctic amplification, is characterized by rapid warming rates that significantly exceed the global average. The extensive melting of snow and ice in the region is a direct consequence of warming [3]. These alterations influence global ocean circulation and accelerate sea level rise, among other significant consequences [4]. Researchers can gain significant insights into the evolving Earth's climate system and its broader impacts on ecosystems and human societies by studying glaciers.

The Vestre Brøggerbreen glacier, which is situated on the Brøggerhalvøya peninsula in Ny-Ålesund, Svalbard, is the subject of this study [4]. The northernmost human settlement in Ny-Ålesund offers a unique perspective on Arctic climate processes [5]. Due to its geographic location and ease of access, this area is especially important for climate research and serves as a center for global scientific cooperation. Vestre Brøggerbreen is one of the most significant glaciers in this area. The glacier is about 3.5 km in length, making it a perfect location for in-depth research due to its small size and high sensitivity to temperature changes [6]. Important clues about the wider effects of climate change on Arctic glaciers can be gained from the glaciers' reaction to shifting environmental conditions.

The cryospheric dynamics of Vestre Brøggerbreen are examined in this study during the 2018–2024 ablation seasons. The study examines connections between temperature, precipitation, surface melt, and surface ice flow (SF) velocity. Higher temperatures and solar radiation cause glaciers to melt at the surface during the ablation season, which lasts the entire summer. This study attempts to measure the effects of changing precipitation patterns and warming temperatures on glacier behavior, specifically about surface melt and ice flow dynamics, by examining data collected over six years.

The objective of the study is to utilize advanced remote sensing techniques to analyze the glacier behavior. High-resolution radar imagery from the Sentinel-1 mission, which is part of the European Union Copernicus program, forms the backbone of the data collection process. Even in an environment of severe Arctic weather, Sentinel-1 radar capabilities allow for reliable year-round monitoring of glacier dynamics [7]. This is especially crucial in an area like Svalbard, where the cloud cover and severe weather can hamper conventional observation techniques. Sentinel-1 SAR technology enables the analysis of surface melt and velocity changes in great detail, which offers important information about the seasonal and annual variations of the glacier.

To measure surface velocity, the study uses the interferometric synthetic aperture radar (InSAR) technique [8]. The InSAR is a potential remote sensing technique that calculates the ground displacements by examining phase differences between several SAR images taken at various times [9]. This method can help measure glacier movement, precisely exposing ice flow patterns and their relationship to climate variables. The InSAR offers a thorough understanding of glacier dynamics over time by detecting disparity in surface velocity. The study also delves into the dynamics of surface melt studies using microwave SAR technology. Synthetic aperture radar is useful for tracking melt events in the Arctic due to its exceptional ability to penetrate through cloud cover and collect data in all weather conditions [10].

Vestre Brøggerbreen and other Svalbard glaciers are contributing freshwater to the Arctic Ocean, which is disrupting regional ocean circulation and influencing global climate systems as they continue to lose mass. Globally, coastal ecosystems, infrastructure, and communities are at serious risk from the ensuing sea level fluctuations. This study improves models for predicting how glaciers will react to climate change by measuring the connections between surface melt, ice flow, temperature, and precipitation.

11.2 Study Area

Svalbard, one of the world's northernmost regions, is an archipelago in the Arctic Ocean that extends latitudes from 74° to 81°N [11]. This high latitude region is characterized by its unique Arctic ecosystems, rich wildlife, and extensive glaciers that cover approximately 60% of its land area [5]. Among the islands in the archipelago, Spitsbergen is the largest, and it hosts Ny-Ålesund, one of the world's northernmost research settlements. Situated on the coast of Kongsfjorden, Ny-Ålesund serves as a major hub for Arctic climate and environmental research, with its strategic location enabling easy access to glaciers and facilitating long-term monitoring [5].

Ny-Ålesund shows a high Arctic climate that includes continuous severe winters and short cool summers. The region's annual mean temperature typically varies between -5°C and -8°C, showing significant seasonal variation. With a mean of 200 to 400 mm per year, precipitation is relatively low and mostly falls as snow. The region's climate is influenced by the West Spitsbergen Current, which brings warmer Atlantic water masses that moderate temperatures in comparison to other Arctic regions at similar latitudes. Ny-Ålesund is the optimal location to study the impacts of climate change in the Arctic since it establishes a relationship between oceanic and atmospheric dynamics.

Ny-Alesund is surrounded by various glaciers, including Vestre Brøggerbreen, Austre Broggerbreen, Midtre Lovenbreen, Austre Lovenbreen, Edithbreen, Pedersenbreen, and Botnfjellbreen. These glaciers have various morphological and dynamic characteristics formed by climatic and geological influences. Among these, Vestre Brøggerbreen is the primary focus of this research [12]. Vestre Brøggerbreen is a small valley glacier located southwest of Ny-Ålesund. It is considered a representative system for understanding glacier behavior in response to climatic changes in the Arctic [5].

The Vestre Brøggerbreen Glacier spans roughly 4.63 km^2 with an elevation of approximately 50 meters at its lowest point to over 700 meters at its highest point. This topographic variation is bound with multiple climatic zones from the accumulation area to the ablation zone. Although the glacier is highly sensitive to fluctuations in temperature and precipitation. Previous studies have documented the glacier's consistent retreat over recent decades, with notable ice losses linked to enhanced surface melt and decreased snow accumulation [13]. The glacier's location near Ny-Ålesund makes it ideal for continuous monitoring using satellite imagery, ground-based surveys,

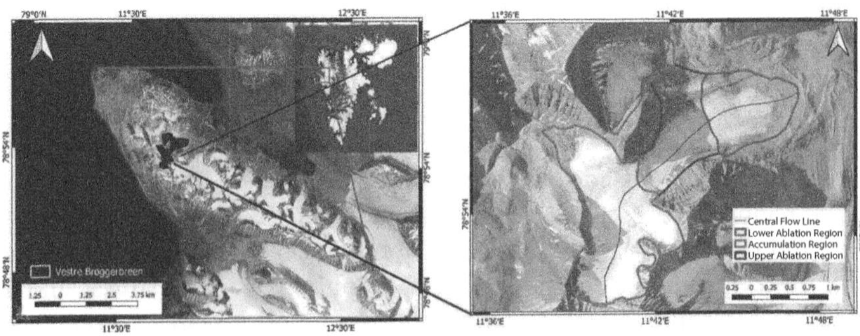

Figure 11.1 Study area map of the Vestre Brøggerbreen Glacier, located near Ny-Ålesund in the Svalbard archipelago at approximately 78.9°N and 11.9°E. The map is referenced in the WGS84/UTM Zone 33X coordinate system for accurate spatial representation.

and unmanned aerial vehicle (UAV)-based remote sensing technologies. Its proximity to Kongsfjorden provides an opportunity to study the interactions between the glacier and fjord ecosystems.

The ablation zone of Vestre Brøggerbreen is shown in Figure 11.1. A lot of surface melt occurs in the summer, resulting in runoff into the nearby fjord system. Examples of glacier dynamics that provide insight into the effects of the local climate include variations in surface velocity and crevasse formation [5].

11.3 Data Used

The dataset used in this study includes Sentinel-1 SAR imagery and meteorological data from the Norwegian Centre for Climate Services (NCCS). The SAR data set includes both ground range detected (GRD) and single look complex (SLC) products from Sentinel-1 missions covering the period from 2018 to 2024. Specific scene IDs for each acquisition are not included here due to extensive data; instead, 71 GRD scenes and 82 SLC scenes are provided. Table 11.1 includes the ablation season for each year along with the corresponding data acquisition time interval. The SAR data were acquired at a frequency of every 12 days during the respective ablation

Table 11.1 Sentinel-1 SLC and GRD data acquisition schedule for the Vestre Brøggerbreen Glacier during ablation months (2018–2024).

Year	Sentinel-1 SLC data		Sentinel-1 GRD data	
	Date		Date	
	From	To	From	To
2018	May-28	Oct-07	June 03	Sep-19
2019	June 04	Oct-02	June 04	Sep-20
2020	May-29	Oct-08	June 04	Sep-26
2021	May-24	Oct-03	June -05	Sep-21
2022	May-31	Sep-28	June -12	Sep-30
2023	May-26	Oct-05	June 07	Sep-23
2024	June 01	Sep-29	June 01	Sep-29

seasons, providing constant coverage of the glacier dynamics throughout each study year.

The high temporal resolution of this data provides variations in surface melt extent and SF velocity during the ablation season. By maintaining this consistent temporal frequency, the study offers detailed knowledge into short-term changes in the glacier surface and provides reliable data for tracking glacier movement and melt patterns over several years [14]. The temperature and precipitation data were important for understanding the climatic conditions that influence glacier behavior. The air temperature variations can be depicted from temperature data, which highly influence surface melt, while precipitation data helps to show snow accumulation trends and the effects on glacier mass balance. Temperature and precipitation data were combined with SAR data to correlate climatic variables with variations in glacier dynamics and surface melt. The combination of datasets enables the analysis of how the Vestre Brøggerbreen Glacier responds to climate conditions during the ablation season. The relationship between weather and glacier activity in the Arctic region improves the accuracy of melt and flow dynamics, which can be enhanced by combining SAR data with temperature and precipitation. This combination of satellite and meteorological data serves as the foundation of the study, contributing both high spatial and temporal resolution to analyze glacier behavior in seasonal and multi-year climate variation.

11.4 Methodology

The present study utilized Sentinel-1 SAR data to analyze the surface melt extent and glacier flow velocity of the Vestre Brøggerbreen Glacier during the ablation seasons from 2018 to 2024. Continuous observation of glacier dynamics under diverse Arctic climate conditions was made possible by the weather-independent characteristics of SAR data. Figure 11.2 illustrates the two main parts of the workflow: surface melt monitoring and glacier velocity analysis. Detection of surface melt was conducted using Sentinel-1 GRD data [15]. Radiometric calibration was carried out to ensure the backscatter values were unchanged for all images, and orbit correction was applied to fix faults in the satellite position. In addition, speckle filtering was employed to reduce noise and improve the ratio of signal to noise. To achieve spatial accuracy, the Arctic digital elevation model (DEM) was used to adjust the terrain and consider the glacier complex topology. After initial processing, backscatter intensity histograms were generated to sort the surface melt, ice, and snow zones. This allowed

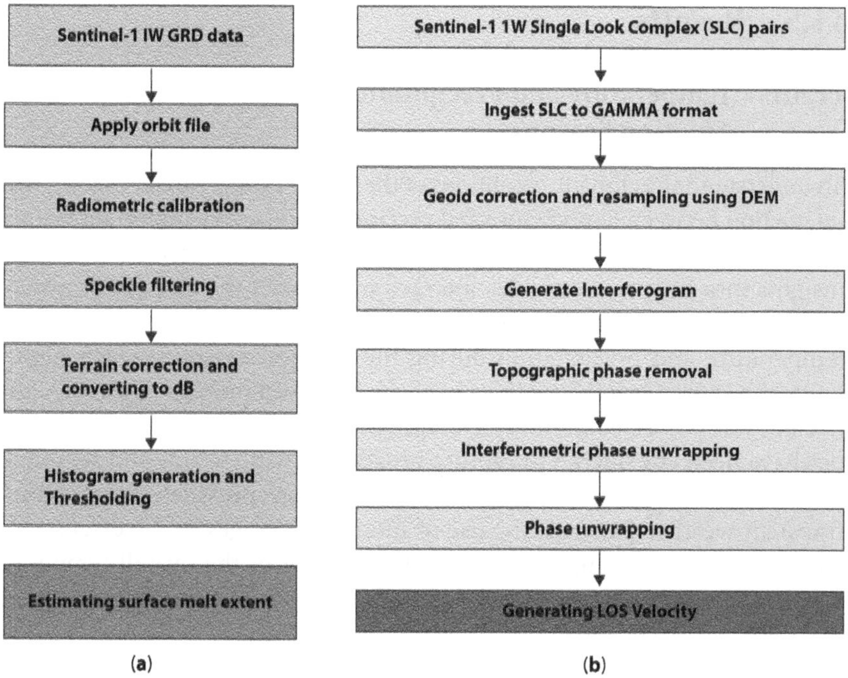

Figure 11.2 Methodological workflow for the Vestre Brøggerbreen Glacier; (a) Surface melt extent estimation using Sentinel-1 GRD data; and (b) SF velocity analysis from Sentinel-1 SLC data.

for the mapping of glacier melt extents, providing an in-depth assessment of how surface melt shifts over time and space during the timeframe. Only the line-of-sight (LOS) displacement from Sentinel-1 SLC data [8, 16] was used. Radiometric calibration, terrain correction with Arctic DEM, and precise co-registration of SAR image pairs to align successive acquisitions were all part of the pre-processing stages. To find the LOS displacement that shows the way the glacier surface shifts into the direction of the radar observation, the phase difference between two SAR images taken at different times was calculated.

The technique yielded measurements of glacier velocity while preserving high temporal resolution to track seasonal variations over the ablation phase. We analyzed the temperature and precipitation data from the NCCS along with the calculated surface melt extents and LOS velocities. The comprehensive study revealed the relationships between glacier flow behavior, surface melt dynamics, and climatic parameters, providing key insights into the impact of climate variability on the Vestre Brøggerbreen Glacier.

11.5 Results

11.5.1 Temperature and Precipitation Dynamics During Ablation Seasons

To understand the climatic conditions of the Vestre Brøggerbreen Glacier, the interaction between temperature and precipitation is a very important factor.

Analyzing data from 2018 to 2024 for the ablation months provides insights into how two variables interact and impact the glacier's seasonal and interannual dynamics. Figure 11.3 illustrates the daily variations in temperature and precipitation during the ablation seasons, which highlights the short-term fluctuations over the observed months. However, for the analysis presented monthly average temperature and precipitation values have been considered to capture broader climatic trends and seasonal characteristics. While the daily data in the figure provides insights into transient weather patterns, the use of monthly averages in the discussion allows for a more comprehensive understanding of the overall temperature-precipitation dynamics and their influence on ablation processes.

11.5.1.1 Temperature and Precipitation Interaction

From 2018 to 2024, a discernible pattern emerges where higher average temperatures during the ablation months are often associated with reduced

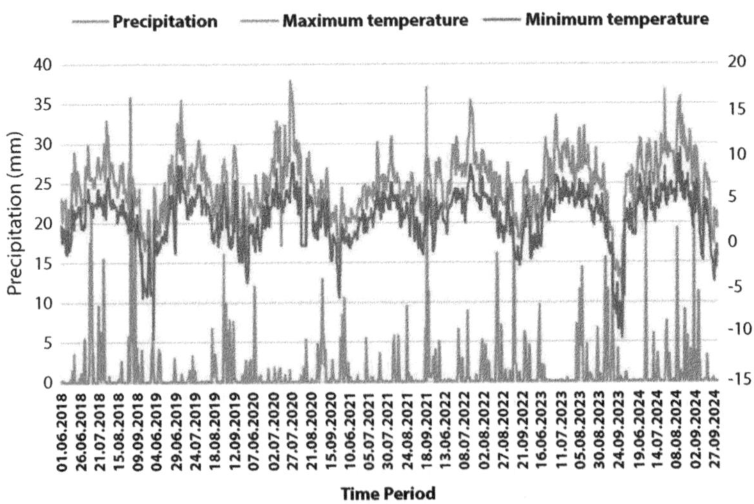

Figure 11.3 Temperature and precipitation trends during the ablation months (June–September) for Vestre Brøggerbreen Glacier from 2018 to 2024.

precipitation, particularly in the peak summer months such as July. For example, July 2020 and July 2023 recorded some of the highest average temperatures (7.55°C and 7.6°C, respectively) and also exhibited relatively low precipitation levels (18.1 mm and 7.9 mm). The inverse relationship suggests that rising temperatures during peak summer months are linked to a decrease in precipitation, potentially driven by shifts in atmospheric moisture dynamics and enhanced evaporation. Conversely, months with cooler temperatures, such as September 2018 (0.5°C), recorded relatively higher precipitation levels (76.7 mm), indicating that cooler conditions may foster enhanced atmospheric stability and greater moisture retention, thereby supporting higher precipitation [17].

11.5.1.2 Seasonal Trends in Temperature and Precipitation Dynamics

On a seasonal basis, June consistently shows moderate temperatures, even though precipitation varies annually. For instance, in June 2018, with an average temperature of 3.4°C, precipitation was relatively low at 8.6 mm, while June 2022 showed an average temperature of 5.7°C recorded similar precipitation levels at 14.1 mm. This highlights that the early summer months do not always show a clear correlation between temperature and precipitation. The onset of melting and evaporation processes, which drive energy fluxes, may vary annually, potentially influenced by different atmospheric dynamics [18]. July, being the warmest month of the year in most instances, shows a stronger inverse correlation between temperature and precipitation. For example, in July 2019, a high average temperature of 7.3°C coincided with minimal precipitation of 8.7 mm, whereas July 2018, with a slightly cooler average temperature of 6.2°C, experienced significantly higher precipitation of 99.4 mm. This highlights that higher temperatures during the peak of the summer melt season may defeat precipitation, likely due to changes in atmospheric moisture dynamics and enhanced evaporation. As the ablation season progresses into August and September, the dynamics of temperature and precipitation develop. August often shows a phase of equilibrium where moderate warming typically coincides with relatively stable precipitation. In August 2021, the average temperature was 5.2°C, and precipitation was 32.4 mm, whereas in August 2018, with a slightly warmer average of 5.6°C, precipitation was much higher at 50.5 mm.

The temperatures start to decrease in September, and there is an incompatible relationship between precipitation and temperature. For example, although there was a lot of precipitation (73 mm) in September 2023, the average temperature fell to -0.5°C, which is significantly lower than typical summer temperatures. This illustrates that the usual September temperature drop that indicates the beginning of winter rarely coincides with precipitation. The atmosphere keeps on moisture, which would lead to snow instead of rain. Significant precipitation in September shows the sign of rising atmospheric moisture retention in colder climates, which could cause snow to accumulate instead of melt and further affect the mass balance and long-term dynamics of the glacier.

11.5.1.3 Climatic Impacts on Glacier Stability

The Vestre Brøggerbreen Glacier dynamics are influenced by the complex climatic impact, which is reflected in the relationships between temperature, precipitation, and glacier stability. In July 2020, which recorded an average temperature of 7.55°C and low precipitation of 8.8 mm, likely contributed to significant melt acceleration during the peak summer period, thereby lowering the glacier ice volume [19]. The relatively warm conditions combined with reduced moisture may result in less snow accumulation and overall ice preservation, increasing the glacier's susceptibility to thinning. The cooler temperatures combined with higher precipitation were observed in other periods, fostering snow retention and mitigating melt. For example, September 2019, which had a low average temperature of 0.5°C and high precipitation of 61.7 mm, demonstrates a scenario in which the cooler conditions contributed to enhanced snow accumulation rather than melt, as snow accumulation insulates the glacier surface and prevents further melting. The increased precipitation in September, particularly in the form of snow, raised the glacier albedo, reflecting more solar radiation and thereby reducing energy available for melt processes. September noticed greater amounts of precipitation, especially snow, which increased the glacier's albedo, reflecting more sunlight and lowering the energy available for melt processes. Likewise, in September 2023, the low temperatures averaged -0.5°C, but showed heavy precipitation of 73 mm, following the trend of snowfall under lower temperatures. The higher precipitation probably helped the glacier retain snow, therefore insulating the ice and slowing down the melt rate [20]. These periods of cool temperatures and significant snowfall contribute positively to the glacier's mass balance, enabling the preservation of ice and providing a buffer against melt-induced ice loss.

According to the findings, high temperatures recorded in July 2023 averaged a temperature of 7.6°C, when the amount of surface melt peaked at 1.38 km^2, certainly accelerating glacier retreat and thinning. However,

September 2023 showed colder temperatures. More snow helps stabilize glaciers by encouraging snow accumulation and slowing down the rate of melting. The long-term stability of the glacier depends on the delicate balance between temperature-driven melt and precipitation-driven snow accumulation. These seasonal dynamics highlight the Arctic's glacier susceptibility to warming trends, which could change the balance between melt and snow accumulation and accelerate ice loss in the future.

11.5.2 Surface Melt Dynamics

The surface melt extent of the Vestre Brøggerbreen Glacier has notable interannual and seasonal variations. These variations underscore the glacier response to climatic forcing, particularly temperature and precipitation during the ablation season. Melt extent was quantified using Sentinel-1 GRD backscatter thresholding algorithms that provide robust spatial and temporal data.

11.5.2.1 Temporal and Seasonal Variations in Melt Extent

Surface melt extent was analyzed for the entire study period (2018–2024) using Sentinel-1 SAR data, as shown in Figure 11.4. For illustrative purposes, the surface melt map for the 2024 ablation season is presented in Figure 11.5 as it represents the most recent and pronounced melt conditions observed during the study. Moderate melt extents were observed in

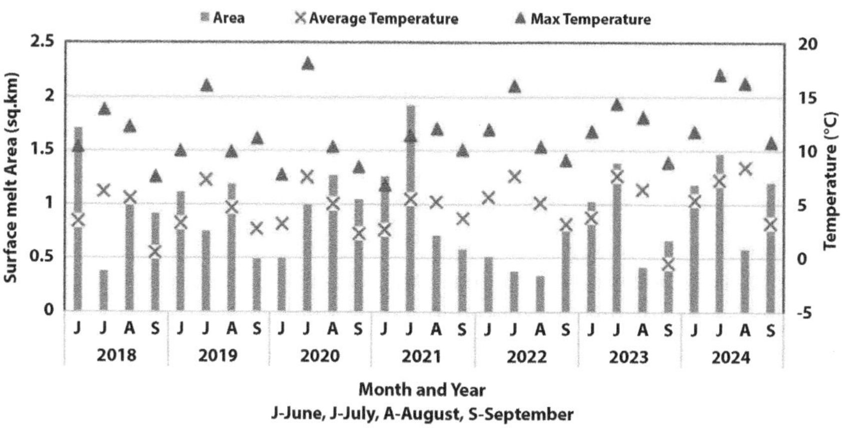

Figure 11.4 Correlation between surface melt extent and temperature during the ablation months (June–September) for the Vestre Brøggerbreen Glacier from 2018 to 2024.

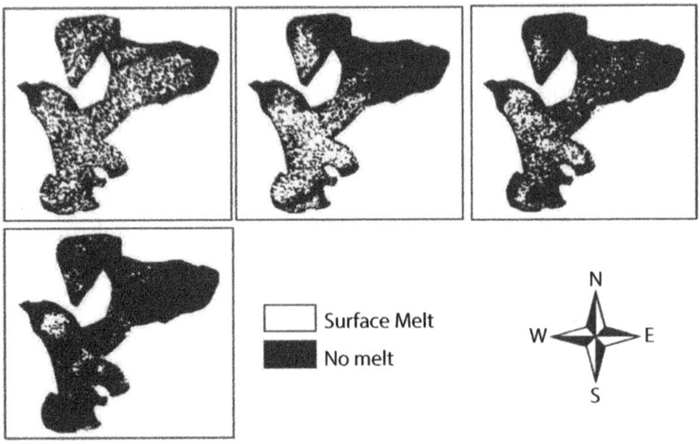

Figure 11.5 Surface melt extent of Vestre Brøggerbreen Glacier during the ablation season of 2024, presented for representation purposes. The analysis comprises the entire study period from 2018 to 2024.

2018 and 2019 with annual averages of 1.02 km^2 and 0.89 km^2, respectively, reflecting relatively stable conditions that serve as baseline references for subsequent years. These years were marked by consistent temperature and precipitation patterns that led to relatively steady melt conditions. In contrast, 2020 exhibited suppressed melt dynamics, particularly in June (0.49 km^2) and September (1.04 km^2), due to cooler temperatures and higher precipitation. For example, September precipitation in 2020 reached 61.7 mm, likely leading to fresh snow accumulation, which enhanced surface albedo and reduced the available energy for melting, resulting in lower melt extent.

The year 2021 marked an increase in melt extent, averaging 1.12 km^2 annually, driven by warmer-than-average conditions in July and August. This was a precursor to the continued warming trends observed in 2024, where elevated melt extents were recorded. Particularly, 1.47 km^2 of surface melt was recorded in June 2024, indicating a sustained warming trend and prolonged solar exposure in recent years, contributing to a higher volume of meltwater [21]. Conversely, 2022 recorded the lowest annual melt extent, especially in August (0.34 km^2), attributed to cooler conditions and significant precipitation events that enhanced albedo, limiting the extent of melt. However, the melt extent rebounded strongly in 2023 with the highest value recorded in July (1.38 km^2) as extreme temperatures intensified surface melting, reflecting a continued upward trend in melt conditions during the mid-summer period. Seasonally clear distinctions emerged across months. June, marking the onset of the ablation season, exhibited moderate melt extents, averaging 1.23 km^2, influenced by rising temperatures and persistent polar daylight. Peak melt extents consistently occurred in July, driven by the season's highest temperatures, such as 7.6°C recorded in July 2023, which facilitated extensive melting. August, as a transitional month, generally showed declining melt extents due to cooling temperatures, reduced solar energy, and shifts in atmospheric conditions. By September, the melt extents reached their lowest values, such as 0.49 km^2 in September 2019, primarily due to cooler conditions and higher precipitation, which often led to snow accumulation and suppressed further melting.

11.5.2.2 Implications of Climatic Drivers on Surface Melt

Temperature and precipitation significantly influenced the temporal and spatial dynamics of surface melt, which acted as key drivers of variability. Elevated temperatures are the primary factor that increases the surface energy absorption and intensifies melting. In July 2023 (Figure 11.4), an average temperature of 7.6°C drove significant melt with an extent of 1.38 km^2. This reflects the direct impact of warmer temperatures in accelerating surface melting. Conversely, in cooler years such as 2022, there were suppressed melt extents due to reduced temperature-driven energy input, with an average surface melt extent of 0.38 km^2 in July. This demonstrates the direct relationship between thermal conditions and melt dynamics, where cooler temperatures lead to less extensive melting.

Precipitation plays a more important role in shaping melt behavior. Rainfall can accelerate melting by reducing surface albedo and delivering energy directly to the glacier. For example, in July 2020, 18.1 mm of precipitation contributed to enhanced melt with a total melt extent of 0.99 km^2. In contrast, snowfall increases albedo, which limits energy absorption and reduces melt. For instance, in September 2019, 61.7 mm of precipitation, primarily in the form of snow, increased albedo, which limited the melt extent [22] to 0.49 km^2. Moreover, September 2022 showed 48.5 mm of precipitation, resulting in less melt than in previous years. These interactions underscore the glacier's sensitivity to climatic variability and the complex nature of melt processes, where both temperature and precipitation through direct and indirect effects combine to determine the extent of surface melt.

11.5.3 Surface Ice Flow Velocity

The SF velocities were analyzed for the entire study period (2018–2024) using Sentinel-1 SAR data, with velocities calculated as absolute magnitudes derived from the modulus of LOS values. This approach highlighted the intensity of glacier flow while neglecting directional components.

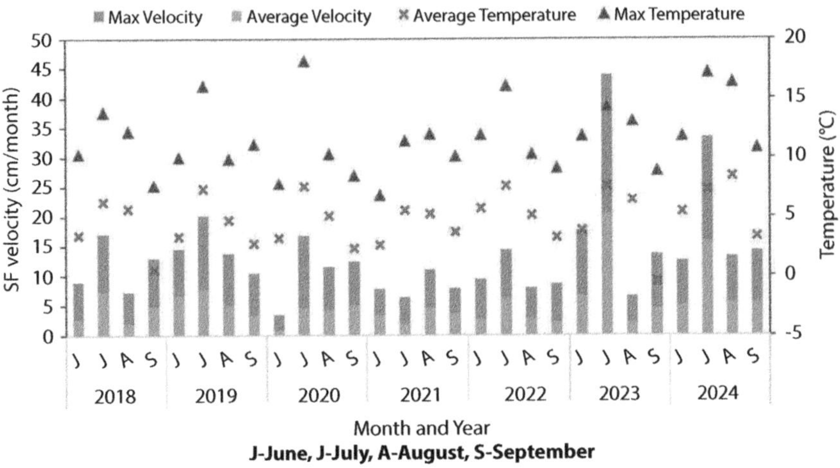

Figure 11.6 Correlation between SF velocity and temperature for the Vestre Brøggerbreen Glacier during ablation Months (2018-2024).

Figure 11.6 shows the annual trends in SF velocity, and for representation purposes, the SF velocity map for the 2024 ablation season is presented in Figure 11.7. The map showcases the most recent and dynamic conditions observed during the study.

11.5.3.1 Seasonal Variability in Surface Velocity

The glacier SF velocity exhibited notable interannual and seasonal variability as reflected in the absolute values. Thesc values capture the magnitude of flow, offering critical insights into the glacier response to seasonal temperature and precipitation disparities; however, they ignore changes in directional flow.

In 2018, velocities were relatively stable with an annual average of 4.23 cm/month. Peak velocity occurred in July (7.3 cm/month), corresponding with increased meltwater availability due to a maximum temperature of 13.8°C. In August, a decrease in velocity to 2.05 cm/month coincided with lower temperatures and lesser meltwater input. Yet, September increased to 4.87 cm/month, possibly related to sudden freezing conditions delaying basal water drainage. In 2019, the velocities had a modest increase with an annual average of 6.6 cm/month in June and July, where peak velocity reached 7.76 cm/month. This development corresponded with elevated temperatures and augmented subglacial hydrological activity. August velocities (5.01 cm/month) indicated equilibrium in subglacial

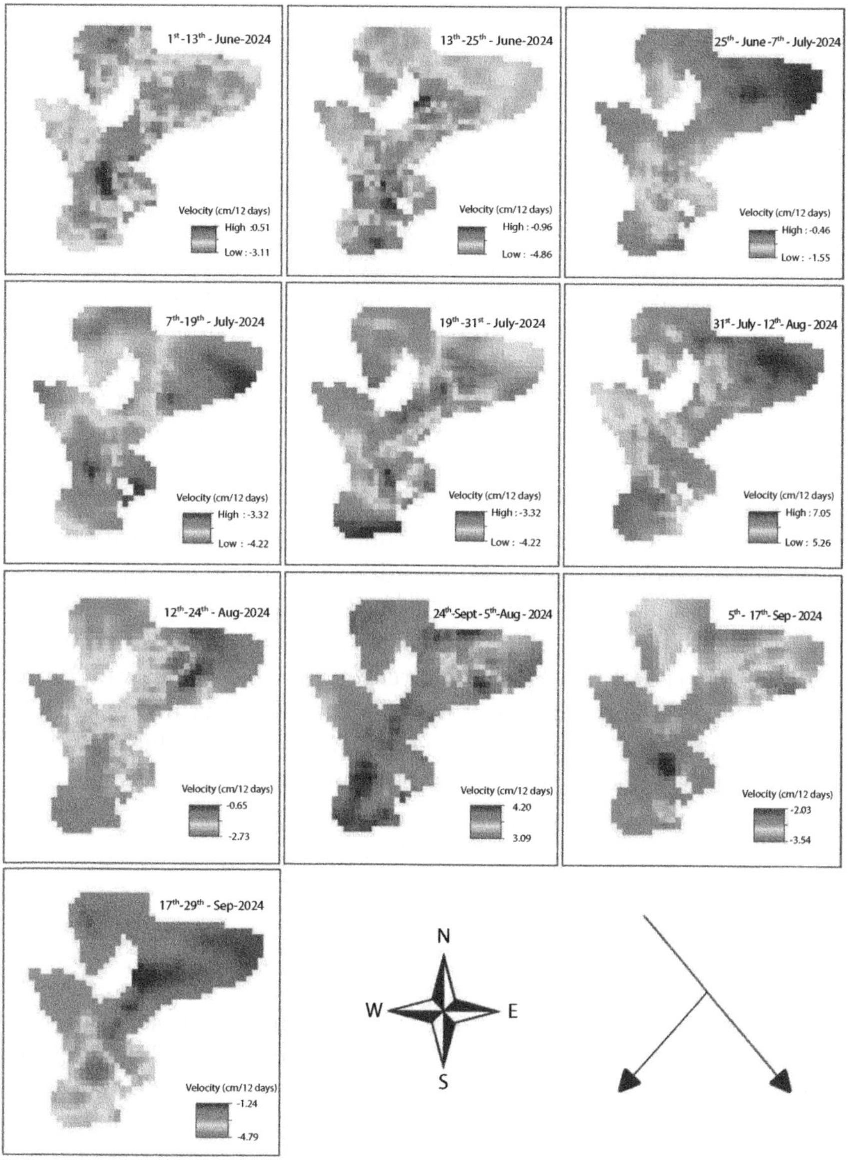

Figure 11.7 The SF velocity of Vestre Brøggerbreen during the ablation season of 2024. The map is for representation purposes, with an analysis covering the entire study period (2018–2024).

water systems; however, September velocities (3.36 cm/month) exhibited a decrease as melting declined. The 2020 ablation season recorded the lowest velocities due to cooler circumstances. June recorded a low velocity of 0.85 cm/month, indicating decreased meltwater input, whereas July demonstrated a slight recovery of 4.6 cm/month. Peak velocities of 5.05 cm/month in September highlighted the glacier's response to late-season precipitation events. In 2021, seasonal variability reemerged with a June velocity of 3.29 cm/month indicated significant meltwater inflow. Peak flow occurred in August at 4.56 cm/month, but July showed diminished values of 1.85 cm/month due to reduced melting conditions. The shift to freezing temperatures in September (3.68 cm/month) corresponds with less hydrological activity, with distinct trends observed from 2022 to 2024. In 2022, velocities were moderate, reaching 2.66 cm/month in June, peaking at 6.04 cm/month in July, and afterwards falling to 2.75 cm/month in August and 2.19 cm/month in September.

In contrast, 2023 demonstrated a noticeable spike in velocity during July (20.4 cm/month), the highest recorded value driven by high temperatures (7.6°C) and minimal precipitation (1.38 sq.km surface melt area). June velocities were also elevated at 6.53 cm/month, while August (2.12 cm/month) marked a steep decline, reflecting reduced meltwater availability. The 2024 ablation season maintained elevated velocities, with an annual average of 5.43 cm/month. Peak flow occurred in July (15.8 cm/month) under high temperatures (7.3°C) and extensive meltwater input (1.46 sq.km surface melt area). Previous years' September (5.3 cm/month) velocities demonstrated a seasonal decline.

11.5.3.2 Interplay of Temperature, Precipitation, and Glacier Hydrology

The interplay between temperature, precipitation, and glacier hydrology profoundly influenced SF velocity. Temperature serves as the primary driver of surface melt by producing meltwater that reduces basal friction and accelerates glacier flow. For instance, July 2023 recorded the highest SF velocity of 20.4 cm/month, corresponding to elevated temperatures (7.6°C) and minimal precipitation. The absence of precipitation facilitated increased surface melt, diminishing subglacial friction, and encouraging basal slide [23]. In July 2024, persistent warm conditions (7.3°C) resulted in a high velocity of 15.8 cm/month propelled by extended meltwater generation. Precipitation has a diverse impact on SF velocity, relying upon its phase and timing. Precipitation, especially during the ablation season, complements basal lubrication by rising meltwater influx into subglacial systems, thus accelerating glacier movement. In July 2019, a notable surface ice flow velocity of 7.76 cm/month, along with substantial precipitation, may have increased base water pressures and enhanced flow. Meanwhile, snowfall increases surface albedo, delays meltwater generation, and leads to reduced velocity. The receding velocity of 3.36 cm/month in September 2019 resulted from precipitation in the form of dominant snowfall and restrained surface melting and meltwater-driven flow.

The interaction between temperature and precipitation considerably influenced seasonal SF velocity patterns. Years characterized by elevated temperatures and restricted precipitation, such as 2023, have accelerated basal sliding owing to the substantial availability of meltwater and reduced damping effects from snowfall. The dominance of snowfall increased surface reflectivity and decreased the generation of meltwater required to produce basal sliding, leading to decreased velocities in colder and wetter years, like 2022. These trends highlighted the glacier's vulnerability to climatic fluctuations. It is clear from the absolute velocity measurements that the glacier is responsive to the amount of meltwater available at different times of the year. During the warmer and drier months, the flow is significantly higher, and during the cooler, wetter months, the flow is very low. The persistent interaction of temperature, precipitation, and glacier hydrology underlines its importance of long-term monitoring to enhance awareness and forecasting of glacier dynamics across many climatic scenarios.

11.5.4 Influence of Surface Melt Extent on Glacier Movement

The impact of surface melt on the sliding velocity of the Vestre Brøggerbreen Glacier during the ablation period emphasizes the essential role of meltwater dynamics in controlling glacier movement. The relationship shows the impact of temperature and precipitation when regulating meltwater generation and the subsequent effects on subglacial hydrology and ice flow mechanisms. The temporal analysis found notable anomalies and patterns related to disparity in surface melt climate anomalies and overall atmospheric conditions.

11.5.4.1 Climatic Events and Their Impact on Velocity Trends

The amount of surface melt that occurred during the ablation months had a direct impact on the glacier SF velocity, particularly during the most intense summer months. With a maximum SF velocity of 23.38 cm/month, the highest melt extent of 1.38 km^2 was recorded in July 2023. In comparison to July 2020, when the maximum SF velocity was 12.2 cm/month indicates a notable 91.5% increase in SF velocity. The spikes demonstrate what caused heat waves, along with other extreme temperature events speed up surface melt, producing vast amounts of meltwater that lubricate the glacier bed and promote basal sliding.

Lower glacier velocities and considerably lower surface melt are features associated with the years 2019 and 2020. In fact, the average SF velocity in June 2020 was only 0.85 cm/month, that were significantly lower than the 2.71 cm/month SF velocity in June 2018. Cooler temperatures (mean of 3.4°C) and steady precipitation are to blame for the decrease, having boosted snow accumulation and inhibited melt. The comparatively low 2020 velocities can also serve as a result of global atmospheric shifts brought on by the COVID-19 pandemic, which momentarily decreased industrial emissions and boosted ozone recovery, thereby reducing solar radiation and surface melt. This supports research that links fewer melt anomalies during the pandemic to lower levels of air pollution [24].

Conversely, periods of elevated surface melt, often associated with heatwaves and warmer-than-average summers, showed corresponding increases in SF velocity. In July 2020, the maximum temperature reached 18.1°C, which occurred with a significant increase in SF velocity to 12.2 cm/month. In June 2024, the surface melt extent reached a peak of 1.47 km^2, which produced a maximum SF velocity of 7.47 cm/month. These results agree with the other studies, highlighting the impact of extended heatwaves on the acceleration of Arctic glacier dynamics [25].

The magnitude of meltwater is affected by fluctuations in snowflake velocity. In August 2024, a significant melt extent of 0.58 km^2 was observed, corresponding to a surface flow velocity of 5.43 cm/month, which highlighted the seasonal influence of temperature fluctuations and meltwater availability on melt-induced flow. The findings highlight the glacier dynamic response to surface melting and climatic conditions. The interaction of temperature, precipitation, and melt extent revealed a complicated and consistent pattern in which more surface melt enhances glacier velocity while reduced melt corresponds to diminished flow. Research highlights the requirement for extended monitoring of surface melt dynamics and glacier movement to predict the future behavior of Arctic glaciers in relation to changing climate conditions.

11.5.4.2 Periodic Variability and Its Effect

The seasonal transition from June to September demonstrated that surface runoff affects SF velocity in various contexts. Melting usually starts in June. For instance, in June 2024, the surface melt extent was 1.47 km^2, indicating that the average SF velocity was 4.97 cm/month. It occurred because meltwater from the surface started to flow into subglacial channels. The month of July typically experiences the most glacier movement. In July 2023, there was the biggest surface melt covering an area of 1.38 km^2. It coincided with the month with the fastest SF speed at 23.38 cm/month. This clearly shows that more meltwater was available and basal lubrication was better.

At the end of ablation season, the SF velocities slowly dropped in August and September because the temperatures decreased and less meltwater was produced. In August 2024, for example, the melt extent dropped to 0.58 km^2, resulting in a moderate SF velocity of 5.43 cm/month. The surface melt extent decreased to 0.49 km^2 by September 2019, the lowest level of the season.

A comparative comparison of wet (high precipitation) and dry (low precipitation) summers elucidates the intricate interplay between surface melt and glacier dynamics. In July 2023, although only 7.9 mm of rain was measured, a lot of surface melt (1.38 km^2) caused fast speeds (23.38 cm/month). July 2022 saw more rain (89.1 mm) than July 2023, but the surface melt extent was smaller (0.38 km^2) because the snow raised the surface albedo. This caused the SF velocity to be 74% lower (6.04 cm/month) than in July 2023. This contrast shows that precipitation can have two different effects. Low precipitation speeds up the process of meltwater production, while heavy precipitation might slow it down to higher albedo production.

The relationship between temperature and precipitation shows changing glacier patterns. Increased surface melt is linked to faster SF velocity, which is caused by meltwater acting as a lubricant under the glacier. Years that are cooler and wetter slow down glacier flow by lowering surface melt. Atmospheric anomalies, such as heatwaves or global emissions reduction during the COVID-19 pandemic, further complicate these dynamics. These results are in line with and reinforce the glacier's sensitivity to changes in the weather, which shows that long-term monitoring is necessary to forecast how the glacier will behave in the future [26].

11.5.5 Spatial Analysis of Dynamics Across Vestre Brøggerbreen

To better understand the behavior of the glacier, spatial and temporal analyses were conducted for 2024, focusing on surface melt and SF velocity dynamics across three distinct zones of the Vestre Brøggerbreen Glacier: the Accumulation Zone, Upper Ablation Zone, and Lower Ablation Zone. The analyses indicate zone-specific patterns influenced by regional elements, such as topography, albedo, and meltwater drainage, which provide important insights into how glaciers adapt to climatic conditions.

11.5.5.1 Surface Melt Zonal Patterns

During the ablation period, surface melt displayed evident temporal and spatial shifts across the glacier zones. The Accumulation Zone constantly possessed the greatest surface melt extent, reaching a peak of 1.027 km^2 in July 2024 and then falling to 0.1682 km^2 in September. Moderate melt extents were observed in the Upper Ablation Zone, peaking at 0.1517 km^2 in July and falling to 0.0269 km^2 in September. Due to its limited exposure to solar radiation and ongoing snow cover, the Lower Ablation Zone experienced minimal melt of 0.0017 km^2 in July and 0.0026 km^2 in September (Figure 11.8). Such trends highlighted how albedo and temperature influence melt variability. July experienced the highest melting caused by elevated summer temperatures (17.5°C), but in September, reduced melting resulted from a transition to cooler temperatures, decreased daylight, and more precipitation.

11.5.5.2 Spatial and Temporal Variability in Surface Ice Flow Velocity (June–September 2024)

The analysis focuses on the 2024 ablation season as it reflects the most recent and pronounced glacier dynamics under current Arctic warming trends. This year captures region-specific variations in melt and velocity,

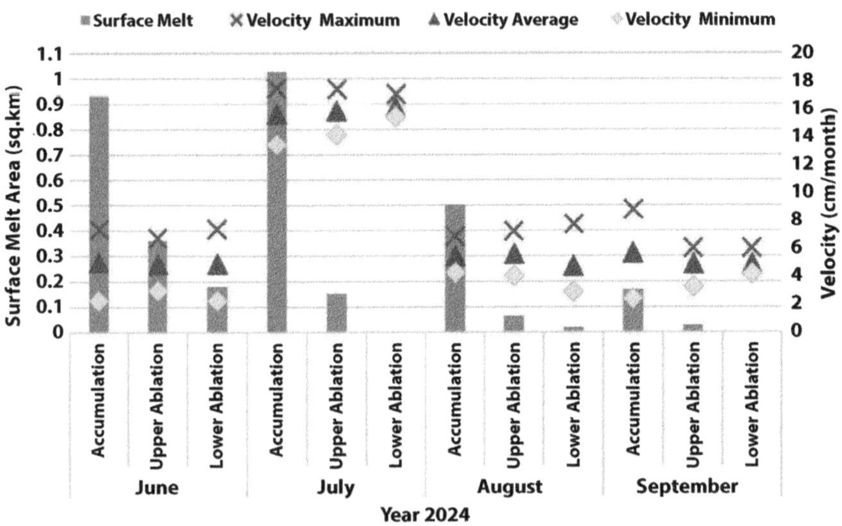

Figure 11.8 Spatial analysis of surface melt extent and surface flow velocity for Vestre Brøggerbreen glacier in 2024 ablation months.

with mid-summer showing peak meltwater-driven flow. Analyzing 2024 data enables a clearer understanding of elevation-dependent and zonal responses to climatic and hydrological conditions. The SF velocity of the Vestre Brøggerbreen Glacier during the 2024 ablation season exhibited significant spatial and temporal variability across its three zones, like the Accumulation Zone, Upper Ablation Zone, and Lower Ablation Zone. These variations show the combined influence of surface melt dynamics, glacier topography, and climatic conditions. In June 2024, the accumulation zone recorded a maximum velocity of 7.4 cm/month and an average velocity of 5.07 cm/month and corresponding to a surface melt extent of 0.9287 km^2. The upper ablation zone exhibited slightly lower velocities with a maximum of 6.7 cm/month and an average of 4.9 cm/month, consistent with its smaller surface melt extent of 0.36 km^2. Similarly, the lower ablation zone showed similar dynamics with a maximum velocity of 7.4 cm/month, an average of 4.9 cm/month, and a surface melt extent of 0.1791 km^2. The basal sliding has begun to influence the onset of melt processes with meltwater availability [27].

By July 2024, velocities across all zones peaked, driven by maximum surface melt and extended high temperatures. The accumulation zone experienced a maximum velocity of 17.5 cm/month, while its average velocity increased to 15.69 cm/month, supported by a melt extent of 1.027 km^2. The Upper Ablation Zone demonstrated a similar peak velocity of 17.4 cm/month and an average velocity of 15.9 cm/month with a lower surface melt extent of 0.1517 km^2. Despite only 0.0017 km^2 of surface melt, the Lower Ablation Zone also displayed high velocities reaching a maximum of 17.1 cm/month and an average of 16.2 cm/month. Particularly in the middle to lower elevations, where basal sliding is most active, high velocities suggested a connection between surface meltwater and glacier movement [28].

In August 2024, velocity values decreased, in together with the decrease in surface melt. With a surface melt extent of 0.5006 km^2, the accumulation zone displayed a maximum velocity of 6.9 cm/month and an average of 5.5 cm/month. Because of its steeper slopes and effective meltwater drainage, the upper ablation zone recorded slightly greater velocities with a maximum of 7.2 cm/month and an average of 5.6 cm/month. The lower ablation zone exhibited moderate dynamics as surface melt decreased to 0.0194 km^2 with a maximum velocity of 7.7 cm/month and an average velocity of 4.8 cm/month.

By September 2024, the ablation season displayed the lowest velocities were caused by cooler temperatures and less meltwater available. The upper ablation zone displayed a maximum velocity of 6.03 cm/month and an average of 4.9 cm/month, while the accumulation zone recorded a maximum velocity of 8.8 cm/month and an average of 5.7 cm/month. With a minimum surface melt extent of 0.0026 km^2 and a maximum velocity of 6.04 cm/month, and an average of 4.9 cm/month, the Lower Ablation Zone exhibited the most reduced changes. The transition to freezing temperature, raising, snow accumulation, and the termination of meltwater-driven flow are all variables that led to these significant reductions.

11.5.5.3 *Influence of Meltwater on Glacier Dynamics*

The connection between surface melt and glacier velocity was highest during the hottest months of summer. Substantial surface melting in July 2024, specifically 1.027 km^2 in the accumulation zone, enhanced subglacial lubrication and accelerated flow. Conversely, reduced velocities across all zones were associated with reduced meltwater availability in September, highlighting the role of meltwater in regulating glacier dynamics.

The interaction between climate and local characteristics, which includes terrain and hydrology, accounts for the variations in SF velocity across different areas. The maximum velocity was frequently observed in the Upper Ablation Zone, distinguished by steeper slopes and heightened meltwater discharge. The Lower Ablation Zone exhibited reduced mobility due to its flat topography and persistent snow cover. The findings aligned with a detailed study on glacier dynamics that highlights the influence of local and climatic variability on zonal melt and flow patterns. Predicting future glacier dynamics in response to changing climatic circumstances demanded consistent monitoring of these temporal and geographical trends.

11.5.5.4 *Relationship between Elevation and Glacier Ice Flow Velocity*

The analysis of elevation against SF velocity was performed for July 2024 (Figure 11.9) for illustrative purposes and to facilitate the analysis. Selecting the most recent year ensures that the results accurately represent the contemporary glacier dynamics influenced by existing environmental conditions. July 2024, as the apex ablation month, provides the most lucid and pertinent insights regarding elevation-dependent velocity disparities, making it a perfect instance for our study. The investigation required digitizing the glacier's major flow line and extracting point-specific velocity and elevation data. The velocity datasets obtained from Sentinel-1 InSAR data indicate the monthly average surface velocity for July 2024, expressed in centimeters per month. The results revealed a clear negative correlation between elevation and velocity that displays considerable spatial heterogeneity along

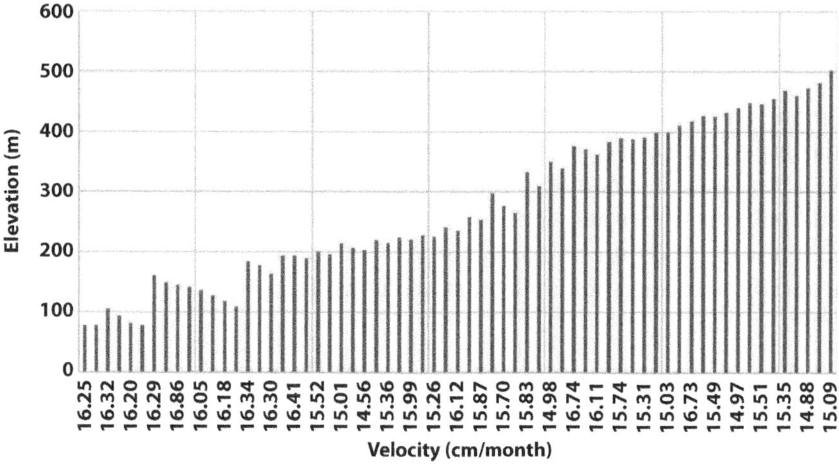

Figure 11.9 The SF velocity as a function of elevation for the Vestre Brøggerbreen Glacier in July 2024.

the flow line. At elevations below 200 meters, the DEM velocities peaked, ranging from 16.73 cm/month to 17.10 cm/month. The increased velocities resulted from heightened meltwater generation at lower altitudes during the ablation season, which improved basal lubrication and accelerated ice movement. The terminus region frequently undergoes intense ice flux due to converging flow lines, whereas intensified interaction with subglacial debris may marginally diminish flow velocities near the glacier edge.

In the mid-elevation range, around 370 meters DEM, the glacier exhibited its highest recorded velocity of 17.10 cm/month. Peak summer melting occurred in this zone, which acts as a transition between the accumulation and ablation areas. In the present study region, the meltwater drainage is especially effective, which helps to improve basal sliding. In this zone, faster flow dynamics are driven by optimal subglacial hydrological activity and adequate meltwater input. Ice flow velocities dramatically dropped above 400 meters DEM at higher elevation, stabilized at 14.51 to 15.51 cm/month. Colder temperatures at higher elevations inhibited surface melting and reduced the amount of basal meltwater available for lubrication tends to be the main cause of this velocity decrease. Moreover, internal deformation rather than basal sliding is the primary force governing ice movement at high altitudes. The elevated regions exhibit a steeper gradient, which restricts gravitational force and reduces ice flow.

The identified inverse correlation between velocity and elevation aligns with a previous study that focuses on the significant influence of elevation-dependent climatic and glaciological parameters on glacier dynamics [6]. The geographical analysis of the DEM velocity profile for the Vestre Brøggerbreen Glacier underscores the dynamic interaction among surface melt, elevation, and ice movement. This study highlights the importance of elevation in influencing glacier velocity, with the highest velocities recorded at mid to low altitudes, indicative of enhanced ice flow and potential meltwater contributions to downstream hydrology. Regular monitoring of elevation velocity is important to know the glacier's response to Arctic warming and its effect on the deduction and global sea level rise.

11.5.6 Climatic Influences on Glacier Dynamics

The analysis of the 2024 ablation season shows the important relationship between surface melt, surface ice flow velocity, and climatic influences, showing the intricate interaction of temperature, precipitation, and glacier hydrology. Peak velocities like 17.5 cm/month were recorded in the accumulation zone during July 2024, which coincided with the maximum melt extents of 1.027 km^2. It emphasizes the significant role of meltwater in driving basal sliding and ice flow. Similar patterns were noticed in the higher ablation zone, where velocities reached 17.4 cm/month, showing the correlation between surface melt and subglacial hydrology. These observations align with the present study [29].

Spatial variability further displayed the impact of glacial topography and climate factors. The lower ablation zone, although it showed minor melt extents of 0.0017 km^2 in July, rather high velocities of 16.2 cm/month, likely influenced by steeper slopes and effective meltwater drainage. In contrast, greater melt extents in the accumulation zone led to continuous ice movement across the season with average velocities of 5.07 cm/month in June and 5.7 cm/month in September. The spatial dynamics highlighted the influence of elevation, slope, and surface energy balance of glacier movement on different climatic conditions [30].

The fluctuation in temperature data highlights the temporal evolution of the melt and velocity pattern in July always displayed the highest temperatures and peak velocities throughout all zones due to increased meltwater production. In September, the temperature declined around 3.3°C, and an increase in precipitation of 14.2 mm showed snow accumulation enhanced the albedo and reduced basal lubrication. The seasonal change in energy balance indicated the transition from dynamics driven by active melt to conditions dominated by stable freezing. Comparisons with previous years reinforce the glacier's sensitivity to climatic variability. In 2023, characterized by comparable temperatures of 7.6°C in July but reduced precipitation, there was a more significant melt and accelerated velocities compared with wetter years such as 2022, where snowfall slowed melt dynamics. These observations align with findings from the dual role of temperature and precipitation in modulating Arctic glacier dynamics [29].

The 2024 findings highlight the potential long-term effects of climate change on glacier stability. Extended warming trends are anticipated to augment meltwater generation, accelerate ice movement, and intensify ice loss. Furthermore, alterations in precipitation patterns, particularly diminished summer snowfall, may intensify these impacts by constraining albedo-driven feedback mechanisms. Understanding these interactions is crucial for predicting the future behavior of Arctic glaciers, particularly in light of their contributions to global sea level rise and the broader cryosphere system.

11.6 Discussion

The study delivers an accurate assessment of the Vestre Brøggerbreen Glacier SF velocity, surface melt extent, and interaction of climatic variables during melt seasons of 2018 to 2024. The results offer important new information about how sensitive glaciers are to changes in the climate and how this affects the stability of the cryosphere using multi-year observations. In the Arctic, the temperature turns out to be the most important factor influencing surface melt and SF velocity. Maturilli *et al.* reported that the observed warming trend in Ny-Ålesund was +1.3 ± 0.7 K per decade for 1994–2013, which is consistent with the increasing melt extents during the study period [31]. An illustration of the accelerated basal sliding caused by meltwater production in July 2023, with a peak temperature of 7.6°C, recorded the highest melt extent of 1.38 km^2 and a maximum SF velocity of 23.4 cm/month. The role of mid-summer temperature peaks in influencing glacier dynamics was further highlighted by July 2024, which showed significant melt extent (1.027 km^2) and sustained warming (7.3°C). In cooler years, the temperature-driven variability was also noticeable. For example, in 2020 observed smaller melt extents (0.4937 km^2 in June) and SF velocities (0.9 cm/month) despite having comparatively lower temperatures (3.16°C in June). This demonstrates the glacier's sensitivity to variations in interannual temperatures as reduced thermal input inhibits the production of meltwater and restricts basal hydrological activity.

The relationship among precipitation, surface melt, and SF velocity was challenging. Rainfall events during warmer months amplified meltwater production and basal lubrication, as seen in July 2018, where high precipitation (99.4 mm) contributed to elevated SF velocities (9.8 cm/month) despite moderate melt extents (0.3759 km^2) [32]. Conversely, the snowfall during cooler periods increased surface albedo, reducing meltwater availability and suppressing SF velocity. For example, September 2019 with snowfall-dominated precipitation (46.2 mm), recorded minimal melt extents (0.49 km^2) and correspondingly low velocities (3.4 cm/month). The interannual variability in precipitation also modulated ablation dynamics in 2022, with higher-than-average precipitation during the ablation season (89.1 mm in July), showing subdued melt extents (0.3752 km^2) and lower SF velocities (6.04 cm/month), emphasizing the suppressive effect of increased snow cover on melt processes.

Seasonal progression and zonal variability: The seasonal transition from June to September revealed distinct patterns in surface melt and SF velocity. June, marking the onset of melting, exhibited moderate melt extents (e.g., 0.9287 km^2 in 2024) and velocities (e.g., 5.07 cm/month in the Accumulation Zone). July consistently recorded the highest melt extents and velocities, driven by peak temperatures and prolonged daylight. For instance, July 2023 showed the highest SF velocity of 23.4 cm/month, corresponding to the largest melt extent (1.38 km^2) [6]. The spatial variability among the glacier zones emphasized the impact of topography and hydrology. Due to effective basal lubrication, the Accumulation Zone, which has larger melt extents (e.g., 1.027 km^2 in July 2024), continuously recorded significant velocities. However, its steep slopes and ice exposure of the Lower Ablation Zone showed notable velocities (16.2 cm/month) despite having minimal melt extents (e.g., 0.0017 km^2 in July 2024). The importance of glacier geometry and subglacial drainage networks in regulating the surface dynamics was highlighted by the zonal disparity.

The results are consistent with regional patterns noted by Sadiq *et al.* and Geetha Priya *et al.* [33], which highlight how long-term warming and changes in precipitation accelerate the dynamics of Arctic glaciers. The Vestre Brøggerbreen Glacier's long-term stability is investigated by the extension of melt seasons in 2023 and 2024, which lasted until September. Long-term warming trends and less summer snowfall may intensify surface melt and promote basal sliding, which could hasten glacier retreat and change the hydrology downstream. The integration of multi-year satellite observations and zonal analysis in this study provides a robust framework for understanding the intricate relationship between climatic drivers and glacier dynamics. These findings underscore the requirement of continuous monitoring to assess the implications of extended Arctic warming on glacier stability and sea level rise.

11.6.1 Field Validation with Unmanned Aerial Vehicle Multispectral Survey

To validate the surface melt extent and enhance the accuracy of remote sensing observations, UAV-based multispectral surveys were conducted over a segment of the ablation region of the Vestre Brøggerbreen Glacier in September 2019. The survey facilitated by the National Centre for Polar and Ocean Research (NCPOR), Ministry of Earth Sciences, Government of India, was a crucial part of a scientific expedition carried out between September 2 and September 30, 2019, in Ny-Ålesund, Svalbard.

A multispectral sensor system that offered high-resolution pictures in multiple bands (RGB) was used in the UAV surveys [33]. The surveys, which were conducted at a flight altitude of 50 meters above ground level, found a spatial resolution of roughly 5 cm/pixel, making it possible to precisely detect exposed ice and surface melt features. Short-term changes in surface conditions were recorded by the chosen surveys, which were carried out on September 14 and September 17 of 2019, and they offered crucial temporal resolution for evaluating glacier dynamics in the late ablation season (Figure 11.10).

High spatial and temporal resolution of UAV data provided an efficient platform to validate surface melt assessments derived from satellites. The ability to record the dynamics of the glacier surface in real-time during the

Figure 11.10 Surface features of the Vestre Brøggerbreen Glacier from UAV Surveys on (a) September 14th, 2019, and (b) September 17th, 2019.

late ablation season provided important information about how the glacier responded to climatic factors like temperature variations and precipitation variability. By bridging the scale gap between *in situ* measurements and regional-scale remote sensing observations, the integration of UAV and satellite data improves the reliability of the study. The main objective was to record variations in the glacier surface between the two survey dates to provide vital information for confirming surface dynamics estimates obtained by remote sensing. High spatial and temporal resolution provided by the UAV data improved the accuracy of surface melt assessments. These observations significantly facilitated understanding short-term shifts in surface conditions and validation of surface melt extent across diverse glacier regions.

Field validation demonstrated the significance of UAV technology in cryosphere research. The ability to monitor Arctic glacier systems in a warming climate is enhanced by the thorough mapping of surface conditions, which helped improve the calibration and validation of datasets derived from satellites and to understand glacier melt dynamics.

11.7 Conclusion

This study undertakes a comprehensive assessment of the cryosphere dynamics of the Vestre Brøggerbreen Glacier from 2018 to 2024 during the ablation season, focusing on surface melt, surface velocity, and their association with temperature and precipitation. The results estimated considerable seasonal variability in these metrics, showing the impact of climatic factors on glacier activity. The integration of temperature data showed a significant impact on surface melt dynamics and the surface ice flow velocity. The summer months from June to September showed average temperatures between 5.6°C in June and 8.4°C in August, with July reaching a maximum temperature of 7.6°C. The surface melt extent in July was almost 141% larger than that reported in June, illustrating the sensitivity of melt dynamics to temperature variations. The dynamics of surface melt were analyzed spatially across three zones: the Accumulation Zone, Upper Ablation Zone, and Lower Ablation Zone. The Accumulation Zone experienced the highest surface melt extent in July at 1.027 km^2, which was a 110% increase from June. The magnitude of surface melt in the Lower Ablation Zone diminished from June to July. These underscore the influence of elevation, albedo, and regional climate on the spatial diversity of melt processes. The surface ice flow velocity assessment revealed the temporal relationship between temperature and surface melt. This study analyzes the cryosphere dynamics of the Vestre Brøggerbreen Glacier throughout the ablation season from

2018 to 2024, concentrating on surface melt, snowflake velocity, and their correlations with temperature and precipitation. The peak surface velocities were observed in July in the Accumulation Zone and the Upper Ablation Zone. The impact of increased meltwater availability on basal lubrication and glacier dynamics was highlighted by estimating that these velocities were almost 209% higher than the average velocities documented in June. Between 2019 and 2020, there was a marked decrease in surface melt and ice flow rates. The global pandemic likely influenced this by decreasing industrial emissions, thereby facilitating the healing of the ozone layer, which in turn lowered temperatures and diminished radiative forcing. Major insights into the complex relationships that govern the glacier dynamics in the Arctic can be derived from the quantitative analysis of geographical and temporal changes. The present study underscores the importance of monitoring and evaluating the impacts of climate change on Arctic glaciers and their impacts on hydrological systems and global sea levels.

Acknowledgment

The authors express their sincere gratitude to the Director of CIIRC, Jyothy Institute of Technology, Bengaluru, and PTICL & TITCL for their invaluable support. The logistical arrangements for access to Ny-Ålesund necessary for this scientific research were made possible by the National Centre for Polar and Ocean Research, Ministry of Earth Sciences, Government of India, Goa. The authors extend their sincere thanks to the Copernicus program and the NCCS for respectfully sharing the invaluable dataset, which was essential to the success of this study. Their input greatly enhanced the research caliber and scope, leading to more thorough analysis and insightful conclusions.

AI disclosure statement: The author acknowledges the use of AI-based tools for minor language editing or grammatical correction. However, all intellectual contributions, analyses, and conclusions presented in this chapter are entirely the author's own.

References

1. Luis, A.J. and Singh, S., High-resolution multispectral mapping facies on glacier surface in the Arctic using WorldView-3 data. *Czech Polar Rep.*, 10, 1, 23–36, 2020.

2. Polyak, L., Alley, R.B., Andrews, J.T., Brigham-Grette, J., Cronin, T.M., Darby, D.A., Dyke, A.S., Fitzpatrick, J.J., Funder, S., Holland, M., Jennings, A.E., Miller, G.H., O'Regan, M., Savelle, J., Serreze, M., St. John, K., White, J.W.C., Wolff, E., History of sea ice in the Arctic. *Quat. Sci. Rev.*, 29, 15, 1757–1778, 2010.
3. White, J.W.C., Alley, R.B., Brigham-Grette, J., Fitzpatrick, J.J., Jennings, A.E., Johnsen, S.J., Miller, G.H., Steven Nerem, R., Polyak, L., Past rates of climate change in the Arctic. *Quat. Sci. Rev.*, 29, 15, 1716–1727, 2010.
4. Ren, G., Wang, J., Lu, Y., Wu, P., Lu, X., Chen, C., Ma, Y., Monitoring changes to arctic vegetation and glaciers at ny-Ålesund, svalbard, based on time series remote sensing. *Remote Sens.*, 13, 19, 2021.
5. Dharwadkar, A. and Kumar, P., Integrated Monitoring of Vestre Broggerbreen Glacier, Savalbard Arctic. *Geol. Surv. India*, 2018.
6. Sadiq, M., Dutta, S., Kumar, P., Jat, S., Gajbhiye, D.Y., Dharwadkar, A., Ice dynamics of Vestre Brøggerbreen glaciers, Ny-Ålesund, Svalbard, Arctic. *J. Earth Syst. Sci.*, 131, 1, 2022.
7. Aschbacher, J. and Milagro-Pérez, M.P., The European Earth monitoring (GMES) programme: Status and perspectives. *Remote Sens. Environ.*, 120, 3–8, 2012.
8. Rott, H., Advances in interferometric synthetic aperture radar (InSAR) in earth system science. *Prog. Phys. Geogr. Earth Environ.*, 33, 6, 769–791, 2009.
9. Gabriel, A.K., Goldstein, R.M., Zebker, H.A., Mapping small elevation changes over large areas: Differential radar interferometry. *J. Geophys. Res. Solid Earth*, 94, B7, 9183–9191, 1989.
10. Nagler, T., Rott, H., Hetzenecker, M., Wuite, J., Potin, P., The Sentinel-1 mission: New opportunities for ice sheet observations. *Remote Sens.*, 7, 7, 9371–9389, 2015.
11. Østby, T.I., Vikhamar Schuler, T., Ove Hagen, J., Hock, R., Kohler, J., Reijmer, C.H., Diagnosing the decline in climatic mass balance of glaciers in Svalbard over 1957-2014. *Cryosphere*, 11, 1, 191–215, 2017.
12. Shashank, B. and Geetha Priya, M., Monitoring of Melting Glaciers of Ny-Ålesund, Svalbard, Arctic Using Space-Based Inputs. *Lect. Notes Electr. Eng*, 966 LNEE, 381–390, 2023.
13. Nuth, C., Moholdt, G., Kohler, J., Hagen, J.O., Kääb, A., Svalbard glacier elevation changes and contribution to sea level rise. *J. Geophys. Res. Earth Surf.*, 115, 1, 1–16, 2010.
14. Strozzi, T., Caduff, R., Jones, N., Barboux, C., Delaloye, R., Bodin, X., Kääb, A., Mätzler, E., Schrott, L., Monitoring rock glacier kinematics with satellite synthetic aperture radar. *Remote Sens.*, 12, 3, 1–24, 2020.
15. Luckman, A., Elvidge, A., Jansen, D., Kulessa, B., Kuipers Munneke, P., King, J., Barrand, N.E., Surface melt and ponding on Larsen C Ice Shelf and the impact of föhn winds. *Antarct. Sci.*, 26, 6, 625–635, 2014.
16. https://sentiwiki.copernicus.eu/web/s1-mission.

17. Førland, E.J., Isaksen, K., Lutz, J., Hanssen-Bauer, I., Schuler, T.V., Dobler, A., Gjelten, H.M., Vikhamar-Schuler, D., Measured and modeled historical precipitation trends for svalbard. *J. Hydrometeorol.*, 21, 6, 1279–1296, 2020.
18. Førland, E.J., Benestad, R., Hanssen-Bauer, I., Haugen, J.E., Skaugen, T.E., Temperature and Precipitation Development at Svalbard 1900–2100. *Adv. Meteorol.*, 2011, 1–14, 2011.
19. Lefauconnier, B. and Hagen, J.O., Glaciers and climate in Svalbard: statistical analysis and reconstruction of the Broggerbreen mass balance for the last 77 years. *Ann. Glaciol.*, 14, 148–152, 1990.
20. Dobreva, I.D., Bishop, M.P., Bush, A.B.G., Climate-glacier dynamics and topographic forcing in the Karakoram Himalaya: Concepts, issues and research directions. *Water (Switzerland)*, 9, 6, 20–26, 2017.
21. Spolaor, A., Scoto, F., Larose, C., Barbaro, E., Burgay, F., Bjorkman, M.P., Cappelletti, D., Dallo, F., de Blasi, F., Divine, D., Dreossi, G., Gabrieli, J., Isaksson, E., Kohler, J., Martma, T., Schmidt, L.S., Schuler, T.V., Stenni, B., Turetta, C., Luks, B., Casado, M., Gallet, J.-C., Climate change is rapidly deteriorating the climatic signal in Svalbard glaciers. *Cryosph.*, 18, 1, 307–320, 2024.
22. Van Pelt, W.J.J., Oerlemans, J., Reijmer, C.H., Pohjola, V.A., Pettersson, R., Van Angelen, J.H., Simulating melt, runoff and refreezing on Nordenskiöldbreen, Svalbard, using a coupled snow and energy balance model. *Cryosphere*, 6, 3, 641–659, 2012.
23. Sundal, A.V., Shepherd, A., Nienow, P., Hanna, E., Palmer, S., Huybrechts, P., Evolution of supra-glacial lakes across the Greenland Ice Sheet. *Remote Sens. Environ.*, 113, 10, 2164–2171, 2009.
24. Steinbrecht, W., Kubistin, D., Plass-Dülmer, C., Davies, J., Tarasick, D.W., Von Der Gathen, P., Deckelmann, H., Jepsen, N., Kivi, R., Lyall, N., Palm, M., Notholt, J., Kois, B., Oelsner, P., Allaart, M., Piters, A., Gill, M., Van Malderen, R., Delcloo, A.W., Sussmann, R., Mahieu, E., Servais, C., Romanens, G., Stübi, R., Ancellet, G., Godin-Beekmann, S., Yamanouchi, S., Strong, K., Johnson, B., Cullis, P., Petropavlovskikh, I., Hannigan, J.W., Hernandez, J.L., Diaz Rodriguez, A., Nakano, T., Chouza, F., Leblanc, T., Torres, C., Garcia, O., Röhling, A.N., Schneider, M., Blumenstock, T., Tully, M., Paton-Walsh, C., Jones, N., Querel, R., Strahan, S., Stauffer, R.M., Thompson, A.M., Inness, A., Engelen, R., Chang, K.L., Cooper, O.R., COVID-19 Crisis Reduces Free Tropospheric Ozone Across the Northern Hemisphere. *Geophys. Res. Lett.*, 48, 5, 1–11, 2021.
25. Blanchard-Wrigglesworth, E., Cox, T., Espinosa, Z.I., Donohoe, A., The Largest Ever Recorded Heatwave—Characteristics and Attribution of the Antarctic Heatwave of March 2022. *Geophys. Res. Lett.*, 50, 17, 1–11, 2023.
26. Trusel, L.D., Das, S.B., Osman, M.B., Evans, M.J., Smith, B.E., Fettweis, X., McConnell, J.R., Noël, B.P.Y., van den Broeke, M.R., Nonlinear rise in Greenland runoff in response to post-industrial Arctic warming. *Nature*, 564, 7734, 104–108, 2018.

27. Benn, D., II and Evans, D.J.A., *Glaciers & Glaciation*, Routledge, London, 2015.
28. Pimentel, S. and Flowers, G.E., A numerical study of hydrologically driven glacier dynamics and subglacial flooding. *Proc. R. Soc. A Math. Phys. Eng. Sci.*, 467, 2126, 537–558, 2011.
29. Joughin, I., Das, S.B., King, M.A., Smith, B.E., Howat, I.M., Flank of the Greenland Ice Sheet. *Sci. (80-.)*., 320, 5877, 781–783, 2008.
30. De Woul, M. and Hock, R., Static mass-balance sensitivity of Arctic glaciers and ice caps using a degree-day approach. *Ann. Glaciol.*, 42, Table1, 217–224, 2005.
31. Maturilli, M., Herber, A., König-Langlo, G., Surface radiation climatology for Ny-Ålesund, Svalbard (78.9°N), basic observations for trend detection. *Theor. Appl. Climatol.*, 120, 1–2, 331–339, 2015.
32. Sundal, A.V., Shepherd, A., Nienow, P., Hanna, E., Palmer, S., Huybrechts, P., Melt-induced speed-up of Greenland ice sheet offset by efficient subglacial drainage. *Nature*, 469, 7331, 521–524, 2011.
33. Geetha Priya, M., Venkatesh, K., Shivanna, L., Devaraj, S., Detecting short-term surface melt over Vestre Broggerbreen, Arctic glacier using indigenously developed unmanned air vehicles. *Geocarto Int.*, 37, 11, 3167–3178, 2022.

12

Legal, Ethical, and Policy Aspects of SAR in Disaster Management

Amit Bhadula[1,2]

[1]*NHPC Limited, Faridabad, India*
[2]*MSc, Geography Department, Durham University, Durham, United Kingdom*

Abstract
Synthetic aperture radar (SAR) technology has become an indispensable tool in disaster management due to its all-weather, day-and-night imaging capability with better temporal and spatial resolution. The advent of various SAR data public domain has been critical in integrating SAR data with disaster management plans. The SAR data has facilitated quick mapping of the impact of potential earthquakes, tsunamis, hurricanes, and subsidence. This unique advantage allows responders and policymakers to monitor, assess, and react to both natural and man-made disasters with increased efficiency. Despite a large-scale understanding of SAR data and its significance over the last decade, various legal, ethical, and policy challenges remain to be addressed. This paper reviews various acts, policies, and frameworks governing the dissemination of SAR data, identifies potential gaps, and provides possible suggestions and steps to be taken. The two schools of thought, realism and liberalism, are two approaches to assessing future potential collaborations. With more commercial players, data availability has become easier but costlier.

***Keywords*:** SAR, disaster management, ethical frameworks, data-sharing policy, sentinel data policy, Gaofen-3 satellite system, capacity building

12.1 Introduction

A disaster can be defined as a sudden event that causes loss of life and property. The United Nations Office for Disaster Risk Reduction (UNDRR) defines a disaster as a severe disruption in the functioning of a community or society

Email: amit.bhadula@gmail.com

Pooja Joshi, Ashutosh Bhatt, Neelam Sharma and Kapil Joshi (eds.) Synthetic Aperture Radar for Disaster Management: Bridging Science and Solutions, (337–362) © 2026 Scrivener Publishing LLC

at any scale, caused by hazardous events interacting with conditions of exposure, vulnerability, and limited coping capacity. Such disruptions often result in significant adverse consequences, including loss of human lives, damage to property and infrastructure, financial setbacks, and environmental and biodiversity degradation. The magnitude of the disaster is shaped by the degree to which people, assets, and systems are exposed and how vulnerable or resilient they are to the hazard. Disasters can overwhelm local resources and capacities, requiring external support and long-term recovery efforts [1, 2]. The Sendai Framework (2015–2030) targets reducing disaster risk through four major priorities, i.e., understanding risk, strengthening disaster risk governance, investing in disaster risk reduction for resilience, and enhancing disaster preparedness. Remote-sensing, especially SAR, plays an important role in achieving all four, which are described further.

Before delving deeper into synthetic aperture radar (SAR) application for disaster risk vis-à-vis various policies, it is imperative to recognize the fact that the frequency of hydrological hazards, i.e., extreme floods, hurricanes, tornadoes, glacial lake outburst floods, and landslide lake outburst flood (LLOF) due to climate change, is expected to increase in the future. Flash floods in the Himalayas in 2013 and 2021 in Uttarakhand and 2024 in Sikkim are some examples. The coastal area in India alone has faced more than 25 cyclones in the last 25 years, with the largest loss being in 1999, wherein more than 10,000 people died. However, with the advent of the Doppler radar, effective risk communication, satellite data from the National Remote Sensing Centre (NRSC), the development of resilient infrastructure, and people's participation, the loss of life has been significantly reduced.

The impact of geological risks, i.e., earthquakes, tsunamis, and subsidence, allows for assessing the impact of earthquakes.

Disaster risk can be defined as a product of hazard and exposure. The increasing population with urbanization and infrastructure creation, like roads, railway lines, and ports, has increased the risk in the last century. The need for effective, timely, and precise disaster management has never been greater. The SAR offers significant capabilities for monitoring terrains, infrastructure in remote, inaccessible areas, as well as in adverse weather conditions. Its remote sensing abilities provide critical data to governments, humanitarian agencies, and researchers.

However, the use of SAR data in public and private domains raises legal, ethical, and policy implications, especially in war zones and international borders. Issues surrounding sovereignty, data ownership, access rights, privacy, commercialization, and international coordination challenge the otherwise straightforward adoption of SAR for humanitarian purposes.

Addressing these concerns is vital for ensuring that SAR technology is leveraged responsibly, equitably, and sustainably in disaster management.

12.2 Legal Aspects

12.2.1 International Legal Frameworks

First, the organization for the international governance of space was the International Telecommunication Union, which was later named the International Telegraph Union, established in 1865 [3, 4]. Although the advent of SAR technology came almost a century later, in 1951, with the help of Carl Wiley [5].

Sputnik's 1957 launch spurred the United Nations' (UN) 1962 Outer Space Principles Declaration. The UN resolution adopted on 13th December 1962, "*Declaration of Legal Principles Governing Activities of States in Exploration and Use of Outer Space,*" was one of the first steps towards collaborative and peaceful use of space [4].

It has been a cornerstone for five treaties between 1967 and 1984, during the peak of the Cold War era. The Outer Space Treaty (OST) (1967), Rescue Agreement (1968), Liability Convention (1972), Registration Convention (1976), and Moon Agreement (1984), and the United Nations Office for Outer Space Affairs (UNOOSA), 2015 [6].

The OST 1967 was the first of its kind agreement between 28 countries for peaceful and collaborative use of space technology, focusing on outer space as well as celestial bodies [7]. This treaty evolved during the Cold War and provides guidelines for different countries to act outside their international borders. The above act assumes that the state is the key player in political interaction; it provides an instrument to deal with the realm of the international political system above the complexity arising from individual countries' interests [6]. The acquisition of satellite data traverses international borders, disputed territories, war zones, and countries with limited access. It can create a complex situation, which calls for assessing various international laws, treaties, frameworks, and acts. This treaty has been ratified by more than 100 countries [8].

In the aftermath of the Cold War in 1991, the development of domestic outer space has been given more emphasis, while the international laws governing outer space have taken a back seat [6]. The development of commercial entities and the lack of further international collaborative efforts have created policy gaps.

The United Nations General Assembly's Resolution 41/65, adopted on December 3, 1986, established 14 principles for the remote sensing of the Earth from space. Within the remote sensing framework, several principles emphasize the importance of data sharing and collaboration.

Principles II, V, VI, and VII advocate for international cooperation, equitable participation, and technical assistance. Principles X and XI stress timely data sharing for environmental protection and disaster response. Principle XII ensures states' access to data and analysis on non-discriminatory, reasonable terms. Principle XIII encourages consultation and mutual benefit in remote sensing programs. Together, these principles promote openness, cooperation, and equitable access to remote sensing data, reinforcing global collaboration, particularly in support of developing countries and shared challenges.

The UN Committee on the Peaceful Uses of Outer Space (UNCOPUOS) is a UN body created in 1958. The UNCOPUS has been responsible for registering satellites, which is now supported by UNOOSA.

The UN resolution report underlines the policy of sharing information on remote sensing technologies. The reiteration of the importance of the Sendai Framework for Disaster Risk Reduction (2015–2030), Training in SAR technologies, application of Sentinel-1 data for disaster management, e.g., Typhoon Rai in the Philippines in 2021, underscores the application of SAR data.

Consequent to the Rio Summit in 1992, a greater emphasis has been given on data sharing for Earth Observation. One of the outcomes is the Global Earth Observation System of Systems (GEOSS). Between 2002 and 2015, a series of Earth observation-focused summits and GEO plenaries shaped the GEOSS. These events facilitated transparency in data sharing, infrastructure strengthening, and enhanced collaboration to support sustainable development and disaster risk mitigation. The Twelfth GEO Plenary and Fourth Ministerial Summit in 2015 resulted in the *Mexico City Declaration*, aligning GEOSS with the SDGs, the Sendai Framework, and climate goals, while endorsing the GEO Strategic Plan 2016–2025. Also in 2015, the Eye on Earth Summit emphasized capacity building, technology, and data sharing, particularly in developing regions. The UN Sustainable Development Summit 2015 recognized Earth observation's role in tracking SDG progress. Earlier GEO plenaries supported initiatives like AfriGEOSS, GEOGLAM, and the Global Forest Observation Initiative (GFOI), while promoting open data principles (e.g., GEOSS Data-CORE). High-level summits—EOS I–III—established GEO and the initial 10-Year Implementation Plan. The foundation was laid at the World Summit on Sustainable Development (2002) and Rio+20 (2012), which recognized the importance of space-based and geospatial data for sustainable

Table 12.1 The SAR satellites and data available with ICSMD (ICSMD, 2025).

S. no.	Agency	SAR satellite	Data	Revisit period
1	Bolivarian Agency for Space Activities (ABAE)	None	Only Optical Data	Not Applicable
2	Canadian Space Agency (CSA)	RADARSAT Constellation Mission & RADARSAT-2	C Band data (HH, HV, VH, VV polarization)	4 to 24 days
3	Indian Space Research Organization (ISRO)	None	Only Optical Data	Not Applicable
4	Russian State Space Corporation (ROSCOSMOS)	None	Only Optical Data	Not Applicable
5	Centre National d'Etudes Spatiales (CNES)	None	Only very high resolution Optical Data	Not Applicable
6	German Aerospace Centre (DLR)	TerraSAR-X, TanDEM-X	X-Band, (HH, HV, VH, VV polarization)	4 days
7	Japan Aerospace Exploration Agency (JAXA)	Daichi-2 (ALOS-2)	L-Band, (HH, HV, VH, VV polarization)	14 days
8	United Arab Emirates Space Agency (UAESA)/Mohamad bin Rashid Space Centre (MBRSC)	None	Only Optical Data	Not Applicable
9	China National Space Administration (CNSA)	Gaofen-3	C-Band (HH, HV, VH, VV polarization)	29 days
10	National Commission for Space Activities (CONAE)-Argentina	SADCOM-1A	L-Band, (HH, HV, VH, VV polarization)/ Dual Pol-quad-pol	2 days

(*Continued*)

Table 12.1 The SAR satellites and data available with ICSMD (ICSMD, 2025). (*Continued*)

S. no.	Agency	SAR satellite	Data	Revisit period
11	National Institute of Space Research (INPE)-Brazil	None	Only Optical Data	Not Applicable
12	ICEYE-Finland	ICEYE-X2, X3, X4, X5, X6, X7	HR/VHR	1 day
13	European Organization for Exploitation of Meteorological Satellites (EUMETSAT)	None	Only Optical Data	Not Applicable
14	European Space Agency (ESA)	Sentinel1A/1B	C-Band (HH, HV, HH+HV, VV+VH polarization)	10 days
15	Korean Aerospace Research Institute (KARI)	KOMPSAT-5	X Band (HH, HV, VH, VV polarization)	28 days (nominal)
16	National Oceanic and Atmospheric Administration (USA)	None	Only Optical Data	Not Applicable
17	Planet (USA)	None	Only High-resolution Optical Data	Not Applicable
18	United States Geological Survey (USGS)	None	Only Optical Data	Not Applicable
19	UK Space Agency	None	Only Optical Data	Not Applicable

development. The GEO is a pivotal international platform for coordinating earth observations to inform science-based policy and global action.

The GEOSS supports the propagation of Sentinel-1 SAR data for disaster management. It promotes free availability of satellite data with minimal cost for commercial data and no cost for research and educational purposes [9].

Though participation of non-spacefaring countries, especially developing countries, has been the highlight of the international legal framework and consequent institutional building. The international legal framework under UN bodies is non-binding 'soft laws' [4] remains fragmented and open to interpretation, allowing states and commercial actors to apply these principles in ways that align with their national interests rather than global humanitarian goals.

The International Charter for Space and Major Disasters (ICSMD) was established in 1999 and provides a platform for national disaster management agencies to activate the charter. Activation of the charter is done by National Disaster Authorities on the occurrence of a disaster, i.e., flood, earthquake, volcano, and tsunami. The Charter provides data and maps for measuring the impact of disasters and helping agencies in managing disasters. Various SAR products are tabulated in Table 12.1 and compared in Figure 12.1.

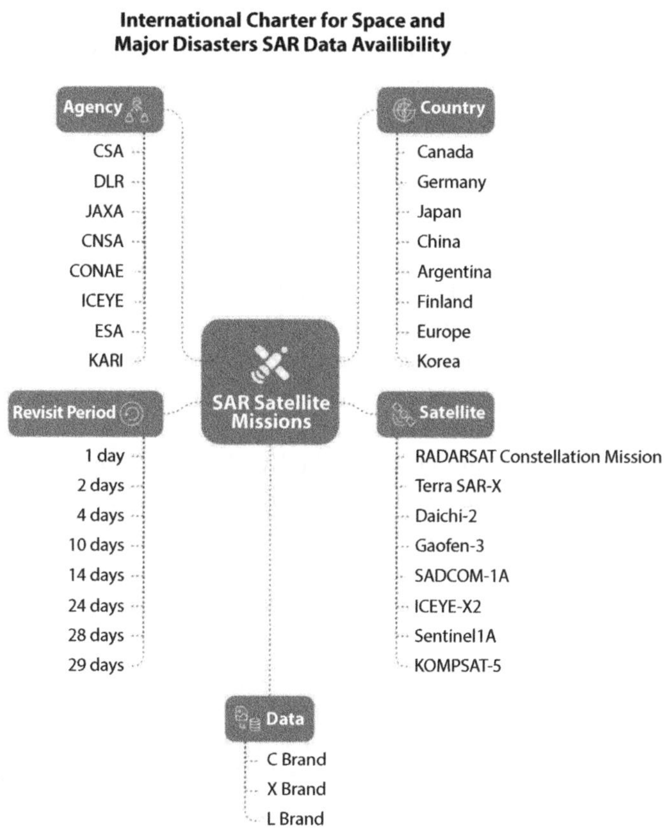

Figure 12.1 The SAR satellites available with ICSMD (ICSMD, 2025).

12.2.2 Country and Regional Policies

12.2.2.1 The United States' Earth Observation Policy and National Plan

The USA's earth observation policy framework may be accessed through the 2013 US Earth Observation Strategy and the 2019 National Plan for Civil Earth Observations, NOAA NESDIS data policy, and Landsat data policies. The Earth Observations 2013, National Plan for Civil Earth Observation policies, NOAA NESDIS, and Landsat data policy all promote full, free, and open access to satellite data, typically at no cost beyond distribution. The Earth Observations 2013 strategy strongly endorses open federal data access without copyright or licensing, explicitly encouraging commercial use. All three policies aim to maximize public benefit, with varying levels of legal complexity and exceptions. The NOAA NESDIS principally supports open data access, but is governed by multiple legal frameworks that allow for data denial in emergencies. Landsat data, managed by USGS, has been freely available online since 2008, with no major restrictions on commercial application.

The US Earth Observation Strategy NSTC subcommittee has recognized four key characteristics of disaster-resilient communities: first, recognition of and understanding of relevant hazards; second, the community understands risk communication protocols; third, individuals at risk are safe in their homes and workplace; fourth, communities must have minimum disruption to life and property. While the 2019 national plan for civil earth observations focuses on support and balance of the portfolio of earth observations, engaging in the earth observation enterprise, and improving the impact of earth observations. But no specific comment on SAR data and its use has been reported.

The above strategies are closely linked with the US Open Data Policy and the US Global Change Research Program Data Policy. US Open Data Policy promotes the default assumption of openness for all data produced by the US Federal government and its agencies. It defines open data as publicly accessible, machine-readable, and interoperable, ensuring usability by end users [13]. The policy mandates providing modifiable, open formats with descriptions, open licenses, granularity, promptly available, and managed by a designated contact person. Exceptions to data openness may be due to privacy, security, confidentiality, trade secrets, or contractual obligations. However, these exceptions are clearly documented and reviewed in consultation with the Office of General Counsel or its equivalent authority.

The US Open data initiative includes NASA's Earth data, Alaska Satellite Facility (ASF), Advanced Collaborative Connections for Earth System Science (ACCESS), and Commercial Small Sat Data Acquisition. Data availability of NISAR in the near future is one of the efforts of the US Open Data policy for wider spread of SAR data in collaboration with ISRO, India.

The US Global Change Research Program's data policy emphasizes full and open access to high-quality data for global change research. It mandates early and sustained efforts in data collection, validation, maintenance, and dissemination. All essential data must be preserved, with designated archives and clear criteria for acquisition, retention, and purging, including systems to prevent data loss. Archives must provide detailed, user-friendly metadata, including quality assessments and guidance for access. Data should conform to national and international standards and be provided at minimal cost, ideally just covering user-request expenses. Any exclusive-use periods must be clearly defined and kept short to ensure broad availability.

12.2.2.2 The National Oceanic and Atmospheric Administration's National Environmental Satellite Data and Information Service Data Policy

The National Oceanic and Atmospheric Service (NOAA) and National Environmental Satellite Data and Information Service (NESDIS) data policy supports the full and open exchange of environmental satellite data, metadata, and products. However, they stop short of clearly defining what "full and open" means. It references 12 legal and policy documents—such as the US Paperwork Reduction Act, OMB Circular A-130, World Meteorological Organization (WMO) Resolution 40, and the GEOSS data-sharing principles—which generally advocate low-cost access, limited to covering the cost of fulfilling user requests [9].

The policy also allows for data denial in cases of war or national security emergencies. Access to data is ultimately governed by international and national laws, organizational policies, agreements, and resource availability [9].

The NOAA NESDIS utilizes 'Sentinel 1A', 'Sentinel 1-B', 'RADARSAT-2', and 'RADARSAT Constellation Mission RCM' (RCM-1, 2, 3) for various SAR-based products, i.e., ocean surface wind, tropical cyclone wind products, sea and lake monitoring, and flood & inundation mapping. Figure 12.2 summarizes the key objectives and applications identified by the NOAA NESDIS Water Science Team for utilizing SAR data in hydrological and flood monitoring systems.

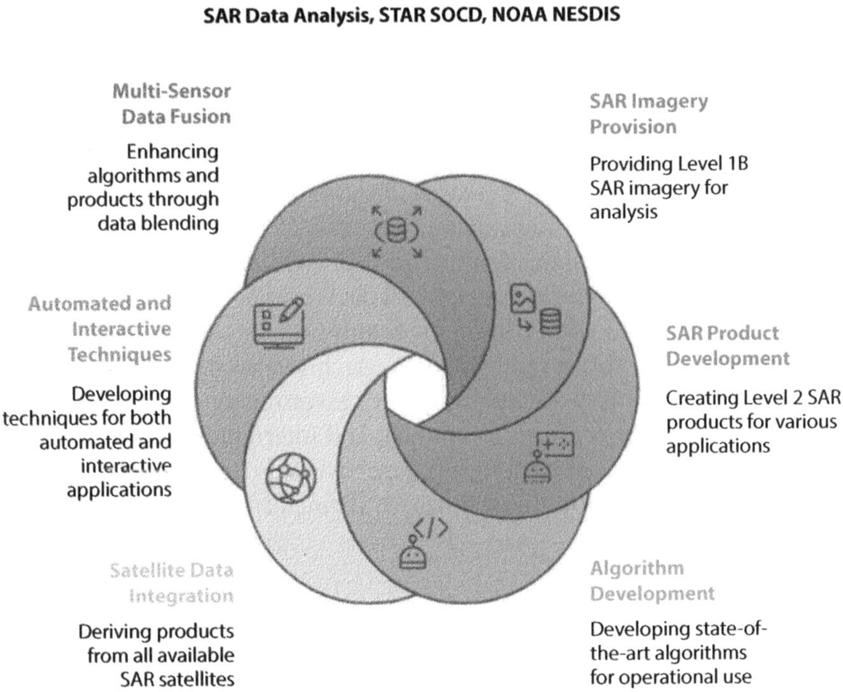

Figure 12.2 The NOAA NESDIS's water science team's objective for the application of SAR data.

12.2.2.3 Landsat Data Policy

The Landsat Data Distribution Policy ensures full, open, and free access to all Landsat data, aligning with US policies like the Earth Observations Strategy and Open Data Policy. Since 2008, the USGS has made Landsat data freely available online, enabling broad use across public and private sectors. The policy aims to support a wide range of applications and foster commercial markets for unenhanced and value-added services. Generally, there are no restrictions on the commercial use of Landsat data, though copyright or licensing applies only if the original data are significantly altered or combined with other datasets to create new products. The RadarSat-2 based operational SAR-derived wind product acquired on 2 April 2025 is depicted in Figure 12.3. The data provided includes only optical imagery; SAR data is not included within its mandate.

The European data policies may be summarized as the Copernicus data policy, ESA data policy, European Organization for the Exploitation of Meteorological Satellites (EUMETSAT) data policy, Sentinel data policy,

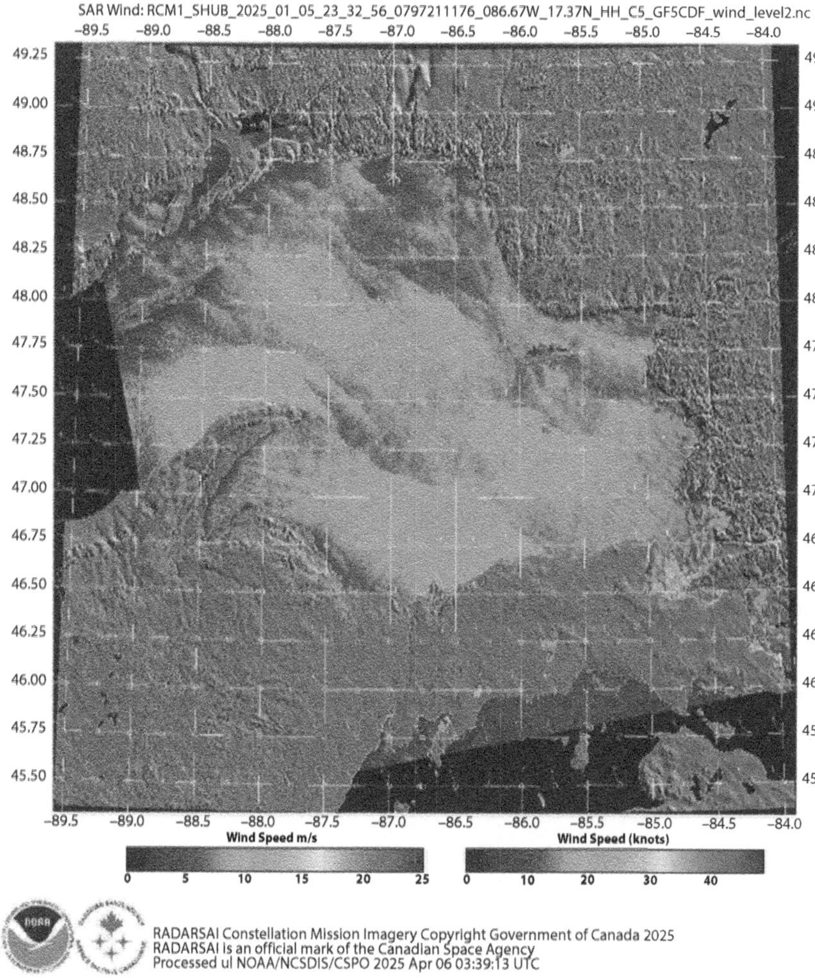

Figure 12.3 The RadarSat-2 data based on 2nd April 2025, operational SAR-derived wind products, NOAA, 2025 [10].

and SPOT (Centre national d'études spatiales or CNES) data policies. The Copernicus and Sentinel data policies both guarantee full, open, and free access to data globally for any user, supporting commercial, public, and scientific use. ESA's policy aligns with Sentinel's but distinguishes between free and restrained datasets, with the latter requiring approval. The Copernicus data policies mentioned in Figure 12.4. The EUMETSAT offers essential data freely to all users, but imposes fees and licenses for other or commercial uses. The SPOT data is free only for non-commercial research use and is limited to older data, requiring licenses. Overall, Copernicus

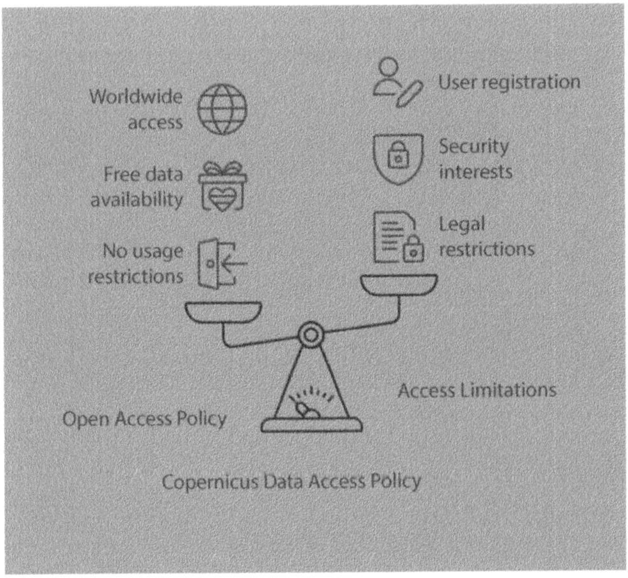

Figure 12.4 Copernicus data policy [9].

and Sentinel promote the most open access, while ESA, EUMETSAT, and SPOT apply varying degrees of restriction. The details are mentioned below.

12.2.2.4 Copernicus Data Policy

The Copernicus Data Policy, established by EU Regulation No. 911/2010 and implemented through Regulation No. 1159/2013, mandates that Copernicus data and information be made available fully, openly, and free of charge. This was reaffirmed by the 2014 Copernicus Regulation (EU No. 377/2014) [9].

12.2.2.5 European Space Agency Data Policy

Data access and usage policy of the European Space Agency (ESA) and the EUMETSAT depicted in Figures 12.5 and 12.6 respectively. The ESA Data Policy for Earth Resource Satellites (ERS-1, ERS-2), Envisat, and Earth Explorer missions aligns with the Sentinel data policy and promotes open and egalitarian access to maximize the use of Earth observation data for scientific, public, and commercial applications. The data obtained from the satellite are segregated into a publicly available and controlled set. Free

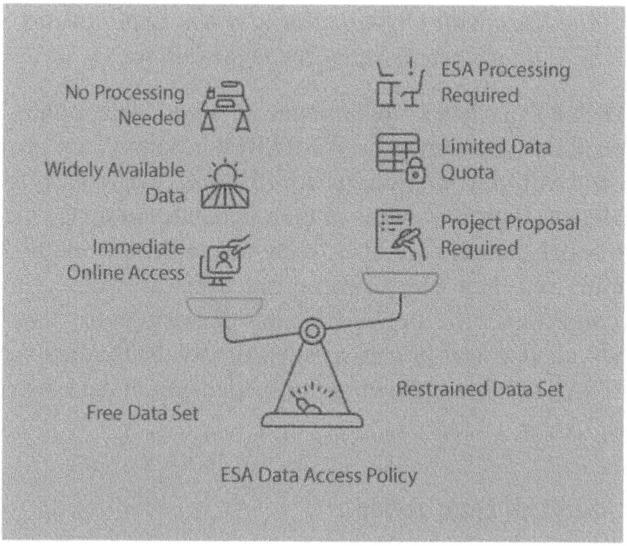

Figure 12.5 The ESA data policy [11].

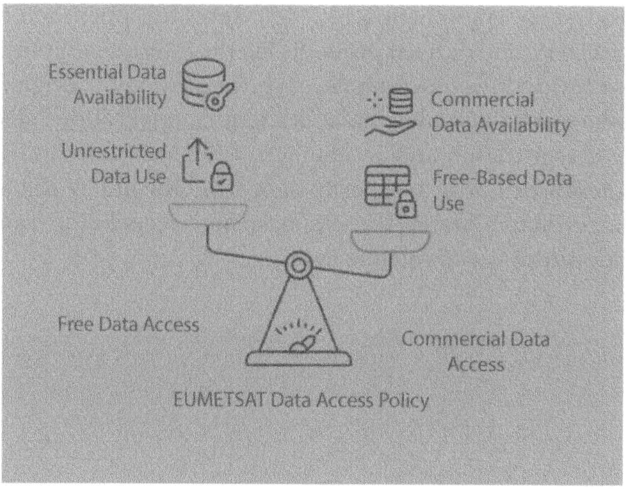

Figure 12.6 The EUMETSAT data policy [12].

data are accessible upon online registration and acceptance of the ESA terms. Restrained data, mainly large SAR datasets from ERS and Envisat, require a project proposal for access due to resource demands. Most datasets, including all Earth Explorer data, are free of charge, with only restricted sets requiring additional approval and processing effort.

12.2.2.6 The European Organization for the Exploitation of Meteorological Satellites Data Policy

The EUMETSAT partially implements an open data policy. Essential data and products, as defined by WMO Resolution 40, are provided free of charge and without restrictions to all users, particularly supporting National Meteorological Services of member states in their official duties. Some additional datasets are also freely available, especially if they are over 24 hours old. However, commercial users are subject to fees and licensing conditions. The policy includes provisions for data denial in emergencies, such as war or crisis, especially for NOAA instrument data on EUMETSAT satellites, allowing the suspension of data distribution in such cases.

12.2.2.7 Sentinel Data Policy

The Sentinel Data Policy, created by the ESA and the European Commission, governs data from Sentinel Earth observation missions under the Copernicus Programme. Sentinel data are provided under an open access framework in Figure 12.7. It ensures full, open, and free access to all Sentinel data and derived products *via* the internet, supporting global use without restrictions based on user type or geographic origin. This policy promotes public, commercial, and scientific applications alike. Access requires user registration and acceptance of the license terms. Exceptions to open access are limited to security concerns. The policy aligns with the broader Copernicus Data Policy, emphasizing accessibility, transparency, and the widespread use of satellite data.

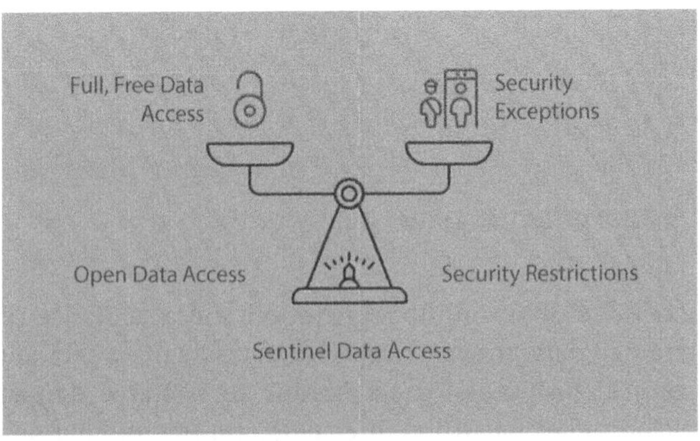

Figure 12.7 Sentinel data policy [11].

12.2.2.8 The Centre National D'études Spatiales (SPOT) Data Policies

In January 2014, CNES announced that SPOT satellite data older than five years and spatial resolution coarser than 10 m would be made freely available to non-commercial users for research purposes. Users are required to sign a license agreement confirming non-commercial research use. Although the SPOT data are not classified as per Panton Principles, the policy reflects a move toward open data. Although this initiative marks a shift towards data accessibility for scientific and educational purposes but the dataset is devoid of SAR data.

12.2.2.9 Inspire Directive (European Union)

Directive 2007/2/EC of the European Parliament and Council, dated 14th of March 2007, which sets up the Infrastructure for Spatial Information in the European Community (INSPIRE), was officially published in the Official Journal on 25th April 2007. It officially entered into force on 15 May 2007. The directive enables the sharing of 34 thematic data sets, but none of them is directly related to SAR data.

12.2.2.10 European Union's Horizon 2020 Program

This program funded various projects on SAR, including the ICEYE project for launching microsatellites for:

12.2.2.11 G-8 Open Data Charter

The comparison of SAR satellite missions under the G-8 framework is depicted in Figure 12.8. The G-8 Open Data Charter, signed in 2013, states, "*We also recognize the benefits of open data can and should be enjoyed by citizens of all nations. In the spirit of openness, we offer this Open Data Charter for consideration by other countries, multinational organizations, and initiatives*". The agreement was signed between Canada, Russia, France, Germany, Italy, Japan, the United Kingdom, and the United States, among which Russia was suspended in 2014. Among these, Canada, Germany, Japan, and Italy have the capability of SAR data acquisition, which is summarized below:

12.2.2.12 Chinese Data-Sharing Policy

The Ministry of Science and Technology of China initiated the scientific data sharing program in 2002 to improve access to meteorological data for national research projects.

SAR Satellite Missions

Characteristic	RADARSAT	TerraSAR	COSMO-SkyMed	ALOS
Country	Canada	Germany	Italy	Japan
Primary Use	Marine, ice, disaster monitoring	Civilian, scientific	Civil, military	Forest, disaster monitoring
SAR Band	C-band	X-band	X-band	L-band
Generations	3	2	2	3

Figure 12.8 G-8 data policy [10].

The Beijing Declaration on Research Data was issued in 2019, following a workshop in Beijing jointly organized by the International Science Council's Committee on Data and the Chinese Academy of Sciences. The declaration called for global collaboration to make public research data as open and accessible as possible.

China's Gaofen-3 satellite system is one of the SAR-based satellites that provide data similar to Sentinel-1 [13, 14].

12.2.2.13 Indian Data-Sharing Policies

The specifications and data access policies of Indian SAR satellites are summarized in Table 12.2, including details of sensor bands, revisit periods, spatial resolutions, and mission lifetimes. The Government of India has promulgated a policy on liberalization of geospatial data on the 28th

Table 12.2 Indian SAR satellites.

S. no.	Satellite	Band	Revisit period	Resolution	Baseline mission duration	Data product and Access
1	RISAT-2B	X – 3.1 cm	3 days	1-0.5 m	5 years	Restricted
2	RISAT2BR-1	X – 3.1 cm	90 min	0.5 – 8 m	5 years	Restricted
3	EOS-01	X	90 min	0.5 – 8 m	5 years	Restricted
4	EOS-04	C-5.5 cm	90 min	1 – 50 m	5 years	Restricted
5	NISAR (Upcoming)	L- 24 cm S-band: 9 cm	12 days	3–10 m mode-dependent	3 years	Free & Open

of December 2022, based on 14 goals of the UN-GGIM. It focuses on the development of sectoral data in collaboration with different ministries, agencies, and private agencies. It envisages developing Geo-spatial infrastructure and manpower development through Public-Private partnership in three stages by 2035. Disaster management and High-resolution digital elevation model creation are some of the key areas that would require the development of SAR data acquisition. The NISAR data acquisition is expected to be one of the key activities in 2025.

12.2.2.14 World Meteorological Organization Resolution 40

The WMO Resolution 40 promotes open global exchange of elementary meteorological data, including satellite Earth observation data. This kind of data dissemination provides access in a non-discriminatory manner with minimal reproduction and delivery fees, especially at the time of disaster. While the overall aim is to support full, open, and free data sharing, commercial meteorological interests have led to more restrictive policies. As a result, only "essential" data—crucial for protecting life, property, and public well-being—are guaranteed to be freely accessible. These are detailed in an annex to the resolution. In response to growing climate data needs, the Sixteenth World Meteorological Congress in 2011 requested a review of Resolution 40 to enhance international climate data exchange and improve access.

The coordination group managing Meteorological Satellites outlines key satellite-derived products used in weather forecasting, environmental monitoring, and climate studies. Vertical temperature and humidity profiles are retrieved using infrared and microwave sensors; they contribute significantly to numerical weather prediction, especially through data from instruments like ATOVS, AIRS, and IASI. GPS radio occultation adds further profile accuracy. Sea surface temperature and land surface temperature are obtained *via* infrared and microwave imagery, with regional and global products aiding in monitoring temperature variations. Snow and ice coverage is tracked using visible, infrared, and microwave sensors. Ocean surface observations, such as sea level, wave height, and wind vectors, are gathered *via* active microwave instruments and are critical for ocean modeling and cyclone forecasting.

12.3 Ethical Frameworks

12.3.1 National Sovereignty and Jurisdiction

The principle of sovereignty remains a critical concern in SAR operations. Many countries regard remote sensing—especially high-resolution SAR imagery—as intrusive, potentially violating national security or internal confidentiality.

Unlike traditional aerial surveillance, SAR systems can observe foreign territories without physical entry, prompting debates about permissible limits. While customary international law does not prohibit such activities, tensions often arise when data is gathered without prior consent, especially during politically sensitive crises.

Moreover, the jurisdiction over SAR data collected by satellites often lies with the launching country or commercial operator. This further complicates any attempts by affected states to regulate how data over their territory is collected, processed, or distributed.

12.3.2 Privacy and Surveillance

Synthetic aperture radar systems, although not optical, can produce detailed images that reveal patterns of human activity, structural layouts, and sometimes even movement through advanced processing. While beneficial in disaster assessment, this raises significant privacy concerns.

For instance, in post-disaster urban areas, SAR can be used to monitor displaced populations, refugee camps, or damaged infrastructure. Without strict controls, such information could be exploited for purposes beyond humanitarian relief—such as surveillance, insurance discrimination, or political coercion.

This is especially sensitive in authoritarian regimes or conflict zones, where SAR data might be used to suppress dissent or monitor civilian movements under the guise of disaster response.

12.3.3 Equity in Data Access

An ethical challenge lies in unequal access to SAR data and processing tools. High-resolution SAR systems are predominantly operated by technologically advanced countries and private corporations, while developing nations often depend on external sources.

Often, countries with the least resources for space technology are impacted by extreme events, i.e., coastal floods and hurricanes. Small Island Developing States like the Maldives, Tuvalu, the Marshall Islands, and Fiji are expected to be affected most by which often have limited access to SAR and meteorological data.

12.3.4 Consent and Community Involvement

While carrying out ethical research and humanitarian work, obtaining informed consent is paramount. While SAR data collection does not involve direct human interaction, communities impacted by SAR-based interventions should still be informed.

For instance, if SAR data is used to justify resettlement or infrastructure changes post-disaster, affected populations should have input in the decision-making process. Lack of engagement can lead to mistrust, resistance, and ineffective implementation.

Integrating local knowledge with SAR insights also enhances data relevance and ethical acceptability, making communities stakeholders rather than mere subjects.

12.4 Policy Challenges

12.4.1 Lack of Standardization

One of the foremost policy issues is the absence of a standardized global framework governing SAR data use. Different countries have different data-sharing policies, processing standards, and access restrictions, making global collaboration difficult.

For instance, while the European Union promotes open-access policies through the Copernicus program, countries like China or India maintain more restrictive regimes. Without harmonized standards, interoperability and timely response can be severely affected.

Developing international protocols—perhaps under the auspices of the UN—could enhance coordination, especially for transboundary disasters.

12.4.2 Balancing Security and Humanitarian Needs

National security concerns often dictate SAR data access. Governments may withhold or delay releasing critical SAR data during disasters for fear of exposing infrastructure, military assets, or politically sensitive areas.

While security is important, overly stringent restrictions can hamper humanitarian operations and delay relief efforts. Policymakers must develop risk mitigation strategies that allow data sharing during emergencies while protecting national interests—such as temporary access licenses or anonymized data processing.

12.4.3 Commercialization vs. Public Interest

The SAR sector is increasingly dominated by private companies offering high-resolution, near-real-time data services. While this enhances technological capacity, it introduces conflicts between profit motives and humanitarian needs.

During disasters, private operators may prioritize clients who pay for expedited access, sidelining emergency responders from low-income countries or non-profit organizations.

Public-private partnerships, guided by policy incentives and social responsibility frameworks, can bridge this gap. Governments and non-governmental organizations can negotiate service-level agreements that ensure free or subsidized SAR access during declared emergencies.

The introduction of private players in this field has complicated the principle with a commercial angle to conquer the space [15], like the US Commercial Space Act 2015, and the INSpace agency in India.

12.5 Case Studies

12.5.1 Lohnak Flash Flood (2023)

The South Lohnak Flash Flood occurred on 3rd–4th October 2024. The NRSC has provided its assessment through Risat-1A. It was reported that the glacier lake area decreased by 101 ha after the flash flood on 17th September 2023 and 4^{rth} October 2013. When compared with the planet data, it was reported as 50 Ha for 27th September 2023 and 6th October 2023. Figure 12.9 presents the RISAT-1A pre- and post-event SAR data for the South Lohnak Lake region, highlighting the affected zones and changes in water extent caused by the outburst (NRSC, 2023).

Figure 12.9 The RISAT-1A data pre-post scenario for South Lohnak Lake (NRSC, 2023).

12.5.2 Myanmar Earthquake (2025)

The United Nations Satellite Centre (UNOSAT) assesses damages during various hazardous events, a 7.1 Richter scale earthquake on 28th March 2025, and its impact in Sagaing, Myanmar, is illustrated in Figure 12.10. The SAR data, corroborated with high-resolution optical data, have enabled rapid assessment of earthquakes, as well as violent conflict zones.

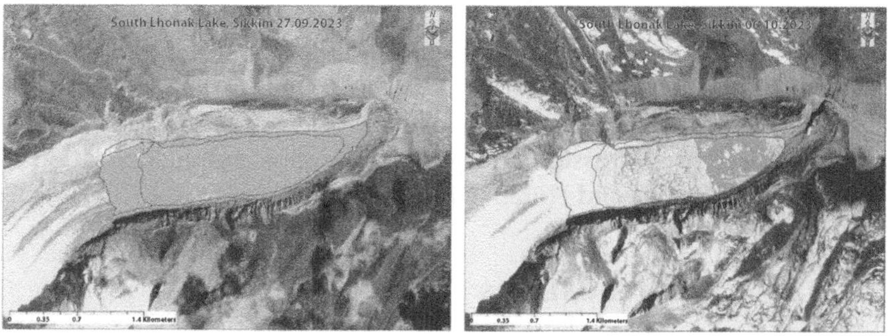

Figure 12.10 Planet Data 3 m resolution pre-post flash flood event (Planet Labs, 2023).

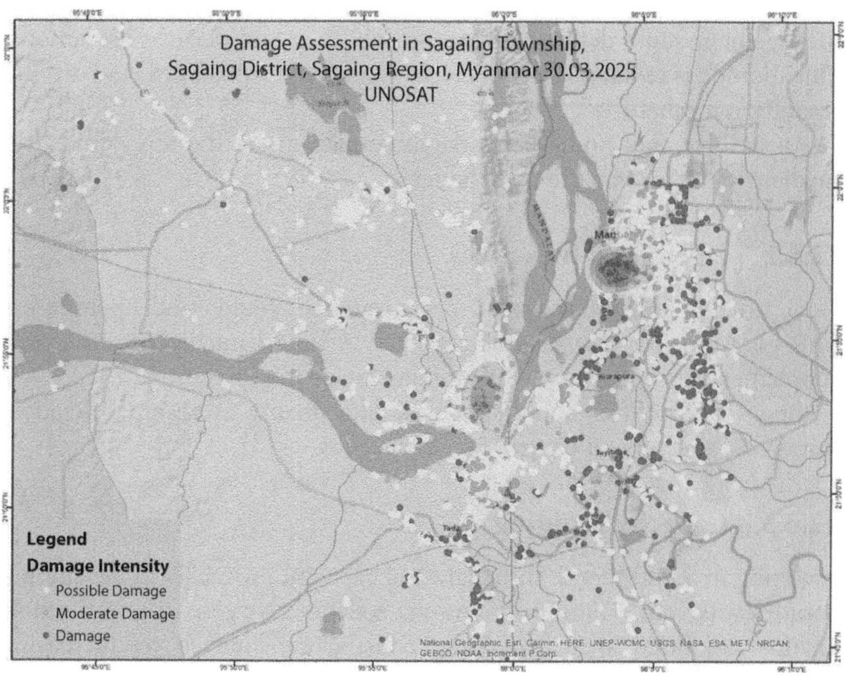

Figure 12.11 Damage assessment with Sentinel-1 in Sagaing, Myanmar. Source: UNOSAT.

It can provide high geometric precision in an uncensored manner [16]. Availability of Sentinel-1 data in the public domain has enabled various such studies to assess [17]. As shown in Figure 12.11, Sentinel-1 imagery was utilized by UNOSAT to assess post-disaster damage in the Sagaing region of Myanmar, underscoring the capability of SAR data for timely and reliable crisis mapping.

12.5.3 Ukraine War (2022–to Date)

The UNOSAT has identified 65 incidents of major damage in various parts of Ukraine since 24th February 2022. Most of the data used is high-resolution optical data to assess the damage to the urban area. Sentinel-1 data have been used to assess damage to urban areas and their zonation in cities of Irpin, Bucha, and Horenka [18].

12.6 Recommendations

12.6.1 International Legal Instruments

A legally binding treaty or protocol focused on remote sensing for disaster management could define roles, responsibilities, and permissible actions. This should cover data collection, sharing, privacy protections, and dispute resolution mechanisms.

An enforcement body or oversight council under the UN could help mediate issues and ensure compliance.

12.6.2 Open Data Initiatives

Governments and international organizations should fund open-access SAR platforms for disaster management. Artificial intelligence tools can supplement these platforms to automate image interpretation.

Open-data policies should include data quality assurance, metadata standards, and version control for consistency.

12.6.3 Capacity Building

Investing in SAR literacy and analytical capacity, especially in developing countries, is vital. Training programs, collaborative research, and scholarships can build local expertise, reducing dependence yet increasing resilience.

Technical assistance can be incorporated into development aid or disaster preparedness grants.

12.6.4 Ethics Committees and Oversight

Establishing ethics review boards for remote sensing projects can prevent misuse and ensure community engagement. These committees should evaluate data collection, usage protocols, and the potential impacts of SAR-based interventions.

Transparency reports can also build public trust and accountability.

12.6.5 Multi-Stakeholder Platforms

Multi-stakeholder coordination involving governments, private companies, non-governmental organizations, and academia should be institutionalized. These platforms can develop common standards, share best practices, and provide peer accountability.

Emergency data-sharing protocols should be negotiated in advance to avoid delays during real crises.

12.7 Conclusion

Synthetic aperture radar has become a cornerstone of modern disaster management, offering unmatched capabilities in data collection and terrain analysis. However, these advantages come with substantial legal, ethical, and policy implications that must be proactively addressed.

International cooperation, transparent policies, ethical foresight, and inclusive decision-making are essential to harness the full potential of SAR for humanitarian goals. As global disasters become more frequent and severe, preparing a robust governance framework is not just an academic exercise—it is a moral and practical imperative.

In general, it can be said that the OST intended to protect the global public interest in space and, for its first few decades in existence, it very successfully achieved this goal.

Space governance today is a combination of international agreements adopted during the space race of the 20th century, together with non-binding principles and guidelines that were created to deal with specific issues as they arose, through several international organizations [4].

With the ever-increasing scope of global governance, the complexity and challenges are also going northwards. The space industry in the past few decades has transformed from a bipolar to a multipolar system as rapidly many new countries are entering into space industry. In addition to that, besides government agencies, many private agencies have developed commercial stakes in the space industry, especially related to disaster management, climate change-related adaptation and mitigation activities, as well as agriculture, forestry, and insurance sectors. This calls for a more polycentric system [4].

The challenges that need to be considered are the commercialization of space activities, the rapidly increasing number of stakeholders, and the lack of an appropriate framework to cater to their management. Launch of rapid satellites like Cube-Sat will further make the registry of space objects difficult in the absence of a new framework and systems for strengthening the OST.

AI disclosure statement: The author acknowledges the use of AI-based tools for minor language editing or grammatical correction. However, all intellectual contributions, analyses, and conclusions presented in this chapter are entirely the author's own.

References

1. Murray, V., The UNDRR/ISC Hazard Definition and Classification Review and Hazard Information Profiles and Links to the Sendai Framework, the SDGs, and the Paris Agreement. *Prehosp. Disaster Med.*, *38*, S1, s70–s71, 2023.
2. Balgah, R.A. and Kimengsi, J.N., Glossary of disaster risk management-related terms, in: *Disaster Management in Sub-Saharan Africa: Policies, Institutions and Processes*, pp. 225–265, Emerald Publishing Limited, Leeds, United Kingdom, 2022.
3. Jakhu, R., International law governing the acquisition and dissemination of satellite imagery. *J. Space L.*, *29*, 65, 2003.
4. Ram, J., International Law Governing the Acquisition and Dissemination of Satellite Imagery, in: *Commercial Satellite Imagery and United Nations Peacekeeping*, pp. 11–30, Routledge, London, United Kingdom, 2017.
5. Lasswell, S.W., History of sar at lockheed martin (previously goodyear aerospace), in: *Radar Sensor Technology IX*, pp. 1–12, vol. 5788, SPIE, 2005, May.
6. Beaver, M., Current Space Law Limitations and Its Implications on Outer Space Conflicts, E-International Relations, 2015.
7. Dembling, P.G. and Arons, D.M., The evolution of the outer space treaty, 1967.
8. Johnson, C.D., The Outer Space Treaty, in: *Oxford Research Encyclopedia of Planetary Science*, 2018.
9. Harris, R. and Baumann, I., Open data policies and satellite Earth observation. *Space Policy*, *32*, 44–53, 2015.

10. National Environmental Satellite, Data, and Information Service (NESDIS), *NESDIS Information and Services*, 2025
11. European Space Agency (ESA), *Updated ESA Earth Observation Data Policy (Simplified Version)*, ESA, 2023.
12. EUMETSAT (European Organisation for the Exploitation of Meteorological Satellites), *EUMETSAT Data Policy*, EUMETSAT, 2021.
13. Meyer, F.J., Hogenson, K., Kennedy, J.H., Lewandowski, A.F., Albright, R.W., Short, G., Flores-Anderson, A.I., Rosen, P.A., Facilitating the Golden Age of Synthetic Aperture Radar: New tools, services, and training to make synthetic aperture radar data more accessible. *IEEE Geosci. Remote Sens. Mag.*, 2025.
14. Zhao, L., Zhang, Q., Li, Y., Qi, Y., Yuan, X., Liu, J., Li, H., China's Gaofen-3 satellite system and its application and prospect. *IEEE J. Sel. Top. Appl. Earth Obs. Remote Sens.*, *14*, 11019–11028, 2021.
15. Dula, A., Private Sector Activities in Outer Space. *Int. Lawyer*, 19, 159, 1985.
16. Brunner, D., Lemoine, G., Bruzzone, L., Earthquake damage assessment of buildings using VHR optical and SAR imagery. *IEEE Trans. Geosci. Remote Sens.*, *48*, 5, 2403–2420, 2010.
17. Plank, S., Rapid damage assessment by means of multi-temporal SAR—A comprehensive review and outlook to Sentinel-1. *Remote Sens.*, *6*, 6, 4870–4906, 2014.
18. Udovychenko, V. and Petrovskyi, I., War-Related in Ukraine Urban Landscapes Damage Assessment Using Radar Data Change Detection Algorithm, in: *International Conference of Young Professionals «GeoTerrace-2024»*, pp. 1–5, vol. 2024, European Association of Geoscientists & Engineers, 2024, October.

Index

Printed and bound by CPI Group (UK) Ltd, Croydon, CR0 4YY

15/09/2026

14962411-0005